T0136236

Sea Turtles of the Eastern Pacific

Arizona–Sonora Desert Museum Studies in Natural History

SERIES EDITOR

Richard C. Brusca

Sea Turtles of the Eastern Pacific
Advances in Research and Conservation

Edited by
Jeffrey A. Seminoff and Bryan P. Wallace

Foreword by Peter C. H. Pritchard

THE UNIVERSITY OF
ARIZONA PRESS

TUCSON

THE UNIVERSITY OF ARIZONA PRESS

www.uapress.arizona.edu

Library of Congress Cataloging-in-Publication Data
Sea turtles of the eastern Pacific : advances in research and conservation / edited by Jeffrey A. Seminoff and Bryan P. Wallace ; foreword by Peter C. H. Pritchard.
    p. cm. — (Arizona-sonora desert museum studies in natural history)
    Includes bibliographical references and index.
ISBN 978-0-8165-1158-7 (cloth :alk. paper)   1. Sea turtles—Pacific Ocean.   2. Sea turtles—Conservation—Pacific Ocean.3. Sea turtles—Research—Pacific Ocean.   I. Seminoff, Jeffrey A. (Jeffrey Aleksandr)   II. Wallace, Bryan P.
   QL666.C536S44   2012
   597.92'809164—dc23
                                                                                          2011051602
Publication of this book is made possible in part by a grant from NOAA—National Marine Fisheries Service.

17   16   15   14   13   12   6   5   4   3   2   1

Contents

Part III: Field-Based Conservation and Signs of Success

Foreword

Nearly five hundred years ago, in 1554, a short, stout, bearded, affable French savant *bon vivant* and former monk by the name of Guillaume Rondelet published a large-format book, ostensibly about marine fishes. The title—*Libri de Piscibus Marinis, in quibus verae Piscium expressae sunt*—was a confusing one. While the title refers to fishes, this must be construed as "fish" in the medieval sense of aquatic creatures in general. Thus, to this day, the papal decree establishing turtles as "fish," and therefore available for consumption during Holy Week, continues to confound the efforts of Venezuelan tortoise conservationists (and Archie Carr's celebrated 1967 book *So Excellent a Fishe* is about turtles, not fish). Rondelet clearly had a broad concept of the word "fish," his book offering not only a droll engraving of a freshwater mammal (a beaver) but also a salamander, a crocodile, and several renderings of bizarre and fanciful hybrids between fish and humans, including the Bishop Fish ("De Pisce Episcopi hanitu") and the King Fish ("De Pesce monachi habitu"), although he avoids committing to whether he believed in the existence of the latter monstrosities himself.

Rondelet spent much of his life near his birthplace at Montpellier, in the south of France, and his book also included a pair of rather Gothic but clearly recognizable engravings of the two species of marine turtles most likely to be found in Europe: the loggerhead (*Testudo corticata*) and the leatherback (*Testudo coriacea* or *T. mercurii*). The engraving of the leatherback was not the first ever published—Candidus (1460) had illustrated the species a century earlier—but Rondelet's was a superior piece of art. Each engraving was accompanied by a lengthy text, summarizing the published statements of the author's predecessors on the subject of turtles. As it happened, the predecessors were ancient indeed—Pliny, Aristotle, Strabo, Oppian, Hesychius—and most of them had little to say about turtles that was not just inaccurate hearsay or, even worse, unchecked imagination. Rondelet did his homework, and his great contribution was to examine the statements of these authors, to quote them (his text flips smoothly from Latin to Greek and back again, to make it comprehensible to all "Men of Learning"), but not to accept what the ancients said unless it could be confirmed. To provide this confirmation, he methodically dissected the carcasses of deceased sea turtles, including leatherbacks, that came his way, and he reported upon their anatomy in great detail. This put him in a higher, more modern scientific category than such fantasy as the observation of Pliny that turtle flesh, mixed with frog flesh, was an excellent remedy for "salamander bites."

Following the excellent, although idiosyncratic, work of Rondelet, a couple of centuries were to pass before significant further steps in sea turtle science were to come about. On a recent visit to the Old Library of my alma mater Magdalen College, Oxford, I found that the collection, established in the 1400s, included only two turtle books, apart from copies of my own work that I had donated. One was Sowerby and Lear's *Tortoises, Terrapins and Turtles* (1871), a sort of abridged version, now quite valuable, of Bell's *Monograph of the Testudinata*, which had never been published in full; and the other was a precious original copy of Rondelet's masterwork, half a millennium old. Grubby-fingered undergraduates are not normally allowed into the Old Library, and these two irreplaceable works were both in mint condition. It is noteworthy that more than three centuries passed between their years of publication.

Actually, there was a work that neatly spanned the interval between the publication of *Libri de Piscibus* and that of *Tortoises, Terrapins and Turtles*—Edward Topsell's *The History of Serpents*. Topsell's book, published in 1658, features a wonderful engraving on the title page of a hugely corpulent snake (with a very thin tail) in the act of swallowing a young child, but the scope of his book, matching that of Rondelet a century earlier, included, among other things, a summary of what the ancients said about turtles. Some of Topsell's comments about sea turtles have great charm, and not a little accuracy—witness the description of the jaws of the green turtle: "They have no teeth, but in the brims of their beaks or snouts are certain eminent divided things like teeth very sharp, and shut upon the under lip like as the cover of a box, and in the confidence of the sharp prickles, and the strength of their hands and backs, they are not afraid to fight with men." Also:

They live in rocks and the sea-sands, and yet they cannot live together in the water, or on the lands, because they want breathing
and sleep, both of which they perform out of the Water: yet Pliny
writeth, that many times they sleep on the top of the water, and his
reason is, because they lie still unmoveable (Except with the water)
and snort like any other creature that sleepeth, but the contrary ap-
peareth, seeing they are found to sleep on the land, and the snorting
noise they make is but an endeavour to breath, which they cannot
well do on the top of the water, and yet better there then in the
bottom.

And:

And for as much as they cannot by Nature, nor dare for accident
long tarry upon the land: they set certain marks with their feet upon
the place where they lay their Egs. Whereby they know the place
again, and are never deceived. Some again say, that after they have
hid their Egs in the earth forty days, the female cometh the just
fortieth day, not failing of her reckoning, and uncovereth her Egs
wherein she findeth her young ones formed, which she taketh out
as joyfully as any man would do Gold out of the earth, and carryeth
them away with her to the water, They lay sometimes an hundred
Egs, and sometimes they lay fewer, but ever the number is very
great.

There are inaccuracies here perhaps, but who would have thought that marine turtle philopatry had been commented upon by Plutarch two thousand
years ago?

In 1758, Carl von Linné published the tenth edition of his *Systema
Naturae*. In this synopsis of the entire living creation, he included two sea
turtles, *Testudo caretta* (the loggerhead) and *Testudo mydas* (the green turtle). Soon to follow, in the twelfth (1766) edition, was the hawksbill, *Testudo
imbricata*, and the leatherback, *Testudo coriacea*, which had recently been
described by the Italian naturalist Domenico Vandelli.

In the earliest comprehensive reviews of turtles of the world, including those of Johann Julius Walbaum (1782), Bernard Germain de La-

Cépède (1788), Johann David Schoepff (1792), and François Marie Daudin (1802), the sea turtles already featured in high profile, generally with detailed accounts of external morphology, and especially of economic (principally gastronomic) value. Walbaum's monograph is an odd one, with lengthy, meticulous accounts of the morphology of the *Americanischen Caret-Schildkröte* and the *Grossfussiger Meerschildkröte* in Gothic German, followed by Latin accounts of *Testudo* (*caretta*), *Testudo imbricata*, and *Testudo macropode*. The book concludes with an engraving not of a sea turtle but of the shell of the world's smallest tortoise, *Homopus signatus*.

Whereas these early writers recognized only a small fraction of the freshwater turtle species known today, four of the eight species of sea turtle now known were featured in these accounts of more than two centuries ago—each, in fact, already denoted by a variety of scientific names, based upon different life stages, or possibly simply a result of lack of due diligence when proposing new names. Thus, LaCépède recognized just twenty-four species of chelonians altogether—tortoises, terrapins, and turtles—but he gave the green turtle (*La Tortue franche*) pride of place. It was the very first account in the "quadrupeds ovipares" section of his *Histoire naturelle*, and also one of the longest accounts, occupying a full twelve pages of text. Even at that early time, the species had already received several names—*Tortue franche* or *Tortue verte* (Cuvier), *Testudo mydas* var. *beta* (Linnaeus), *Testudo viridis* (Schneider), and *Caretta esculenta* (Merrem)—the last name translating to "the edible loggerhead." He gave possible credence to one more name—*La Tortue ecaille-verte*, or "the green-scaled turtle," which he accepted upon the word of Monsieur de Bomare. The verbose trend continued, and by 1889 the green turtle had received no fewer than forty-eight different scientific names and combinations of names, according to the most recent synonymy (Uwe Fritz and Peter Havaš's 2006 *Checklist of Chelonians of the World*), with just five more added since—and three of these really just typographical errors.

Some of the other sea turtles known today—the loggerhead, hawksbill, and leatherback—were recognized, each with various synonyms, by LaCépède, although his account of the leatherback (or *luth*) was relatively short and concentrated on individual adult specimens that had washed up dead, here and there, from time to time.

It was a while before the other valid species of sea turtles were identified. The first of this second wave of novelties was the olive ridley, *Lepido-chelys olivacea*, named by Johann Friedrich von Eschscholtz in 1829, followed by the black turtle (*Chelonia agassizii*) by Marie Firmin Bocourt in 1868; and the flatback (*Natator depressa*) and Kemp's ridley, *Lepidochelys kempii*, similar turtles from opposite parts of the world, both named by Samuel Garman in 1880.

Apart from these genuine novelties, the nineteenth century and early twentieth century were a sort of second dark ages for sea turtle science, with a steady output of useless synonymous names proposed, no fieldwork done, and a tendency among both popular and scientific writers to ignore perfectly valid taxa, including the flatback turtle and the two ridleys. No end was seen to this fruitless epoch until P. E. P. Deraniyagala, working alone in Ceylon, published his monumental *Tetrapod Reptiles of Ceylon* in 1939, and Archie Carr published his important *Notes on Sea Turtles* in 1942. These authors straightened out the classification of the extant species of marine turtle, and Carr's work led to a steady output of original sea turtle science, with various students and colleagues as coauthors, right up to the time of his death in 1987. This switch of focus from Carr's original work with the freshwater turtle genus *Pseudemys* was motivated in large part by his recognition, when he was examining the published turtle literature in preparation for writing his 1952 magnum opus *Handbook of Turtles*, that almost no information was available on the biology of sea turtles.

Today the output of published information in the field of marine turtles is enormous, especially if one includes the annual production of "gray literature," because peer-reviewed contributions are now a distinct minority among the important sources in this field. Within this copious torrent of literature, a new genre is emerging: monographs of the sea turtles— the actual books, as opposed to the scientific papers, scholarly notes, and raw data compilations. The books are still relatively few, and most do not cover sea turtles on a global basis or all aspects of turtle life. Those with wide coverage include Blair E. Witherington's *Sea Turtles* (2006), David Gulko and Karen L. Eckert's *Sea Turtles: An Ecological Guide* (2004), and James R. Spotila's *Sea Turtles: A Complete Guide* (2004). Others (e.g., Carl Safina's *Voyage of the Turtle* [2006] and Gary Paul Nabhan's *Singing the Turtles to Sea* [2003]) offer a mystical or personal approach to turtles. James Jerome Parsons's *The Green Turtle and Man* (1962) was the first of the single-species compendia, and others had to wait nearly half a century before enough data had accumulated to make book-length monographs feasible—*Biology and Conservation of Ridley Sea Turtles* (2007), edited by Pamela T. Plotkin, and *Loggerhead Sea Turtles* (2003), edited by Alan B. Bolten and Blair E. Witherington.

Jack Frazier's *Marine Turtles as Flagships* (2005) is also multi-authored and offers some sociological and culture-themed papers on the special role that marine turtles may play as symbols of their ecosystems for conservation purposes. The two-volume *The Biology of Sea Turtles*, edited by Peter L. Lutz, John A. Musick, and Jeanette Wyneken (1997, 2003) presents a series of rigorous scientific contributions on many aspects of sea turtles and their lives. But in addition to these, there is a new subgenre of sea turtle books—those that describe the marine turtles of a specified, restricted region of the world, such as Jacques Fretey's *Biogeography and Conservation of Marine Turtles of the Atlantic Coast of Africa* (2001), Kartik Shanker and B. C. Choudhury's *Marine Turtles of the Indian Subcontinent* (2006), Irene Kinan's *Proceedings of the Western Pacific Sea Turtle Cooperative Research and Management Workshop* (2002), Jacques Fretey's *Les Tortues marines de Guyane* (2005), and Luis Felipe López Jurado and Ana Liria Loza's enigmatically titled *Marine Turtles: Recovery of Extinct Populations* (2007), which covers the turtles of Macaronesia and the Mediterranean.

You now hold in your hands a new and comprehensive work on the sea turtles of the East Pacific, an area of extraordinary latitudinal extent (from Alaska to Chile), where study of marine turtles lagged even as turtles in Caribbean Costa Rica, Australia, and Southeast Asia became the subject of intensive field studies. In this unique zone, the turtles show adaptations to the special environmental conditions, which include the very narrow coastal shelf in most of the area, and the cooling effects of both the Humboldt Current from the south and the Alaska Current and California Current from the north. The combined aspects of this region have produced the smallest adult leatherbacks in the world; the blackest green turtles; dark, small, steep-shelled olive ridleys that float high in the cool waters, trying to catch warming rays of the rising sun on the exposed tops of their shells; immature loggerheads that visit for a while but migrate from and back to Japan and Australia, one of the most prodigious migrations of any sea turtle; and hawksbills that seem to be struggling with a difficult habitat and are rare and marginal throughout the region. It's not just the cold-tolerant leatherbacks that penetrate to the northern and southern extremes of East Pacific waters; the black turtles have been found in Alaska and are regularly seen around Pisco, Perú, and even habitually swim with the penguins in Galápagos waters.

Only now are East Pacific turtle studies reaching the viable, or even vibrant, level justly celebrated in this book. The East Pacific hosts two of the largest ridley nesting arribadas in the world (Escobilla, México, and Ostional, Costa Rica), of which both myself and the entire scientific world were blissfully unaware when my own East Pacific wanderings started in 1967.

Living in Florida, I am close to Caribbean and Atlantic turtle beaches, but the Pacific is more distant, and some early fieldwork opportunities took me not to the eastern, but to the western Pacific—Micronesia, New Caledonia, and Papua New Guinea. It was in 1967 that I first saw the greatest of all oceans. (I celebrated by collecting a gallon of Pacific water at midnight and bringing it back to Gainesville.) I was starting my dissertation work at the University of Florida and needed access to olive ridleys for morphological and other studies. The Kemp's ridley arribadas in Tamaulipas, México, had been brought to public attention in 1963, but the first unconfirmed rumors of olive ridley arribadas were not published until David Caldwell's 1967 essay in the now defunct but fondly remembered *International Turtle and Tortoise Society Journal*, and these stories centered upon the Mexican Pacific state of Jalisco, specifically at Bahía de Banderas.

Investigations by René Márquez and Antonio Montoya later established that the actual arribada site was Playón de Mismaloya, Jalisco. But by all accounts this arribada has since gone extinct, a hint that even undisturbed arribadas may not be permanent—a topic I enlarged upon in my chapter "Arribadas I Have Known" in Pamela Plotkin's book on ridleys. Astonishingly, the arribada at Escobilla in Oaxaca, despite decades of massive-scale industrial slaughter of the adult ridleys, continues in full spate and has increased dramatically since the slaughter was curtailed in 1990 (see chapter 13 in this volume).

My destination on the 1967 trip was the scenically spectacular Gulf of Fonseca in Honduras, where Archie Carr had observed nesting olive ridleys two decades earlier and where Cynthia Lagueux saw them still coming

ashore twenty years after my visit, in 1987. Throughout this forty-year period there was probably almost no survival of ridley eggs in the Gulf of
Fonseca; the *hueveros* patrolled the beaches so thoroughly that every nest
was collected. So how come there were—and are—still turtles nesting
there? In retrospect, the explanation of this seeming miracle was almost
certainly the existence of undocumented small arribadas just a little to the
south, in Nicaragua, and much bigger ones one nation farther south, in
Costa Rica. These may have pumped out enough millions of hatchlings into
local waters each year to keep the Gulf of Fonseca beaches, not too far away,
adequately supplied with breeders.

My two decades with the Florida Audubon Society taught me
much about "real-world" conservation, whereby opposing stakeholders
need to be engaged rather than demonized or ignored. I was party to a dramatic example of such interaction when Kim Cliffton and Carlos Nagel invited me to participate in negotiations with Antonio Suarez, the notorious
proprietor of the frightful turtle slaughterhouse at San Agustinillo, Oaxaca,
at which, every day for several months a year, several hundred adult olive
ridleys met their end. The negotiations were an education to all of us, and
while we never surrendered our own principles, we had to concede that
Suarez made valid points from time to time. He insisted that political and
sociological realities were a vital part of any conservation plan, and also that
his financial support for rigorous protection of nesting turtles and their eggs
gave him the "right" to harvest some number of adult turtles, provided he
operated within quotas. The only uncertainty, to him, was how many. This
did not accord with demographic theory, which considered adjustment of
egg quotas of trivial importance compared to the protection of near-adult
and adult turtles.

But ridleys and their arribadas are confusing things, and the turtles
were still nesting in significant numbers when the slaughterhouse was finally closed by presidential decree in 1990. In the ensuing years the annual
nest totals immediately increased spectacularly, as each year intact cohorts
of newly matured females joined the ranks of the breeders.

My search for insights into olive ridleys continued for many years.
Much of this time was spent looking for, or waiting for, arribadas. Highly
memorable was our first camp at Piedra de Tlalcoyunque in Guerrero,
where we were blown away by a hurricane the first night, then became pro-

foundly sick from drinking the local water, and the next morning witnessed the fatal aftermath of a drug-related shootout in the village square of San Luis La Loma—but we did not witness an arribada.

Thirty years after my interactions with Antonio Suarez in Oaxaca, I paid another visit, with my student Jean Jang, who was doing a thesis on arribadas for her master's degree. Much had changed. The slaughterhouse had given way to a major research center, the Centro Mexicano de la Tortuga, still under construction but already well populated with sea turtles of the world and with a comprehensive live collection of the local freshwater and terrestrial chelonians as well. Strangest of all, the exact site of the slaughterhouse was now occupied by a kindergarten, of all things, and stranger still, the school was named after Sigmund Freud, of all people. In other respects, the place was simply a typical Mexican coastal village.

Many of us have boxes and cases full of unpublished information, which we intend to work up and submit for publication "some time." Until "some time" is reached, the data remain unavailable to the world. Dr. Ernesto Albavera Padilla, the director of the research station, also had a great deal of unpublished information about the Oaxaca ridleys in his files, and many excellent photos, too. What was different was that, unlike many of us, he was more than willing to share his unpublished data. The photocopier and computer labored for hours, and he handed me a two-inch thick stack of documents and photos, all pertaining to the local sea turtles.

It was clear that at least three of the former Pacific coast arribadas—those in Piedra de Tlalcoyunque, Playón de Mismaloya, and Chacahua, in the states of Guerrero, Jalisco, and Oaxaca, respectively—had simply disappeared, or at best had lost 99 percent of their turtles. On the other hand, arribadas were flourishing at Ixtapilla, Michoacán, and at Morro Ayuta and Escobilla in Oaxaca. The Escobilla data were the most interesting, not only because of the extraordinary number of turtles using the beach but also because this was the population exploited on a massive scale in the years preceding 1990. Today, it has a higher level of protection than most turtle beaches in the world, with a military detachment encamped on the beach throughout the season and a requirement that any visitors to the beach carry documents of permission from the authorities in Puerto Angel. Extensive patrols are made on the 17-km beach every day, and estimates are made of the number of turtles nesting, especially on arribada days. These estimates have now been gathered each year since 1973; the number of nests oscillated around 100,000–200,000 during the years of intense commercial exploitation, but after 1990 they really took off, reaching the figure of one million nests (to celebrate the new millennium?) in 2000 and touching 1,400,000 nests in 2007. The only "low" year in the recent past was 1998, with slightly more than 200,000 nestings, but after all, this was an El Niño year. Such extraordinary nesting numbers, together with the disappearance of the other Pacific arribadas during the same period, give one grounds for deep contemplation about the true status of the species *Lepidochelys oliva-cea*, and the difficulty of finding an International Union for Conservation of Nature category (Endangered? Critically Endangered?) that really fits. Unfortunately, the former categories of "Recovering" and "Conservation Dependent" are no longer used.

On another Pacific coast trip, I was on the lookout for ridley skulls, to compare with those of Atlantic ridleys of both species. Clean or dry skulls proved elusive, but there were lots of freshly severed heads lying around in coastal areas of Oaxaca. My strategy for packing these for the long trip home to Gainesville was to half-fill a footlocker with the heads, then top the box up with fresh beach sand and latch the whole thing down until we got home—any kind of actual preservation would make the skulls difficult to clean and prepare. Our plans were somewhat unraveled when we passed through customs in Laredo. Our destination of Gainesville seemed to convince the inspection officer that we were automatically involved in the marijuana business, and he insisted on making a thorough search. He finally came to the footlocker, and against our advice, he opened it. He was greeted by an intolerable stench and the deafening buzz of thousands of flies—metamorphosed maggots—that flew up into his face. Instantly discarding his cool demeanor, the officer ran away in terror and dispatched two junior officials to complete the inspection. Those were the days before the Convention on International Trade in Endangered Species of Wild Fauna and Flora, when there was no objection to importing the skulls into the United States—but sand was illegal (it might harbor nematodes), so the stinking mass of rot-blackened sand had to be excavated by hand and deposited in the waste bins at the customs station. Rules, after all, are rules. But I did keep my skulls.

In addition to the multitudes of olive ridleys, the region is the almost exclusive home of the black turtle, but this form was not recognized as a distinct taxonomic entity until David Caldwell's 1962 paper, despite its bearing Bocourt's name *Chelonia agassizii* from nearly a century earlier. It was also my fate, one season, to be the director of the black turtle conservation project at Maruata and Colola, Michoacán. For many years now this project has been led by Javier Alvarado Díaz, Carlos Delgado Trejo, and their colleagues (see chapter 11), but at the time the field director was the remarkable Kim Cliffton, an American who, among his many distinctions, was being an honorary member of the Comcáac (Seri Indian) community in Northwest México, despite his blond hair and imposing height of six feet four inches. Dedicated to his task, he not only patrolled the beaches at night and caught male turtles at sea by day but also personally set up roadblocks to stop traffic and confiscate live black turtles and their eggs from poachers driving the booty to surreptitious markets to the north. How Kim stayed alive intercepting *hueveros* and desperadoes in this gun-totin' part of the world I do not know.

As the nominal project boss for the season, I visited the Maruata project from time to time, but Kim was not used to being subordinate to anyone, and I was in an unenviable position. We agreed to retire the arrangement at the end of the year. There are some individuals in this world—Jacques Fretey is another, and I think I am one, too—who simply do not fit into hierarchical systems. Kim was so unimpressed by rank that, when Russell Train, administrator of the U.S. Environmental Protection Agency and also chairman of World Wildlife Fund USA (which funded the black turtle project), paid a visit to the project at Maruata, no special arrangements were made, no dinners planned, and within one day Train found himself relegated to the role of babysitter for the Cliffton infant while Kim and his young wife spent the day catching turtles at sea by leaping off the front of their boat.

To facilitate the search for poachers, Kim asked if he could borrow the small Cessna aircraft that my fellow scientist at Florida Audubon used for manatee surveys. This request was duly granted, and Kim came to Florida to fly the plane around the northern Gulf and across the high Sierras of México to Maruata. At the end of the season, we had to bring the plane home, and Kim was not available for this task, so I recruited an adventurous pilot named Gray Bower, and her husband as copilot. Gray and I discussed the best route home, and we agreed that the transmountain route was too dangerous, so we looked at the charts to find an alternative. There seemed to be a low passage called La Ventosa that stretched from the Pacific to the Gulf across the Isthmus of Tehuantepec, so we elected to fly southeast along the Pacific coast 600 miles and then embark upon this new route.

It was a momentous decision that nearly killed us all. The isthmus indeed offered a low passage but was hemmed in by high mountains, with thick clouds and roaring winds that made me belatedly understand what La Ventosa actually meant ("the windy place"). We had been sold jet fuel for our little Cessna by soldiers at Ixtepec military airport, where the authorities were far from happy to have us land at all. The engine carbonized rapidly, and we had cylinder failure twice (once in midair) before we abandoned the mission.

But while flying down the Pacific coast, before we got the bad fuel, we had selected an altitude of about 300 feet (or less) that allowed us to observe and count leatherback turtle nesting tracks. It seemed that there were extraordinary numbers, on beaches from Michoacán to Oaxaca, indicating a huge population. This resulted in my doing a drastic recalculation of the global population of leatherbacks, raising it from between 29,000 and 40,000 adult females to 115,000. This was published in *Copeia* in 1982. As it turned out, this was a vital, although accidental, baseline observation for what was to come, because in the subsequent decades the East Pacific leatherback population took a precipitous drop that has alarmed the entire sea turtle conservation community. There appears to be a very real possibility of extinction of East Pacific leatherbacks nesting in the key nations of México and Costa Rica. The causes have not been unambiguously identified, but mortality on beaches and collection of eggs have been curtailed for enough years now that they are probably not the causes, and the actual culprit is almost certainly incidental catch by longliners and other fishing operations in the open Pacific.

Another East Pacific project, which I remember very fondly, was my involvement with the leatherbacks at Playa Grande, in Costa Rica—a major colony that for some reason was undocumented in the scientific literature until I went there in February 1983. Subsequent monitoring by Nayudel Guadamuz indicated that up to nearly 200 turtles were nesting nightly on this 4-km beach. With my colleague Maria-Teresa Koberg as the chief source of energy and inspiration for the operation, we prepared a management plan for the beach and proposed a marine national park to protect the nesting turtles. This was quite controversial, even within the conservation community, and involved extensive interaction with stakeholders, not only the villagers who had habitually collected the eggs but also the developers who had acquired the land without realizing its importance as a leatherback nesting ground. The park itself, originally called Las Baulas de Guanacaste but now known as Parque Nacional Marino Las Baulas, has had a complex history, steered in the right direction in recent years by Jim Spotila and Frank Paladino. With the reelection of President Oscar Arias, the buyout of the land developers at Playa Grande became a hot issue once again. Teté Koberg, meanwhile, has prepared the text of a marvelous manuscript titled "Against All Odds," in which the politics and personalities of the players in the saga of Las Baulas are described, not without affection but with remarkable objectivity. It is my earnest hope that this truly original literary work will be published soon. Meanwhile, fieldwork at Playa Grande continues under the leadership of Rotney Piedra Chacón and colleagues (see chapter 8).

But it was not just on the mainland of the Americas that I went in search of turtles. In 1970, I first realized my youthful ambition to do fieldwork in the Galápagos Islands, where the sea turtles had never been scientifically studied, although they had been heavily collected in 1905–1906 by the California Academy of Sciences expedition. So for the next four seasons my wife and I spent the early months of each year in the archipelago, she tending our infant son and I tagging sea turtles by night and walking massive stretches of coastline of the larger islands by day, patiently documenting the beaches on which turtles seemed to be nesting and entering the data in the field on detailed outline maps of the islands. There was a lot of nesting, especially along the north coasts of Santa Cruz, Santiago, and San Cristobal islands, as well as on western Baltra and southern Isabela islands. Once on Quinta Playa (Isabela) we tagged seventy-six turtles in one night. Sometimes there was confusion of turtle tracks with those of sea lions, but in the dry conditions prevailing, tracks and nest pits were usually evident and clear for many weeks afterward.

We also talked at length with Carmen Angermeyer and her husband, Fritz. Fritz had lived in Galápagos since the late 1930s, and he and

Carmen knew more about Galápagos sea turtles than anyone else alive. They explained to me with great care and enthusiasm the differences between the two forms of *Chelonia*—known locally, in Galápagos English, as the black turtle and the yellow turtle. Fritz's brother Karl had laboriously constructed an artificial cave out of lava blocks and cement, and he had furnished it with the fruits of a long lifetime of accumulation of treasures of the sea. There were turtle shells of many sizes in there, and it was immediately clear to me that there were two kinds of green turtle—even the foot-long juveniles were clearly divisible into those with intense black pigmentation and distinct narrowing of the shell above the hind limbs, and those with heart-shaped shells and attractive radiating patterns of brown and amber on each scute.

The Angermeyers explained that the yellow turtles, although reaching a good size, were never in breeding condition, neither males nor females. When slaughtered, they yielded a great deal of yellow, sweet-tasting fat that made excellent butter, contrasting sharply with the fishy-tasting oil of the black turtles. The lack of indications of sexual maturity even in the biggest yellow turtles was odd indeed, and I thought of various increasingly unlikely explanations for this. It was thirty years before Peter Dutton clarified the issue using genetic techniques not available in the early 1970s. The yellow turtles turned out to be subadults of the very large green turtles from the far western side of the Pacific Ocean, nesting in such places as Vanuatu, New Caledonia, and French Polynesia. The black turtles were, like the sea lions, an East Pacific specialty, an enclave of a species also found on American Pacific shores. They nested in Michoacán and in Costa Rica and constituted an important resource for the Comcáac and other indigenous peoples in the Gulf of California. Sometimes, our tagged Galápagos turtles showed up on mainland shores from Costa Rica to Perú, but others were probably year-round residents of Galápagos waters. The coexistence of these two distinct forms of *Chelonia* offered a strong argument that they should be recognized as full species. Since my early work on Galápagos turtles, the research has been continued by Derek Green, who studied the turtles in the water as well as on land; then by Mario Hurtado from the Instituto Nacional de Pesca in Guayaquil; and in more recent years by Patricia Zárate, whose work is summarized in chapter 3 of this book.

One could identify many East Pacific sea turtle projects that would be worthwhile, but high on the priority list must be a determination of hawksbill populations throughout the region. In my own visits to the region I have rarely encountered hawksbills, and those individuals I have seen, in Oaxaca, the Galápagos, or Costa Rica, have been scattered juveniles rather than adults. Throughout my travels I encountered hardly any nests, or even rumors of nests.

Perhaps the environment is indeed marginal, with coastal water too deep and too cold. But accumulating information does suggest that this may not be the whole story. The recent global assessment of hawksbill populations by Jeanne Mortimer and Marydele Donnelly gives some tantalizing fragments of information about hawksbills in the East Pacific. These fragments include reports that, in the 1950s, the crew of a small fishing boat in the Gulf of California might expect to catch five to seven hawksbills in one night. Farther south, in Colombia, hawksbills were reported to nest from Guapi south to the border with Ecuador. Most of the other records were very old, and confusion of the hawksbill with other species is always possible. However, a 2011 report by Mike Liles and colleagues in El Salvador identifies several hawksbill nesting beaches in Pacific El Salvador, especially in Bahía de Jiquilisco and Los Cóbanos. Collectively, there were about 300 nests per year. Few nests exceeded ninety eggs, so the reproductive output was markedly lower than in the Caribbean, where nests of 150–200 eggs are commonly seen. This suggests that, like leatherbacks, the species may be constrained in the East Pacific by the oligotrophic conditions and low temperatures and feeding options, which may serve to keep the turtles small and rare, with low clutch sizes. One hopes that the new Eastern Pacific Hawksbill Initiative will further develop this theory and identify areas worthy of further investigation, as described by Alexander Gaos and Ingrid Yañez in chapter 10.

On the other hand, some species, such as olive ridleys and green turtles, have been the focus of conservation efforts for decades and in many cases have shown signs of recovery. For example, although some of the ridley arribadas in eastern México—San Luis la Loma, Piedra de Tlalcoyunque, Bahía de Banderas—have collapsed, the surviving Pacific ones at Escobilla and Ostional are large enough for the overall East Pacific population to be larger than ever, even though bizarrely concentrated (see chapter 13, where Pamela Plotkin and colleagues meditate upon this remarkable phenomenon). *Chelonia* in the Galápagos Islands appears to be secure, and in the Gulf of California there are intensive and successful programs to involve the traditional turtle fishing communities in the conservation effort for black turtles by J. Nichols, Jeff Seminoff, and colleagues. Efforts are also afoot to reduce the very high incidental capture of young *Caretta* in the course of their astonishing migration from Japan to Baja California (see chapter 12), but the nesting populations in Japan have diminished, and there is still much conservation work to be done.

But it is the leatherbacks that we worry about the most, because their population collapse has been regionwide, intense, unresponsive to heroic beach protection efforts, and simply hard to interpret. One would have expected that two decades of good beach protection of adults, eggs, and hatchlings would have stabilized the population even if widespread incidental catch and mortality were a constant stress on the adults. In the Atlantic, incidental catch of leatherbacks is massive in waters around Trinidad, yet the nesting populations there have increased dramatically in recent decades, and other turtle populations with well-protected nesting grounds, such as green turtles at Tortuguero, Costa Rica, have maintained their numbers in the face of a daunting level of directed take in five or six Caribbean nations.

One is reduced to reflecting that the leatherback is a mysterious creature in many ways—the largest of all turtles, but very unturtlelike, even unreptilian, in structure and physiology, and with juvenile and subadult stages still completely unknown and undocumented. Possibly its nesting population fluctuations—dramatic sustained increase in the Atlantic, equally dramatic reduction in the East Pacific, not to mention extinction in Malaysia—reflect events simply above or outside the plane of human activities, positive or negative, such as the El Niño–Southern Oscillation, sunspots, or jellyfish blooms. Leatherbacks can surprise us positively as well as disappoint us—witness the appearance in Guyana in the year 2000 of nesting flotillas that generated not four or five turtles per night, as was customary, but fifty or more, an unexplained tenfold increase. Local fishermen had no explanation, nor did the biologists. Where did they come from?

So, after decades of neglect, much is now being done with sea turtles of the eastern Pacific, and much more is known about their biology and conservation status. There is hope, but there are still challenges. As you read *Sea Turtles of the Eastern Pacific: Advances in Research and Conservation*, many of these stories—both the encouraging and discouraging—come to life through the words of the cadre of biologists working to protect these marvelous creatures throughout the region. We can only hope that their stories, and actions, someday bear the fruits of species recovery.

—Peter C. H. Pritchard

Introduction

JEFFREY A. SEMINOFF AND BRYAN P. WALLACE

Why a book about sea turtles in the eastern Pacific Ocean, you ask? As the editors and authors of this book can (and will) tell you, there's no place on Earth like this region, and its turtles are no exception. Where else would one find the environmental paradox of bone-chilling water temperatures along the equator, which is exactly the situation off the coast of South America, including the "tropical" waters of the Galápagos Archipelago, one of the most biodiverse areas in the world? Where else could marine iguanas, green turtles, and penguins swim the same waters? Where else could ocean productivity fluctuate dramatically between supporting the largest-tonnage fisheries on the planet to being veritable "ocean deserts"? Where else could leatherbacks, green turtles, and olive ridleys shrink? Where else would hawksbills stop being coral reef critters and abandon their migratory ways? Nowhere but the eastern side of the world's largest ocean.

The Pacific coastline of the Americas runs like a backbone through the region, joining a dozen countries in a north–south axis along thousands of kilometers across two hemispheres. This coastline also anchors sea turtle life cycles, providing countless sandy spots for turtles to repeat their ancient reproductive rituals, and backstopping varied marine habitats where five sea turtle species—the leatherback (*Dermochelys coriacea*), the green turtle (*Chelonia mydas*), the hawksbill (*Eretmochelys imbricata*), the olive ridley (*Lepidochelys olivacea*), and the loggerhead (*Caretta caretta*)—swim, eat, grow, mate, live, and die (plates 1–10).

Like our colleagues in the region, we editors have spent our careers studying sea turtles—either by casting and monitoring nets in spectacular bays and lagoons (J.A.S.) or by plodding along kilometers of sandy beaches bathed in moon and starlight (B.P.W.) (see plates 13 and 14). Our first joint research venture turned out to be symbolic of the enigmatic nature of eastern Pacific sea turtles. Curious as to why leatherbacks in the eastern Pacific were inferior in size, reproductive output, and even population size and status to their western Atlantic sisters, we were looking for explanations in a seemingly unlikely place—turtle tissues. We compared telltale signatures of stable carbon and nitrogen isotopes in leatherback eggs from nesting populations in Pacific Costa Rica and the Caribbean to identify potential differences in nutrients that the mother turtles had acquired in their different feeding areas. We expected our analyses to reflect differences in food quality and/or availability that would correspond to the major differences between the two populations. Of course, as with most ecological research, especially of ocean-going animals, the results were not so straightforward. What our "sea turtle CSI" investigation uncovered were subtle clues that put turtles from each ocean basin at their respective scenes of the crime—in other words, those isotopic fingerprints revealed differences in turtle tissues that paralleled differences in oceanographic processes driving food-web functioning in the two feeding areas. Food availability is much less predictable in the eastern Pacific than in the western Atlantic (or in any other region, for that matter), resulting in smaller—and less fecund—leatherbacks. So, same species, different environments, very different results.

As it turns out, this anecdote reflects the broader truth about eastern Pacific sea turtles: because of unique traits of the ocean environment, sea turtles in this region are *different* from their counterparts elsewhere. Specifically, the boom-and-bust cycle of El Niño–Southern Oscillation (ENSO) overlaid upon longer-term oceanographic fluctuations has manifested in differences in growth rates, body sizes, and reproductive outputs among breeding populations of sea turtles that forage in the eastern Pacific relative to populations in other ocean regions. As a result, eastern Pacific sea turtles seem to ignore—or flagrantly violate—what textbooks say they should be or do as members of their species. You read above about how unique eastern Pacific leatherbacks are (and you will read much more about them in this book). Typical species descriptions say that hawksbills are sponge-eating, reef-dwelling, warm-water turtles, and green turtles are big, brightly colored seagrass grazers. Those descriptions would work anywhere else except the eastern Pacific, where such "typical" rules about how turtles look and live simply do not apply (plates 1–10).

And it appears that the uniqueness of the eastern Pacific and its sea turtles is nothing new. Evidence from genetics and geology shows that sea turtle populations in the eastern Pacific have waxed and waned over several long-term cycles, as environmental conditions in this vast region have swung dramatically between Ice Ages. Thus, the populations that exist today in the eastern Pacific Ocean were the last of their respective species to recolonize a previously inhabited region. For a sea turtle, life in the eastern Pacific has rarely been easy (see plates 11 and 12).

This has never been more true than today. All sea turtle populations are depleted relative to historic (i.e., recent decades) levels. While there are some encouraging examples of population recovery, some populations are teetering on extinction's precipice. As in other regions of the globe, the chief causes of these population declines in the eastern Pacific are human activities, including—but not restricted to—consumption of meat, eggs, and other turtle products, as well as incidental mortality in fishing gear (plates 11 and 12). The wild card that eastern Pacific turtles must deal with, however, is high variability and unpredictability of environmental conditions in the region that ultimately make them more susceptible to unsustainable levels of mortality. This one-two punch of environment and anthropogenic pressures complicates efforts by humans as well as turtles to restore populations in the region.

In the face of these challenges, turtles and humans forge ahead. Fortunately, a mosaic of conservation initiatives involving grassroots organizations, governmental wildlife agencies, and international consortia has led to recovery of some of the region's most ecologically and culturally important sea turtle populations (plates 15 and 16). For the first time in the decades-long history of sea turtle conservation in the eastern Pacific, efforts are becoming biologically holistic—not just targeting specific nesting beaches or foraging areas in isolation but taking into account population structures, connectivity among geographically and environmentally distinct habitats, and the influence of oceanography on life history. Furthermore, conservation efforts are being directed at the supply lines and demand for turtle products that historically contributed to the demise of these marvelous marine creatures. Sea turtle conservation is now more multidisciplinary than ever, joining sea turtle biologists, wildlife managers, economists, policy experts, subsistence fishermen, local communities, and many others in the fight to address challenges to persistence of sea turtles and marine biodiversity (plates 13 and 14). This change in fortune has resulted in sea turtles being among the most recognizable conservation "flagships" in the eastern Pacific and has united people and organizations across international boundaries, languages, and traditions.

This book provides not only an assessment of the present-day situation of sea turtle conservation in this singular region, but also a look back in time and a glance toward the future. More than thirty authors have contributed to the book's thirteen chapters, which include photos, figures, and color maps produced using up-to-date nesting data contributed to the State of the World's Sea Turtles (SWOT). We are greatly indebted to each of these contributors, all of whom are experts and colleagues who have shared the front lines in conservation for decades. The stories contained within these pages convey the experiences, setbacks, triumphs, and passions of an intriguing cast of characters. Leading the charge is Peter Pritchard, whose foreword provides a vivid backdrop for sea turtle conservation in the eastern Pacific through decades of research by boat, plane, car, and foot.

The book is organized into three parts: biogeography of eastern Pacific sea turtles, international management approaches, and field-based conservation models and signs of success. In the opening chapter of part I, Jeff Seminoff and colleagues explore the diversity and life styles of sea turtles in the eastern Pacific and detail how—and why—the region's sea turtles are unique relative to other ocean regions. This theme is picked up in the following chapter by Vince Saba, who describes the dynamic oceanography of the region, as well as sophisticated modeling approaches that elegantly elucidate the profound impacts of a fluctuating environment on turtles in the region. In chapter 3, Patricia Zárate guides us on an island-hopping expedition to illustrate the importance of insular sites throughout the eastern Pacific and shares her legacy of work in the Galápagos Archipelago. Rounding off the first section, Scott Benson and Peter Dutton shift the focus northward to U.S. waters, which, although not well known for sea turtles, host one of the most important—and best-studied—populations of foraging leatherbacks in the Pacific. We also learn about the green turtles residing in urbanized waters of southern California.

The second part of this book examines the role of management and international instruments such as treaties, international conventions, and nonbinding multinational accords in sea turtle conservation. Biologists

Mark Helvey and Christina Fahy, both of the U.S. National Marine Fisheries Service (NMFS), the U.S. fisheries management agency, convey first-hand accounts describing the fisheries management framework in the United States. The following chapter, by Martin Hall, Yonat Swimmer, and MariLuz Parga of the Inter-American Tropical Tuna Commission (the regional fisheries management organization in the eastern Pacific) and NMFS, describe their work to reduce sea turtle bycatch in artisanal and industrial fisheries of the region. Hall's team shows that there is no easy answer to the issue of sea turtle bycatch but that many options are available to limit turtle interactions, such as time area closures, innovative gear technology, and better information on where turtles are not present. Part II closes with a summary of the international instruments in the region by famed sea turtle guru Jack Frazier of the Smithsonian Institute. Frazier demonstrates that for any conservation effort to be successful over the long term, some level of top-down legislative framework is critical.

The third part consists of case studies, with personal accounts that bring us into the action, giving us a perspective from the front lines of some of the region's most successful and iconic sea turtle conservation projects. The six chapters in part III illustrate the many faces and facets of a sea turtle conservation effort, including engaging with local stakeholders for protection of turtles at nesting beaches, discerning important sites for nesting and in marine environments, and the business of monitoring populations as they rebound toward recovery. In chapter 8, Bryan Wallace and Rotney Piedra describe the difficult balancing act between human interests and protecting leatherbacks and their nesting beaches in Costa Rica's well-known Parque Nacional Marino Las Baulas. We then follow the coast northward, to México, where Ana Barragán describes the growth in conservation of sea turtle nesting beaches in México. From the initial student trip that led to science's discovery of a major nesting beach to the launch of a cohesive, regionwide leatherback conservation network, we learn about what it takes to make a conservation effort prosper. In chapter 10, Alexander Gaos and Ingrid Yañez detail the compelling evolution of the Eastern Pacific Hawksbill Initiative (Iniciativa Carey del Pacifico Oriental, or ICAPO, in Spanish), which is perhaps the most cohesive sea turtle conservation network in the region. The authors delight us with a compelling account of how hawksbill turtles, once thought to be on the verge of extinction in the eastern Pacific (if not already functionally extinct), are holding on in unexpected ways and places. Carlos Delgado-Trejo and Javier Alvarado-Díaz follow by tracing the past decline of green turtles—known locally as black turtles—in the state of Michoacán, México, due to the country's once-legal turtle fishery, through to today, as the population makes a remarkable comeback to nesting levels not seen in more than two decades. Moving from the beach to the water, chapter 12 by Hoyt Peckham and David Maldonado Díaz details the creative and persistent efforts to reduce the devastating impacts of an artisanal longline fishery in Baja México on loggerhead turtles, a local population that has its nesting beach origins 10,000 km away in Japan. It's a true conservation success story of our efforts to create sustainable fisheries, and this case study underscores how local initiatives with modest beginnings can develop into world-renown conservation models. Conservation success stories continue with the final chapter, by Pam Plotkin, Raquel Briseño-Dueñas, and Alberto Abreu-Grobois, who tell of the olive ridley, the world's most abundant sea turtle, that was long the focus of a harvest bonanza throughout the region. Considering that literally hundreds of thousands of ridleys were hunted for decades, the rebound of the largest mass nesting sites on the planet is a remarkable testimonial of the resilience of sea turtles.

We are confident that this text will be a valuable resource—a reference manual of sorts—for biologists, conservation managers, students, and turtle enthusiasts alike that have an interest in sea turtles, the eastern Pacific, or (we hope) both. For this we owe a debt of gratitude to the contributors to this book, who collectively represent more than 500 years of experience in researching and conserving sea turtles in the eastern Pacific. We have entrusted the development of this book to these people, all of whom are experts in their respective areas of interest and, conveniently and fortunately for the editors, close colleagues and friends. Thanks to all of you for putting up with deadlines, prodding, edits, and our attempts at cat herding—the next round is on us.

We also thank the multitude of agencies and organizations that have facilitated the development of this book: National Oceanic and Atmospheric Administration, Conservation International, and the International Sea Turtle Society. Also, numerous individuals have helped make this book possible, including Allyson Carter, Andrew DiMatteo, Katie Wedemeyer, and Rick Brusca. We give special thanks to Katie Wedemeyer and Trish Watson for their amazing assistance with the final editing of this volume.

And, of course, thanks to the turtles, for the inspiration, for living in such exciting, beautiful, and interesting places for us to visit, and for sharing your world—if only a little bit—with us. We promise to keep working hard.

Lastly, we are incredibly thankful to our wives, Jennifer Gilmore and Jessica Wallace, who have put up with our long hours and our constant travels to the region and have calmed us when our frustrations were boiling over. If it weren't for their best-friendship and unwavering support, this book—and our barnstorming the eastern Pacific—would have never been possible.

PART ONE

Sea Turtle Biology and Human Dimensions

1

Biology and Conservation of Sea Turtles in the Eastern Pacific Ocean

A General Overview

JEFFREY A. SEMINOFF, JOANNA ALFARO-SHIGUETO, DIEGO AMOROCHO, RANDALL ARAUZ, ANDRES BAQUERO GALLEGOS, DIDIHER CHACÓN CHAVERRI, ALEXANDER R. GAOS, SHALEYLA KELEZ, JEFFREY C. MANGEL, JOSE URTEAGA, AND BRYAN P. WALLACE

Summary

The waters of the eastern Pacific Ocean (hereafter EP) host important feeding and nesting areas for four sea turtle species: the leatherback (*Dermochelys coriacea*), the green turtle (*Chelonia mydas*), the hawksbill (*Eretmochelys imbricata*), and the olive ridley (*Lepidochelys olivacea*) (see plates 1–10). In addition, a fifth species, the loggerhead turtle (*Caretta caretta*), feeds in the northern- and southernmost latitudes of the EP but nests on distant beaches in the western Pacific. While green turtles and hawksbills depend on shallow coastal habitats for food resources, leatherbacks, loggerheads, and olive ridleys are more generally tied to offshore pelagic waters for foraging. Because of the unique oceanography of these habitats in the EP, sea turtles in this region have evolved distinct biology, morphology, and behavior compared with their counterparts in other parts of the world.

Through the exploitation of eggs and turtles as food, as well as incidental mortality in fishing gear, sea turtle populations have been subjected to human impacts throughout the EP (plates 11 and 12). While some remote nesting rookeries have withstood these threats and remain relatively intact, most populations have declined to some extent, many severely so. Efforts to recover populations started in the mid-1960s with the initiation of nesting beach conservation camps at a few major nesting beaches in México. By the late 1990s, sea turtle conservation camps were established at hundreds of beaches throughout the EP, and the number of similar operations increases every year. Over the last decade, national and regional sea turtle conservation strategies and action plans have been developed and enacted, and grassroots sea turtle conservation has prospered through increased networking and coordination among various stakeholders (e.g., biologists, wildlife managers, and local communities) throughout the region (plates 13–16). These efforts have contributed to significant recoveries of olive ridley nesting rookeries in the EP and the partial restoration of green turtle rookeries in México. However, in some cases—particularly with leatherbacks and hawksbills—the outlook remains grim. Recovery of these nesting assemblages will require prompt, broad-based action on the part of countries on whose beaches these turtles nest (plates 3 and 7), and a redoubling of efforts to mitigate the impacts of industrial and artisanal fisheries bycatch in the nearshore and offshore waters within which they forage.

Introduction

The eastern Pacific is among the most important areas on Earth for marine turtles (plate 1). The vast region hosts five of the world's seven sea turtle species: the leatherback, the green turtle, the hawksbill, the olive ridley, and the loggerhead (fig. 1.1; plates 2, 4, 6, 8, 10). Over the course of their lives, individual turtles of all five species inhabit broadly separated localities—from marine to terrestrial ecosystems—in the EP. Foraging turtles access a diversity of marine habitats in temperate, subtropical, and tropical zones of the EP, including marine algae and seagrass pastures, coral and rocky reefs, mangrove estuaries, and open water. Nesting turtles haul ashore for egg-laying on sandy beaches (plates 2, 4, 6, 8), primarily in tropical stretches along the mainland coast but also on several of the offshore islands and archipelagos (see chapter 3).

Regardless of the particular habitat, one thing is clear: the EP region is like no other in terms of both oceanography and biogeography. It is a massive region (plate 1), with upwards of 20,000 km (12,500 miles) of coastline and a total area of nearly 60 million square kilometers (about 38

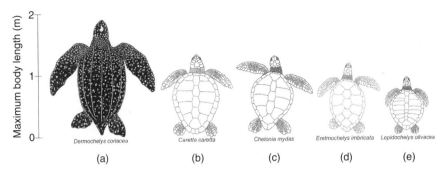

FIGURE 1.1. The five sea turtle species found in the eastern Pacific: the leatherback (*Dermochelys coriacea*), the loggerhead turtle (*Caretta caretta*), the green turtle (*Chelonia mydas*), the hawksbill (*Eretmochelys imbricata*), and the olive ridley (*Lepidochelys olivacea*). Maximum body size is for females at their respective nesting beaches. Illustrations courtesy of Thomas MacFarland. See plates 2, 4, 6, 8, and 10 for photographs of each species.

million square miles). Complex bathymetry and substantial spatiotemporal variability in oceanographic characteristics prevail throughout (Fiedler 1992; Chavez et al. 1999). The lack of an extensive continental shelf and expansive area with deep oceanic waters (to 5,000 meters) in the EP contribute to this system having the unusual combination of relatively low overall nutrient levels (i.e., oligotrophic) but reasonable productivity, which is sparked by mesoscale upwelling events throughout the region (Barber 2001).

The EP has several hotspots of wind-driven upwelling, often aided by ocean currents, which bring cool, nutrient-rich waters from the deepest zones of the water column to the ocean surface (Fiedler et al. 1991). For example, Perú hosts a seasonal tongue of freshly upwelled water (via the Humboldt Current), which extends into equatorial waters (Strub et al. 1995, 1998). Farther north, the coastal shelf is even narrower off the U.S. coast, and in areas like Monterey Bay, California, there is seasonal upwelling so close to shore that many offshore species such as blue whales, orcas, and leatherback turtles enter waters only a few kilometers from the coast (Croll et al. 2005; see chapter 4 in this volume). Not surprisingly, these two marine areas—Perú and central California—have been epicenters of major fisheries over the years, none more notable than the sardine fishery off California—made famous by John Steinbeck's *Cannery Row*—or its southern counterpart, the Peruvian anchovy fishery, whose crash in the 1940s (see Sharp and McLain 1993) led to the discovery and description of the El Niño–Southern Oscillation (ENSO). Current and wind-driven upwelling hotspots are also present near the Galápagos Archipelago (Palacios 2002), the Costa Rica Dome (Fiedler 1992), and the Gulf of California (Alvarez-Borrego 2010). These sites, as well as the coasts of Perú and the United States, are important areas of aggregation for a variety of marine organisms (Worm et al. 2005; Pennington et al. 2006), including sea turtles.

In addition to the spatial patchiness and variable nature of primary productivity in the EP, there is substantial temporal variation in local and regional oceanography. The most notorious of these processes is ENSO, which, although centered along the southeastern Pacific Ocean, relaxes upwelling events throughout the EP, resulting in increased surface water temperatures and lower primary productivity in the entire EP region (Chavez et al. 1999). Also characteristic of the EP are decadal oscillations in mean surface water temperatures that bring distinct warm or "positive" phases and cool or "negative" phases to waters throughout the region. These oscillatory changes occur on different spatial scales, including the Pacific Decadal Oscillation (Mantua et al. 1997) centered north of 20°N, and the Interdecadal Pacific Oscillation (Ghil and Vautard 1991), which broadly manifests throughout tropical and temperate waters of the entire EP.

The tremendous variability in oceanographic conditions through time and space within the EP has profound biological consequences on the region's biota (Chavez et al. 1999). In several cases, sea turtles respond behaviorally and/or demographically to these changes. For example, leatherback turtles may delay reproduction for a season or more because of the limitation in available resources that occurs during ENSO periods (Saba et al. 2007, 2008a, 2008b; see chapter 2). With respect to foraging behavior, leatherbacks off the U.S. West Coast avoid entering their typical nearshore habitats when waters are overly warm (Benson et al. 2007), a habitat use pattern that has been linked to positive phases of ENSO and the Northern Oscillation Index (Schwing et al. 2002). In contrast, green turtles aggregate in greater density along the Peruvian coast during the warm ENSO periods, presumably because of the warming of local waters that are usually too frigid for this species. This trend is reflected in the substantial increase in the local harvest rate of green turtles during ENSO events (Quiñones et al. 2010).

In addition to adapting their behaviors in response to changing oceanography, sea turtles in the EP in many cases possess unique morphological and demographic traits relative to their conspecifics elsewhere. Leatherbacks and green turtles mature at smaller sizes and lay fewer eggs per clutch than do their Atlantic and Indian Ocean counterparts, which appears to be due to lesser productivity of the EP and its impacts on the lives, body development, and reproductive periodicity of these two species (Wallace et al. 2006; Saba et al. 2008b, Suryan et al. 2009, Wallace and Saba 2009). Likewise, although adult hawksbill turtles are strongly tied to coral reefs in the rest of the world (Plotkin 2003), in the EP they often reside in mangrove estuaries (see chapter 10; Gaos et al. 2011), a preference that has likely evolved in response to the rarity of major coral reefs in the EP (Glynn 2001).

Together, these oceanographic features underscore just how dynamic the EP is and how important its waters are for sea turtle populations. However, turtles are not the only beings that have been supported by ocean productivity in the EP; extensive human settlement and the resource extraction that often goes with it have long been part of the region. Clearly, humans have exacted a heavy toll on sea turtles in this region (plates 11 and 12), and all five species have suffered to some extent. As described elsewhere in this volume, the hardest hit of these have been leatherbacks (see chapters 8 and 9) and hawksbills (chapter 10). Loggerheads have also decreased substantially, which is largely based on nesting trends in the western Pacific since no nesting occurs in the EP, but is at least partially due to high bycatch rates off northwest Pacific México (Peckham et al. 2008; see chapter 12). Also sustaining heavy declines in decades past, but now inching upward in population numbers, are green turtles (chapter 11) and olive ridleys (chapter 13).

Despite the apparently dire conservation status for several EP populations, sea turtles have demonstrated time and again that they are incredibly resilient to human impacts—in many cases for the same reasons that they have been so vulnerable and so slow to respond to conservation efforts: broad distribution, delayed maturation, and great longevity. These dichotomous consequences of sea turtle traits suggest that although there is still time to prevent local extinctions, there is no time to spare. Transboundary conservation efforts must emphasize collaborative management of marine activities within the EP, as well as with other regions of the Pacific Ocean, recognizing that sea turtles are indifferent to geopolitical boundaries. Specifically, effective conservation must involve international partnerships and information exchange, and holistic, sustainable marine resource management, thereby enabling the development of protection strategies that encompass the entire life cycles of these ocean connectors (Seminoff and Zárate 2008; Shillinger et al. 2008). In this chapter we describe the diverse life histories of sea turtles in the EP, identify threats to their survival (plates 11 and 12), describe the legacy of sea turtle conservation that has continued for the last forty years (plates 15 and 16), and suggest a number of conservation alternatives that will promote the recovery of local sea turtle populations.

Species Accounts

The waters of the EP host five of the world's seven sea turtle species (fig. 1.1); the only species not found here are the flatback turtle (*Natator depressus*), an Australian endemic, and the Kemp's ridley (*Lepidochelys kempii*), which is quasi-endemic to the Gulf of México. These seven species are distributed among two families: Dermochelyidae, with its sole member, the leatherback, and Chelonidae, which has the remaining six hard-shelled species (Ernst and Barbour 1989). In general, the life history of these turtles follows a fundamental pattern of slow growth, delayed maturity, and broad-scale movement. However, each of the species in the EP is also unique in its own way relative to conspecifics in other ocean regions. In the following accounts, we describe the standout aspects of the biology, threats, and conservation of each species and elaborate on why the EP is like no other place on Earth (see color plates).

Dermochelys coriacea (English: leatherback; Spanish: baula, dorso de cuero, galápago, laúd, garapacho, canal, siete filos, siete quillas, tinglar; plates 2 and 3)

Leatherbacks roaming the oceans today are virtually carbon copies of their ancestors from more than 100 million years ago; they *are* living dinosaurs. By far the largest of all living sea turtles, leatherbacks are easily distinguished from other extant sea turtles by their massive size and lack of a completely fused, bony shell. Leatherbacks can attain a carapace length of nearly two meters (seven feet) and a weight of up to 800 kg (1,760 lbs); they have a teardrop-shaped body and a somewhat flexible, smooth-skinned shell with dorsal and ventral keels. Together, these features give leatherbacks unparalleled hydrodynamics and swimming efficiency among turtles (Jones et al. 2011). In addition, leatherbacks have peripheral insulation, counter-current blood flow, and the ability to utilize heat generated internally during exercise. This suite of adaptations results in a unique thermoregulatory system in leatherbacks (Paladino et al. 1990; Bostrom et al. 2010) that prevents excessive cooling or overheating and enables individuals to access abundant food resources in pelagic waters that are too cold for other sea turtle species (Wallace and Jones 2008; see chapter 4 in this volume).

Leatherback nesting occurs along the Pacific coast of the Americas from México to Ecuador (plate 3), but concentrates in two beach complexes—one in Costa Rica (see chapter 8) and the other in México (see chapter 9)—with a third, lesser nesting assemblage in Nicaragua (Urteaga and Chacón 2007; see plate 3). The number of nesting female leatherbacks in the EP has declined by more than 90 percent over the past twenty years, which is in stark contrast to leatherback populations in the Atlantic Ocean, most of which are either stable or increasing and are much more abundant (U.S. National Marine Fisheries Service and U.S. Fish and Wildlife Service 2007; Saba et al. 2008a; Wallace and Saba 2009). Bycatch in commercial and artisanal and drift net fisheries in distant feeding areas in the southeast Pacific (Hays-Brown and Brown 1982; Frazier and Brito 1990; Eckert and Sarti 1997; Arauz et al. 2000; Spotila et al. 2000; Alfaro-Shigueto et al. 2007, 2011) and decades of comprehensive egg poaching on nesting beaches (Santidrián Tomillo et al. 2007, 2008; Sarti Martínez et al. 2007), occurring against the backdrop of variable resource availability in the EP, have been implicated in the regionwide population crash (Wallace and Saba 2009; see chapters 8 and 9 in this volume). The situation is so dire that some researchers fear that this regional population will become extinct in the near future (Spotila et al. 2000).

Chelonia mydas (English: green turtle, East Pacific green turtle, black turtle; Spanish: tortuga negra, tortuga prieta, tortuga blanca, tortuga verde, tortuga; plates 4 and 5)

EP green turtles have the broadest coastal distribution of any sea turtle in the region; they occur in bays, estuaries, and lagoons from southern California to Chile. They are not exclusive herbivores, as reported for green turtles elsewhere (Bjorndal 1997), and instead rely heavily on invertebrate prey such as sponges, mollusks, and jellies (Seminoff et al. 2002a; López-Mendilaharsu et al. 2005; Amorocho and Reina 2007, 2008; Carrión-Cortez et al. 2010). In addition to their novel diet, green turtles also have several unique behaviors. Most notably, individuals in the higher latitudes become dormant during cold periods and even undertake a hibernation-like behavior known as overwintering during which individuals spend hours, days, or even weeks underwater during the coldest of periods (Felger et al. 1976).

Green turtles in the EP are the most taxonomically controversial species in the Pacific. Compared with green turtles in other ocean regions, those in the EP are smaller (up to 90 cm in shell length and 150 kg in body weight), are uniquely shaped and colored, and lay fewer eggs (Hirth 1990). These traits have led some scientists to declare green turtles in the EP as a separate species, *Chelonia agassizii* (Pritchard 1999), but those citing genetic data argue it is, at most, a subspecies (i.e., *Chelonia mydas agassizii*) and perhaps nothing more than a unique population of green turtles (Karl and Bowen 1999). When considering the tremendous variation in what an EP green turtle looks like, the question of species-level differences is not cut-and-dry: while most individuals in the EP possess melanism distinct from green turtles in other areas, some possess the same countershading and ivory-colored plastron found in the species worldwide (plate 4).

Green turtle nesting ranges from the cape region of México's Baja California peninsula to Ecuador (Wester et al. 2010; plate 5), with major nesting rookeries in Michoacán, México (plate 5; see chapter 11), and the Galápagos Islands (see chapter 3). Lesser nesting sites are found in the Revillagigedo Islands off central México and several small beaches along the Central American coastline, particularly from El Salvador through northwest Costa Rica. While long-term nesting trends are available only for Michoacán, these data indicate an encouraging sign: the population has been on the rise since 2000, and annual nesting levels at this Mexican rookery in the beginning of the 2010s are the largest they have been since the early 1980s. Certainly, much of this recovery can be attributed to the positive impacts of nesting beach protection that have been ongoing since 1979 (see chapter 11), but ongoing grassroots conservation efforts in foraging areas that have presumably cut down on the rate of illegal hunting of green turtles (Nichols et al. 2002; López-Castro et al. 2010) have also likely helped the situation.

Eretmochelys imbricata (English: hawksbill turtle; Spanish: tortuga carey; plates 6 and 7)

Hawksbill turtles are the most poorly understood sea turtle species in the EP. Until recently, no major nesting beaches had been identified in the entire region, and hawksbills were thought to nest only sporadically throughout tropical latitudes of the EP. However, efforts of ICAPO (Iniciativa Carey del Pacifico Oriental [Eastern Pacific Hawksbill Initiative]; see chapter 10) since 2008 have revealed at least three important nesting areas, in El Salvador (three major sites), Nicaragua (Estero Padre Ramos), and Ecuador (Parque Nacional Machalilla), with perhaps a few other important nesting sites in as-of-yet unexplored areas such as Panamá and the Tres Marías Islands off central México (plate 7; Gaos et al. 2010; Liles et al. 2011).

In another major oddity for the biology of sea turtles in the EP relative to elsewhere, it appears that hawksbills may have an entirely different set of habitat preferences than their counterparts in other parts of the world. While coral reefs have been the primary sites for foraging hawksbills worldwide—and indeed, juveniles in the EP have been found in the few coral reefs in the EP—it appears that adult hawksbills prefer to forage in mangrove-lined estuaries or rocky reef areas (Gaos et al. 2011; A. R. Gaos, unpublished data). What's more, despite the paradigm that "all adults migrate," recent satellite telemetry work suggests that at least some hawksbill turtles are nonmigratory and instead appear to live their entire lives within a circumscribed area (A. R. Gaos, unpublished data).

Hawksbills in the EP reach 90 cm in length and 100 kg in weight, and the species is best described by the coffee- and caramel-colored scutes on the carapace and plastron, conspicuously overlaid in imbricate fashion, particularly evident on juveniles and subadults. Hawksbills stand alone as being the only sea turtle species fervently sought after primarily for its shell and only secondarily for its meat and skin. The beauty and workability of the shell's keratinaceous plates, known as tortoiseshell, *carey*, *penca*, or *bekko*, have made the hawksbill turtle the target of a colossal global harvest for artisanal uses (Groombridge and Luxmoore 1989), and the EP is no exception (Gaos et al. 2010). The material is fashioned into combs, pendants, and fine jewelry (plate 12).

Clearly, the harvest of hawksbills for eggs, meat, and tortoiseshell has had an impact on population numbers in the EP. Indeed, the species is rare to absent in many localities. However, it is still not clear if the apparently small size of the hawksbill population is due exclusively to decades—if not centuries—of harvest, or if the EP simply was never a preferred area because of its general lack of coral reefs, which have been long considered the defining habitat of hawksbills (Carr et al. 1966; van Dam and Diez 1996). Certainly both factors contribute to the low abundance of hawksbills in the region.

Caretta caretta (English: loggerhead turtle; Spanish: tortuga amarilla, cabezona, caguama, caretta, javelina, perica; plate 10)

The loggerhead is an enigmatic species. The species does not nest in the EP but instead crosses the Pacific Ocean to remain in EP waters as juveniles, only to leave again upon reaching adulthood, never to return (Kamezaki et al. 1997; Sakamoto et al. 1997). In northern latitudes off Pacific México, an area where loggerheads have been studied for many years, scientists originally speculated that there must have been some undiscovered nesting sites in the EP because it was hard to fathom that a turtle would cross the entire Pacific Ocean, particularly a small juvenile turtle, as are sometimes seen of Baja (Peckham et al. 2008). The first indication that "Mexican loggerheads" were also "Japanese loggerheads" came in the late 1980s via a fortuitous flipper tag recovery from a turtle that was marked in Japan and found in Mexican waters (Uchida and Teruya 1988). Direct evidence has since been gathered through additional genetic analysis (Bowen et al. 1995), flipper tag recoveries (Resendiz et al. 1998), and satellite telemetry (Nichols et al. 2000; Kobayashi et al. 2008), and the story of this North Pacific loggerhead stock has been followed closely by resource managers and the fishers from across the northern Pacific whose livelihoods have been jeopardized by loggerhead bycatch and U.S. fisheries closures (see chapters 5 and 12).

In southern latitudes, off Perú and Chile, the species was not even confirmed as being present until very recently (Kelez et al. 2003; Alfaro-

Shigueto et al. 2004). For many years the presence of this species was obscured by the fact that it was commonly confused with olive ridley turtles (Frazier 1985). Today, however, the presence of loggerheads off Chile and Perú is common knowledge, largely because of increased focused scientific research on the population (Pajuelo et al. 2010; Kelez Sara 2011; Mangel et al. 2011) and from increased reporting of interactions between loggerheads and fishing gear (Alfaro-Shigueto et al. 2004, 2008, 2010, 2011; Donoso and Dutton 2010; Kelez et al. 2007, 2008). Recent information from fisheries interactions has confirmed that loggerheads are also occasionally present in waters of Ecuador (Alava 2008). From these findings and associated research, it has become clear that life history of loggerheads in the South Pacific seems to parallel that of the North Pacific population, where posthatchlings migrate to the coasts of South America from nesting beaches in Australia and New Caledonia, as evidenced by mitochondrial DNA analysis (Boyle et al. 2009). As in the northern EP, fisheries bycatch in the southern EP is very problematic, and the strategies to combat these impacts are less obvious because the industrial and artisanal fisheries operating here are less regulated than those operating in the North Pacific (Alfaro-Shigueto et al. 2010).

Relative to historic abundance, today's loggerhead nesting populations in the North Pacific (Japan) and South Pacific (Australia) are substantially reduced. One ray of hope is that nesting in Japan may be showing early signs of recovery. Although it is much too soon to distinguish a trend, nesting for the species in 2008 and 2009 was the highest it had been in more than a decade (Y. Matsuzawa, pers. commun.). This promising news is likely related to near-complete protection of Japan's remaining nesting beaches, the closure of the North Pacific high-seas drift net fishery, and improved management of the remaining U.S.-based fisheries in the North Pacific. In México, one project that bears mention is the local initiative of ProCaguama, whose scientists work with local fishers to switch fishing gears from highly lethal bottom-set longlines and drift nets to single-line gear to reduce bycatch mortality off Baja (Peckham et al. 2007; see chapter 10). However, tempering this news is the fact that a relatively new pound net fishery has proliferated throughout much of Japan that has been catching—and killing—juvenile and adult loggerheads (S. H. Peckham, pers. commun.).

Lepidochelys olivacea (English: olive ridley turtle; Spanish: tortuga golfina, lora, pico de loro, parlama, paslama; plates 8 and 9)

The olive ridley is by far the most abundant sea turtle species in the EP, and probably in the world. A recent study by Eguchi et al. (2007) using shipboard observations estimated the at-sea population at more than two million turtles! Olive ridleys have been the focus of considerable attention recently because of the recovery of some nesting populations along the Pacific coast (see chapter 13). Olive ridleys are also the smallest sea turtle in the Pacific Ocean, reaching 70 cm in length and 60 kg in body weight. They forage on a variety of invertebrates in both coastal and pelagic habitats of the EP (Kopitsky et al. 2005).

Perhaps most intriguing about the species are its multiple nesting strategies. Olive ridleys are usually solitary nesters, but in several places they come ashore in massive, synchronized aggregations of females called *arribadas* (Spanish for "arrivals"; plate 8). During these *arribada* events thousands of turtles—sometimes hundreds of thousands—come ashore almost simultaneously during the same few nights each month during the nesting season to lay their eggs. While solitary nesting beaches are dispersed throughout the region—olive ridleys have the broadest nesting range of any sea turtle in the EP, on beaches from northwestern México to northern Perú—*arribada* beaches are less widespread and occur at only a few sites, in southern México (Escobilla), Costa Rica (Ostional and Nancite), Nicaragua (Chacocente and La Flor), and Panamá (La Marinera and Isla Caña) (Plate 9).

Sea Turtle Life History in the EP

Sea turtles in the EP, much like elsewhere, have a remarkable life cycle—individuals inhabit broadly separated localities over the course of their lives (Fig. 1.2). From hatchling to adult, sea turtles undergo ontogenetic shifts in habitat use that encompass nesting beaches, juvenile developmental habitats, and adult foraging areas. Upon emerging from their nest, hatchlings depart the nesting beach to begin the oceanic phase of their life cycle, floating passively for a year or more in major current systems (gyres) that serve as open ocean developmental grounds. Not only do these young turtles travel amazing distances, but they also connect the EP with the rest

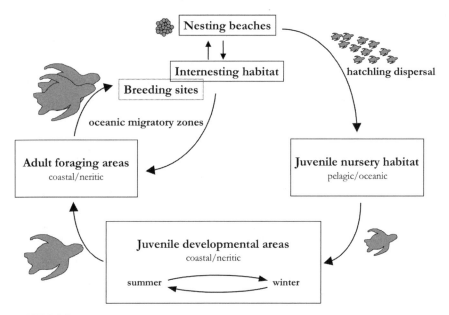

FIGURE 1.2. Generalized life history of the hard-shelled sea turtles. Not included in this group is the leatherback, a species with juvenile developmental areas and adult foraging areas in the oceanic zone.

of the Pacific. Loggerhead turtles provide a prime example of this trans-Pacific phenomenon. After emerging from nesting beaches in Japan and Australia, hatchling loggerheads become entrained in the North and South Pacific gyres, respectively, and ride the strong ocean currents to offshore developmental grounds in the high seas of the central North and South Pacific (Bowen et al. 1995; Boyle et al. 2009). Some juvenile loggerheads ride the eastbound surface currents all the way to the EP and have been found foraging in waters of México, Perú, Chile, and Ecuador (see below). Trans-Pacific current-based transfer also affects green turtles, only in this case it goes both ways: some green turtles from South Pacific rookeries in French Polynesia move eastward and find their way into foraging habitats of the southeastern Pacific (Amorocho 2009; see chapter 3), whereas some green turtles from the EP move all the way to western Pacific waters off Australia and Japan, presumably via westbound equatorial surface currents (Omuta 2005; C. Limpus, pers. commun.).

While living in the high seas, oceanic juveniles commonly associate with offshore rafts of sargassum and other floating debris that aggregate at frontal areas. These "mini ecosystems" provide important food resources and offer protection from predators (Nichols et al. 2001). After several or more years living in oceanic environments, species such as green and hawksbill turtles recruit to neritic (i.e., nearshore) habitats, where they feed and grow to maturity (Seminoff et al. 2002a, 2003; López-Mendilaharsu et al. 2005; Amorocho and Reina 2007, 2008). During these adolescent years, individuals often go through a series of ontogenic, or life-stage-based, shifts in habitat preference. In the case of green turtles, the smallest of neritic juveniles often prefer shallow protected bays or lagoons (López-Mendilaharsu et al. 2005). As they grow in size, they shift to more open, high-energy coastal areas, where they reside within specific home range areas (Seminoff et al. 2002b). For Pacific loggerheads, their developmental migrations cover broader areas: juveniles feed and grow in the EP and then, upon maturity, return to the western Pacific to nest and spend the rest of their lives as adults (Bowen et al. 1995; Boyle et al. 2009). With respect to olive ridleys and leatherbacks, the habitat transitions are less distinct, as these species commonly remain in offshore waters for most of their lives, even as large juveniles and adults (Plotkin 2003; Shillinger et al. 2008). Hawksbill habitat shifts are unknown; whether they follow one of the patterns demonstrated by these other species or exhibit something new remains in question.

Upon reaching maturity, which can take anywhere from 10 and 35 years, depending on the species (Chaloupka and Musick 1997), sea turtles carry out breeding migrations every few years that take them back and forth between their adult foraging grounds and the same beach (or region) from which they originally emerged as hatchlings decades earlier (Lohmann et al. 1997). This biological phenomenon, known as philopatry, is among the most amazing examples of animal imprinting in the natural world and in many cases results in turtles traveling tens of thousands of kilometers over the course of their lives (Meylan et al. 1990; Bowen et al. 2004).

Various migratory strategies are undertaken by adult sea turtles, and in all cases these migrations establish connectivity among disparate habitats within the EP. More coastal species such as hawksbills and green turtles tend to migrate along continental shelf regions, moving hundreds to a thousand or more kilometers along the EP coastline. Green turtles that migrate from the continental rookeries in México and Costa Rica generally stay within a hundred kilometers of the coastline as they move north and south before and after nesting (G. Blanco, unpublished data; J. Nichols, unpublished data). Such is also the case for hawksbill turtles that disperse from continental rookeries in El Salvador, Nicaragua, and Ecuador to continental foraging sites within a few hundred kilometers or less from their nesting beaches (A. R. Gaos, unpublished data).

Adult migrations also bring sea turtles through vast stretches of the high seas. For example, green turtles that nest in the Galápagos, Revillagigedos, and Tres Marías Islands commonly move to continental foraging habitats that are more than a thousand kilometers from their insular nesting sites (Seminoff et al. 2008; P. Dutton, unpublished data). Also, sometimes these offshore regions *are* the migratory destination for turtles. In the case of leatherbacks and olive ridleys, postnesting turtles may simply move into the oceanic realm and stay there, their "foraging area" being less defined by geographic landmarks and hinged more on prevailing oceanography and marine productivity (Plotkin et al. 1995; Eckert and Sarti 1997; Eguchi et al. 2007; Shillinger et al. 2008). However, at least some leatherbacks do indeed migrate from nesting sites to relatively nearshore areas within the EP, only in this case they come from nesting beaches in Indonesia and the Solomon Islands, some 16,000 km from their seasonal foraging grounds of the U.S. West Coast (see chapter 4).

The complex life history of sea turtles, their shifts in habitat preference, adult migrations, and need to nest on terrestrial beaches underscore the fact that virtually any piece of coast or ocean in the EP can play an important role for sea turtles. As we describe later in this chapter, research and conservation programs must therefore focus on multiple habitats in multiple countries, with efforts throughout the region enlisting a diversity of stakeholders (see plates 13–16).

Human Threats to Sea Turtles

Sea Turtle Use in the Eastern Pacific

As is so often the case with many of our natural resources, human uses of sea turtle products that began as sustainable in decades and centuries past have evolved to have devastating negative impacts, largely the result of

TABLE 1.1. Conservation status of sea turtles occurring in the eastern Pacific.

Species	ESA Status	IUCN Red List	CITES	Level in EP Historical	Level in EP Current
Leatherback turtle, *Dermochelys coriacea*	Endangered	Critically endangered	Appendix 1	Common	Extremely rare
Green turtle, *Chelonia mydas*	Endangered*	Endangered**	Appendix 1	Highly abundant	Common
Hawksbill turtle, *Eretmochelys imbricata*	Endangered	Critically endangered	Appendix 1	Uncommon at most sites	Extremely rare
Loggerhead turtle, *Caretta caretta*	Threatened***	Endangered	Appendix 1	Common	Uncommon
Olive Ridley turtle, *Lepidochelys olivacea*	Threatened	Vulnerable	Appendix 1	Highly abundant	Common

Abbreviations: CITES, Convention on International Trade in Endangered Species of Wild Fauna and Flora (also known as the Washington Convention); ESA, U.S. Endangered Species Act of 1973; IUCN, International Union for Conservation of Nature.
*The ESA lists all but the Pacific México population of green turtles as Threatened.
**The IUCN Red List of Threatened Species lists the Hawaiian green turtle subpopulation as Near Threatened.
***The ESA lists the North Pacific and South Pacific loggerhead distinct population segments as Endangered.

the ever burgeoning human population (plates 11 and 12). Traditional human customs and industrial advances have exacted a heavy toll on sea turtle populations in the EP; all are listed as Vulnerable, Endangered, or Critically Endangered on the International Union for Conservation of Nature's Red List and as either Threatened or Endangered on the U.S. Endangered Species Act (Table 1.1). All life phases of sea turtles, from egg to hatchling to juvenile to reproductive adult to senescent adult, are susceptible in one form or another to anthropogenic threats. In earlier times turtles provided a source of nutrition, medicine, and material to human societies. Sea turtles were, and continue to be, harvested primarily for their meat, although other products have served important nonfood uses. Sea turtle oil was for many years used as a cold remedy, and the meat, eggs, and other products have been highly valued for their supposed aphrodisiacal qualities, beliefs that strongly persist in the EP and around the globe (Campbell 2003). These so–called traditions sustain the culture of consumption; turtle meat is the preferred food of many *fiestas*, *quinceñeras*, and (in México) *Semana Santa* (Holy Week) gatherings, and eggs are a common *bocadita* (finger food) in bars and restaurants across the EP. Indeed, the two most profound impacts to sea turtles have come from the collection of eggs and the directed take of turtles from nesting beaches and foraging areas in the region.

Decades of egg harvest have decimated many nesting assemblages in the EP (plate 12). This harvest has taken many forms, from single families collecting eggs for subsistence use, to "professional" egg collectors taking every last egg they can get their hands on to sell at market (see chapters 8 and 9). In some countries and localities, egg harvest has been legal, while in others it is illegal but persistent because of lack of enforcement. In addition to increasing human populations, the high monetary value of eggs, a consistent market demand, and severe poverty in many of the countries in the EP where sea turtles are found are compounding these pressures.

Industrial-scale exploitation of sea turtles in the EP began in the early twentieth century, with captures occurring primarily at sea. These first "fisheries" focused primarily on green turtles and olive ridleys (Cliffton et al. 1982; Green and Ortiz-Crespo 1982; Hays-Brown and Brown 1982; see chapters 4 and 13). By the 1960s harvests had reached a production peak, particularly in countries such as Ecuador and México (see chapters 4, 9, and 13). At the same time, large-scale removal of nesting females and eggs at nesting beaches increased, and, as in other parts of the world, this demand outstripped the ability of these slow-growing animals to regenerate. Consequently, this exorbitant exploitation contributed to sea turtle population declines in the EP.

The Bycatch Problem

Bycatch in the EP has been identified as among the highest conservation priorities for sea turtles globally (Wallace et al. 2010, 2011). This impact can be attributed to two general fishing sectors: industrial fleets and artisanal fleets. Bycatch in coastal areas occurs principally in shrimp trawlers, gill nets, and bottom longlines (e.g., Orrego and Arauz 2004). However, since 1996 and 1997 all countries from México to Ecuador have mandated the use of turtle excluder devices (TEDs) for all industrial fleets, to meet the requirements to export shrimp to the United States under the U.S. Magnuson-Stevens Fishery Conservation and Management Act (see chapter 5). Since then, bycatch has not been thoroughly evaluated, but it is largely known that most fishermen either improperly implement TEDs or remove them entirely from their trawls. As was the case with sea turtle meat and egg collection, an almost total lack of enforcement of bycatch mitigation measures by local authorities only confounds the problem.

Additionally, TEDs are not always a requirement for artisanal shrimping boats, which with today's technology are becoming more "industrial" in ability and have been reported to catch large numbers of sea turtles (Arauz et al. 2000; A. Zavala, pers. commun.). Bottom-set longlines and gill nets, both artisanal and industrial, also interact frequently with sea turtles and can have devastating mortality rates, as has been the case in artisanal fisheries of Baja California, México (Peckham et al. 2007), and Perú (Kelez et al. 2007; Alfaro-Shigueto et al. 2011). In purse seine fisheries, which typically target tuna and other large pelagic fish species, turtle catch rates are highest during "log sets," in which the net is set around natural floating objects, or fish aggregation devices (Hall 1998), because turtles associate with these floating objects, presumably looking for shelter. Pelagic longlines in the EP are used to capture such species as tunas, swordfish, billfishes, mahi-mahi, and sharks. All species of sea turtles that occur in the region interact with longline gear in the EP, but species frequency, bycatch, and mortality rates vary spatially and seasonally (Kelez et al. 2008).

Unfortunately, sea turtle bycatch in small-scale fisheries has not been evaluated for most of the EP. Considering the widespread presence of these fisheries and their potential threats to sea turtles, rigorous assessments of sea turtle bycatch should be at the top of the EP's sea turtle research agenda.

From Top Down to Bottom Up: EP Sea Turtle Conservation in the New Millennium

Today, although protected by strong laws in nations within the EP as well as international treaties, sea turtles, their eggs, and products continue to be exploited and sold in the region, in some cases heavily so (see plates 11 and 12). Governmental authorities and nongovernmental organizations have acted to limit illegal trade and reduce incidental capture of sea turtles; however, the inability to root out sea turtle black market networks has slowed the progress of sea turtle conservation in many areas.

Nevertheless, there have been some important advances in the EP. There are indications that wildlife enforcement branches of local and national governments are stepping up their efforts to enforce existing laws, although successes in stemming sea turtle exploitation through legal channels are few and far between. In addition, a multitude of nongovernmental organizations and conservation networks are raising awareness about sea turtle conservation (plates 15 and 16).

The first of these conservation alliances commenced in 1997, when after years of information exchange about shared populations between the nations of the region, the Central American Regional Network for the conservation of sea turtles was created. The first product that resulted from this collaborative effort was the creation of a national sea turtle network in each country of the region, as well as the development of firsthand tools, such as a regional diagnosis, a ten-year strategic plan, a manual of best practices, and four regional training and information workshops for people in the region (e.g., Chacón and Arauz 2001). This initiative is managed by stakeholders in various sectors (private, nongovernmental, and governmental) across the region. Like many such initiatives, the Central American Regional Network works under the principle that "the benefits and achievements from working in alliance are much higher than those from working alone."

The list of dynamic regional networks does not stop there. Farther north, in México there are two fantastic networks that underscore just how synergistic regional partnerships can be. Perhaps the most active group is Grupo Tortuguero, a grassroots community-based conservation network spanning dozens of the coastal communities in northwestern México. What started in 1999 as an annual meeting of fishermen, biologists, and conservationists has since broadened and now enlists people from all sectors, including marine resource stakeholders, politicians, commercial and sport fishermen, biologists, educators, students, and conservationists (Pesenti and Nichols 2002). Farther south in México, there is the Community Network for the Recovery of the Leatherback in México, which provides a perfect example of how government and community can work together; in this case, the Mexican government works together with community groups in southern México for a common goal of leatherback conservation (see chapter 9). Another collaborative effort, which spans México but includes partners from across the EP region, is ICAPO (Iniciativa Carey del Pacifico Oriental [Eastern Pacific Hawksbill Initiative]; Gaos et al. 2010). This highly cohesive network is a group of scientists and other stakeholders united in the common goal of developing a foundation of scientific knowledge about hawksbill turtles in hopes that this will pave the way for effective conservation (see chapter 10).

The activities of these groups are reducing the demand for sea turtles throughout the region by addressing the social and ecological roots of sea turtle exploitation, transforming sea turtles from a food and consumptive economic resource to a cultural, ecological, and nonconsumptive recreational resource. These activities are having a profound effect in many ways, as demonstrated by the nesting beach protection programs and environmental education programs focusing on sea turtles that are sprouting up throughout the region.

While no single law or treaty can be 100 percent effective at minimizing anthropogenic impacts to sea turtles in these areas, there are several international conservation agreements and laws in the region that, when taken together, provide a framework within which sea turtle conservation advances can be made (see chapter 7). In addition to protection provided by local marine reserves throughout the region, sea turtles may benefit from broader regional efforts, such as (1) the Eastern Tropical Pacific Marine Corridor (CMAR) Initiative supported by the governments of Costa Rica, Panamá, Colombia, and Ecuador, which is a voluntary agreement to work toward sustainable use and conservation of marine resources in these countries' waters; (2) the Eastern Tropical Pacific Seascape Program managed by Conservation International that supports cooperative marine management in the eastern tropical Pacific, including implementation of the CMAR Initiative; (3) the Inter-American Tropical Tuna Commission (IATTC) and its bycatch reduction efforts, which are among the world's finest for regional fisheries management organizations; (4) the Inter-American Convention for the Protection and Conservation of Sea Turtles (IAC), which is designed to lessen impacts on sea turtles from fisheries and other human impacts; and (5) the Permanent Commission of the South Pacific (Lima Convention), which in 2006 updated its action plan for the Regional Program for the Conservation of Marine Turtles in the Southeast Pacific.

Looking ahead, sea turtle conservation in the EP will require successful implementation and greater integration among the region's international accords, and better coordination between these instruments and the on-the-ground conservation movements. New legislation and enforcement of existing laws that curb the flow of turtle products in the region's coastal communities are also necessary, although it is increasingly clear that any such efforts will be effective only if the underlying human social drivers, such as local demand for sea turtle products or increasing fleet sizes despite lower target species catch rates (i.e., overfishing), are also addressed. By implementing both new and existing conservation measures in an integrated manner, management efforts may be more effective at providing habitat protection that extends from nesting beaches and internesting habitats within coastal corridors to nearshore and offshore foraging areas, thereby conserving all life history phases of sea turtles in the EP.

REFERENCES

Alava, J. J. 2008. Loggerhead sea turtles (*Caretta caretta*) in marine waters off Ecuador: occurrence, distribution and bycatch from the eastern Pacific Ocean. Marine Turtle Newsletter 118:8–11.

Alfaro-Shigueto, J., P. H. Dutton, J. Mangel, and D. Vega. 2004. First confirmed occurrence of loggerhead turtles in Peru. Marine Turtle Newsletter 103:7–11.

Alfaro-Shigueto, J., P. H. Dutton, M. F. Van Bressem, and J. Mangel. 2007. Interactions between leatherback turtles and Peruvian artisanal fisheries. Chelonian Conservation and Biology 6:129–134.

Alfaro-Shigueto, J., J. C. Mangel, F. Bernedo, P. H. Dutton, J. A. Seminoff, and B. J. Godley. 2011. Small scale fisheries of Peru: a major sink for marine turtles in the Pacific? Journal of Applied Ecology 48:1432–1440.

Alfaro-Shigueto, J., J. C. Mangel, M. Pajuelo, P. H. Dutton, J. A. Seminoff, and B. J. Godley. 2010. Where small can have a large impact: structure and characterization of small-scale fisheries in Peru. Fisheries Research 106(1):8–17.

Alfaro-Shigueto, J., J. C. Mangel, J. A. Seminoff, and P. H. Dutton. 2008. Demography of loggerhead turtles *Caretta caretta* in the southeastern Pacific Ocean: fisheries-based observations and implications for management. Endangered Species Research 5:129–135.

Alvarez-Borrego, S. 2010. Physical, chemical, and biological oceanography of the Gulf of California. In: Brusca, R. (Ed.), The Gulf of California, Biodiversity and Conservation. University of Arizona Press, Tucson. Pp. 72–95.

Amorocho, D. F. 2009. Foraging ecology and population structure of the green sea turtle (*Chelonia mydas*) in the eastern Pacific coast of Colombia. Ph.D. dissertation, Monash University, Melbourne, Australia.

Amorocho, D. F., and R. Reina. 2007. Feeding ecology of the East Pacific green sea turtle *Chelonia mydas agassizii* at Gorgona National Park, Colombia. Endangered Species Research 3:43–51.

Amorocho, D. F., and R. Reina. 2008. Intake passage time, digesta composition, and digestibility in East Pacific green turtles (*Chelonia mydas agassizii*) at Gorgona National Park, Colombian Pacific. Journal of Experimental Marine Biology and Ecology 360:117–124.

Arauz, R., O. Rodriguez, R. Vargas, and A. Segura. 2000. Incidental capture of sea turtles by Costa Rica's longline fleet. In: Kalb, H. J., and T. Wibbels (Eds.), Proceedings of the Nineteenth Annual Symposium on Sea Turtle Biology and Conservation. NOAA Tech. Memo. NMFS-SEFSC-443. U.S. Department of Commerce, U.S. National Marine Fisheries Service, Miami, FL. Pp. 62–64.

Barber, R. T. 2001. Ocean ecosystems. Encyclopedia of Biodiversity 4:427–437.

Benson, S. R., K. A. Forney, J. T. Harvey, J. V. Caretta, and P. H. Dutton. 2007. Abundance, distribution, and habitat of leatherback turtles (*Dermochelys coriacea*) off California 1990–2003. Fisheries Bulletin 105(3):337–347.

Bjorndal, K. A. 1997. Foraging ecology and nutrition of sea turtles. In: Lutz, P. L., and J. A. Musick (Eds.), The Biology of Sea Turtles, Vol. 1. CRC Press, Boca Raton, FL. Pp. 199–232.

Bostrom, B. L., Jones, T. T., Hastings, M., and Jones, D. R. 2010. Behaviour and physiology: the thermal strategy of leatherback turtles. PLoS One 5(11):e13925; doi:10.1371/journal.pone.0013925.

Bowen, B. W., F. A. Abreu-Grobois, G. H. Balazs, N. Kamezaki, C. J. Limpus, and R. J. Ferl. 1995. Trans-Pacific migrations of the loggerhead turtle (*Caretta caretta*) demonstrated with mitochondrial DNA markers. Proceedings of the National Academy of Sciences of the United States of America 92:3371–3734.

Bowen, B. W., A. L. Bass, S. Chow, M. Bostrom, K. A. Bjorndal, A. B. Bolten, T. Okuyama, B. M. Bolker, S. Epperly, E. LaCasella, D. Shaver, M. Dodd, S. R. Hopkins-Murphy, J. A. Musick, M. Swingle, K. Rankin-Baransky, W. Teas, W. N. Witzell, and P. H. Dutton. 2004. Natal homing in juvenile loggerhead turtles (*Caretta caretta*). Molecular Ecology 13:3797–3808.

Boyle, M. C., N. N. Fitzsimmons, C. J. Limpus, S. Kelez, X. Velez-Zuazo, and M. Waycott. 2009. Evidence for transoceanic migrations by loggerhead turtles in the southern Pacific Ocean. Proceedings of the Royal Society of London, Series B 276:1993–1999.

Campbell, L. 2003. Contemporary culture, use, and conservation of sea turtles. In: Lutz, P., J. Musick, and J. Wyneken (Eds.), The Biology of Sea Turtles, Vol. 2. CRC Press, Boca Raton, FL. Pp. 307–338.

Carr, A., H. Hirth, and L. Ogren. 1966. The ecology and migrations of sea turtles. 6. The hawksbill turtle in the Caribbean Sea. American Museum Novitates 2248:1–29.

Carrión-Cortez, J., P. Zárate, and J. A. Seminoff. 2010. Feeding ecology of the green sea turtle (*Chelonia mydas*) in the Galápagos Islands. Journal of the Marine Biological Association of the United Kingdom 90:1005–1013.

Chacón, D., and R. Arauz. 2001. Diagnóstico regional y planificación estratégica para la conservación de las tortugas marinas en Centroamérica. Red regional para la conservación de las tortugas marinas de Centroamérica, San José, Costa Rica.

Chaloupka, M. Y., and J. A. Musick. 1997. Age, growth, and population dynamics. In: Lutz,

P. L., and J. A. Musick (Eds.), The Biology of Sea Turtles, Vol. 1. CRC Press, Boca Raton, FL. Pp. 233–276.

Chavez F. P., P. G. Strutton, G. E. Friederich, R. A. Feely, G. C. Feldman, D. G. Foley, and M. J. McPhaden. 1999. Biological and chemical response of the equatorial Pacific Ocean to the 1997–98 El Niño. Science 286:2126–2131.

Cliffton, K., D. O. Cornejo, and R. S. Felger. 1982. Sea turtles of the Pacific coast of Mexico. In: Bjorndal, K. A. (Ed.), Biology and Conservation of Sea Turtles. Smithsonian Institution Press, Washington, DC. Pp. 199–209.

Croll, D., B. Marinovic, S. Benson, F. P. Chavez, N. Black, R. Ternullo, and B. R. Tershy. 2005. From wind to whales: trophic links in a coastal upwelling system. Marine Ecology Progress Series 289:117–130.

Donoso, M., and P. H. Dutton. 2010. Sea turtle bycatch in the Chilean pelagic longline fishery in the southeastern Pacific: Opportunities for conservation. Biological Conservation 143:2672–2684.

Eckert, S. A., and L. Sarti. 1997. Distant fisheries implicated in the loss of the world's largest leatherback nesting population. Marine Turtle Newsletter 78:2–7.

Eguchi, T., T. Gerrodette, R. Pitman, J. A. Seminoff, and P. H. Dutton. 2007. At-sea density an abundance estimates of the olive ridley turtle (*Lepidochelys olivacea*) in the eastern tropical Pacific. Endangered Species Research 3:191–203.

Ernst, C. H., and R. W. Barbour. 1989. Turtles of the World. Smithsonian Institution Press, Washington, DC.

Felger, R. S., K. Cliffton, and P. J. Regal. 1976. Winter dormancy in sea turtles: independent discovery and exploitation in the Gulf of California by two local cultures. Science 191:283–285.

Fiedler, P. C. 1992. Seasonal Climatologies and Variability of Eastern Tropical Pacific Surface Waters. NOAA Tech. Rep. NMFS 109. U.S. Department of Commerce, Springfield, VA.

Fiedler, P. C., V. Philbrick, and F. P. Chavez, 1991. Oceanic upwelling and productivity in the eastern tropical Pacific. Limnology and Oceanography 36:1834–1850.

Frazier, J. 1985. Misidentifications of sea turtles in the East Pacific: *Caretta caretta* and *Lepidochelys olivacea*. Journal of Herpetology 19:1–11.

Frazier, J. G., and J. L. M. Brito. 1990. Incidental capture of marine turtles by the swordfish fishery at San Antonio, Chile. Marine Turtle Newsletter 49:8–13.

Gaos, A. R., F. A. Abreu-Grobois, J. Alfaro-Shigueto, D. Amorocho, R. Arauz, A. Baquero, R. Briseño, D. Chacón, C. Dueñas, C. Hasbún, M. Liles, G. Mariona, C. Muccio, J. P. Muñoz, W. J. Nichols, M. Peña, J. A. Seminoff, M. Vásquez, J. Urteaga, B. P. Wallace, I. Yañez, and P. Zárate. 2010. Signs of hope in the eastern Pacific: international collaboration reveals encouraging status for severely depleted population of hawksbill turtles. Oryx 44:595–601.

Gaos, A. R., R. R. Lewison, I. L. Yañez, W. J. Nichols, A. Baquero, M. Liles, M. Vasquez, J. Urteaga, B. P. Wallace, and J. A. Seminoff. 2011. Shifting the life-history paradigm: discovery of novel habitat use by hawksbill turtles. Biology Letters; doi:10.1098/rsbl.2011.0603.

Ghil, M., and R. Vautard. 1991. Interdecadal oscillations and the warming trend in global temperature time series. Nature 350:324–327.

Glynn, P. W. 2001. Eastern Pacific coral reef ecosystems. In: Seeliger, U., and B. Kjerfve (Eds.), Coastal Marine Ecosystems of Latin America. Ecological Studies, Vol. 144. Springer, Berlin. Pp. 281–305.

Green, D., and F. Ortiz-Crespo. 1982. Status of sea turtle populations in the central eastern Pacific. In: Bjorndal, K. A. (Ed.), Biology and Conservation of Sea Turtles. Smithsonian Institution Press, Washington, DC. Pp. 221–233.

Groombridge, B., and R. Luxmoore. 1989. The Green Turtle and Hawksbill (Reptilia: Cheloniidae): World Status, Exploitation and Trade. CITES Secretariat, Lausanne, Switzerland.

Hall, M. A. 1998. An ecological view of the tuna–dolphin problem: impacts and trade-offs. Reviews in Fish Biology and Fisheries 8:1–34.

Hays-Brown, C., and W. M. Brown. 1982. Status of sea turtles in the southeastern Pacific: emphasis on Peru. In: Bjorndal, K. A. (Ed.), Biology and Conservation of Sea Turtles. Smithsonian Institution Press, Washington, DC. Pp. 235–240.

Hirth. H. F. 1990. Synopsis of the biological data on the green turtle, *Chelonia mydas* (Linnaeus 1758). FAO Fisheries Synopsis 85(1):1–19.

Jones, T. T., B. Bostrom, M. Carey, B. Imlach, J. Mickelson, S. Eckert, P. Opay, Y. Swimmer, J. A. Seminoff, and D. R. Jones. 2011. Determining Transmitter Drag and Best-Practice Attachment Procedures for Sea Turtle Biotelemetry Studies. NOAA Tech. Memo. NMFS-SWFSC-480. U.S. Department of Commerce, U.S. National Marine Fisheries Service, La Jolla, CA.

Kamezaki, N., I. Miyakawa, H. Suganuma, K. Omuta, Y. Nakajima, K. Goto, K. Sato, Y. Matsuzawa, M. Samejima, M. Ishii, and T. Iwamoto. 1997. Post-nesting migration of Japanese loggerhead turtles, *Caretta caretta*. Wildlife Conservation Japan 3:29–39.

Karl, S. A., and B. W. Bowen. 1999. Evolutionary significant units versus geopolitical taxonomy: molecular systematics of an endangered sea turtle (genus Chelonia). Conservation Biology 13:990–999.

Kelez, S., X. Velez-Zuazo, and C. Manrique. 2003. New evidence on the loggerhead sea turtle Caretta caretta (Linnaeus 1758) in Perú. Ecologia Aplicada 2:141–142.

Kelez, S., X. Velez-Zuazo, and C. Manrique. 2007. Incidental capture of sea turtles by Peruvian medium-scale longline fisheries. In: Mast, R. B., B. J. Hutchinson, and A. H. Hutchinson (Comps.), Proceedings of the Twenty-Fourth Annual Symposium on Sea Turtle Biology and Conservation. NOAA Tech. Memo. NMFS-SEFSC-567. U.S. Department of Commerce, U.S. National Marine Fisheries Service, Miami, FL. P. 27.

Kelez, S., B. Wallace, D. Dunn, and W. J. Nichols. 2008. Sea turtle bycatch in the Eastern Pacific: a regional review. In: Dean, K., and M. Lopez (Comps.), Proceedings of the Twenty-Eighth Annual Symposium on Sea Turtle Biology and Conservation. NOAA Tech. Memo. NMFS-SEFSC-602. U.S. Department of Commerce, U.S. National Marine Fisheries Service, Miami, FL. Pp. 110–111.

Kelez Sara, S. 2011. Bycatch and Foraging Ecology of Sea Turtles in the Eastern Pacific. Ph.D. dissertation, Duke University, Durham, NC.

Kobayashi, D. R., J. J. Polovina, D. M. Parker, N. Kamezaki, I. J. Chenge, I. Uchida, P. H. Dutton, and G. H. Balazs. 2008. Pelagic habitat characterization of loggerhead sea turtles, *Caretta caretta*, in the North Pacific Ocean (1997–2006): insights from satellite tag tracking and remotely sensed data. Journal of Experimental Marine Biology and Ecology 356:96–114.

Kopitsky, K. L., R. L. Pitman, and P. H. Dutton. 2005. Aspects of olive ridley feeding ecology in the eastern tropical Pacific. In: Coyne, M. S., and R. D. Clark (Comps.), Proceedings of the Twenty-First Annual Symposium on Sea Turtle Biology and Conservation. NOAA Tech. Memo. NMFS-SEFSC-528. U.S. Department of Commerce, U.S. National Marine Fisheries Service, Miami, FL. P. 217.

Liles, M. J., M. V. Jandres, W. A. Lopez, G. I. Mariona, C. R. Hasbún, and J. A. Seminoff. 2011. Hawksbill turtles *Eretmochelys imbricata* in El Salvador: nesting distribution and mortality at the largest remaining nesting aggregation in the eastern Pacific Ocean. Endangered Species Research 14:23–30.

Lohmann, K. J., B. E. Witherington, C. M. F. Lohmann, and M. Salmon. 1997. Orientation, navigation, and natal beach homing in sea turtles. In: Lutz, P. L., and J. A. Musick (Eds.), The Biology of Sea Turtles, Vol. 1. CRC Press, Boca Raton, FL. Pp. 107–136.

López-Castro, M. C., V. Koch, A. Mariscasl-Loza, and W. J. Nichols. 2010. Long-term monitoring of black turtles *Chelonia mydas* at coastal foraging areas off the Baja California Peninsula. Endangered Species Research 11:35–45

López-Mendilaharsu, M., S. Gardner, J. A. Seminoff, and R. Riosmena-Rodriguez. 2005. Identifying critical foraging habitats of the green turtle (*Chelonia mydas*) along the Pacific coast of the Baja California peninsula, México. Aquatic Conservation: Marine and Freshwater Ecosystems 15:259–269.

Mangel, J. C., J. Alfaro-Shigueto, M. J. Witt, P. H. Dutton, J. A. Seminoff, and B. J. Godley. 2011. Post-capture movements of loggerhead turtles in the southeastern Pacific Ocean assessed by satellite tracking. Marine Ecology Progress Series 433:261–272.

Mantua, N. J., S. R. Hare, X. Zhang, J. M. Wallace, and R. C. Francis. 1997. A Pacific interdecadal climate oscillation with impacts on salmon production. Bulletin of the American Meteorological Society 78:1069–1079.

Meylan, A. B., B. W. Bowen, and J. C. Avise. 1990. A genetic test of the natal homing versus social facilitation models for green turtle migration. Science 248:724–728.

Nichols, W. J., K. E. Bird, and S. Garcia. 2002. Community-based research and its application to sea turtle conservation in Bahía Magdalena, BCS, México. Marine Turtle Newsletter 89:4–7.

Nichols, W. J., L. Brooks, M. Lopez, and J. A. Seminoff. 2001. Record of pelagic east Pacific green turtles associated with *Macrocystis* mats near Baja California Sur, México. Marine Turtle Newsletter 93:10–11.

Nichols, W. J., A. Resendiz, J. A. Seminoff, and B. Resendiz. 2000. Transpacific loggerhead turtle migration monitored with satellite telemetry. Bulletin of Marine Science 67:937–947.

Omuta, Y. 2005. A black turtle entangled by fishing net and stranded on Maehama Beach, Yakushima. Umigame Newsletter of Japan 63:2–3.

Orrego, C. M., and R. Arauz. 2004. Mortality of sea turtles along of the Pacific coast of Costa Rica. In: Coyne, M. S., and R. D. Clark (Eds.), Proceedings of the Twenty-First Annual Symposium on Sea Turtle Biology and Conservation. NOAA Tech. Memo.

NMFS-SEFSC-528. U.S. Department of Commerce, U.S. National Marine Fisheries Service, Miami, FL. Pp. 265–266.

Pajuelo, M., K. A. Bjorndal, J. Alfaro-Shigueto, J. A. Seminoff, J. Mangel, and A. B. Bolten. 2010. Stable isotope dichotomy in loggerhead turtles reveals Pacific-Atlantic oceanographic differences. Marine Ecology Progress Series 417:277–285.

Palacios, D. M. 2002. Factors influencing the island-mass effect of the Galápagos Islands. Geophysical Research Letters 29:2134.

Paladino, F. V., M. P. O'Connor, and J. R. Spotila. 1990. Metabolism of leatherback turtles: gigantothermy and thermoregulation of dinosaurs. Nature 344:858–860.

Peckham, S. H., D. Maldonado-Díaz, V. Koch, A. Mancini, A. Gaos, M. T. Tinker, and W. J. Nichols. 2008. High mortality of loggerhead turtles due to bycatch, human consumption and strandings at Baja California Sur, México, 2003 to 2007. Endangered Species Research 5:171–183.

Peckham, S. H., D. Maldonado, A. Walli, G. Ruiz, and L. B. Crowder. 2007. Small-scale fisheries bycatch jeopardizes endangered Pacific loggerhead turtles. PLoS One; doi: 10.1371/journal.pone.0001041.

Pennington, J. T., K. L. Mahoney, V. S. Kuwahara, D. D. Kolber, R. Calienes, and F. P. Chavez. 2006. Primary production in the eastern tropical Pacific: a review. Progress in Oceanography 69:285–317.

Pesenti, C., and W. J. Nichols. 2002. Signs of Success: Fourth Annual Meeting of the Sea Turtle Conservation Network of the Californias (Grupo Tortuguero de las Californias). Marine Turtle Newsletter 97:14–16.

Plotkin, P. 2003. Adult habitat use and migrations. In: Lutz, P. L., J. A. Musick, and J. Wyneken (Eds.), The Biology of Sea Turtles, Vol. 2. CRC Press, Boca Raton, FL. Pp. 225–241.

Plotkin, P. T., R. A. Byles, D. C. Rostal, and D. Owens. 1995. Independent versus socially facilitated oceanic migrations of the olive ridley, *Lepidochelys olivacea*. Marine Biology 122:137–143.

Pritchard, P. C. H. 1999. Status of the black turtle. Conservation Biology 13:1000–1003.

Quiñones, J., V. González Carman, J. Zeballos. S. Purca, and H. Mianzan. 2010. Effects of El Niño-driven environmental variability on black turtle migration to Peruvian foraging grounds. Hydrobiologia 645:69–79.

Resendiz, A., B. Resendiz, W. J. Nichols, J. A. Seminoff, and N. Kamezaki. 1998. First confirmation of a trans-Pacific migration of a tagged loggerhead sea turtle (*Caretta caretta*), released in Baja California. Pacific Science 52:151–153.

Saba, V. S., P. Santidrián-Tomillo, R. D. Reina, J. R. Spotila, J. A. Musik, D. A. Evans, and F. V. Paladino. 2007. The effect of the El Niño Southern Oscillation on the reproductive frequency of eastern Pacific leatherback turtles. Journal of Applied Ecology 44:395–404.

Saba, V. S., G. L. Shillinger, A. M. Swithenbank, B. A. Block, J. R. Spotila, J. A. Musick, and F. V. Paladino. 2008a. An oceanographic context for the foraging ecology of eastern Pacific leatherback turtles: consequences of ENSO. Deep-Sea Research I 55:646–660.

Saba, V. S., J. R. Spotila, F. P. Chavez, and J. A. Musick. 2008b. Bottom-up and climatic forcing on the worldwide population of leatherback turtles. Ecology 89:1414–1427.

Sakamoto, W., T. Bando, N. Arai, and N. Baba. 1997. Migration paths of adult female and male loggerhead turtles, *Caretta caretta*, determined through satellite telemetry. Fisheries Science 63:547–552.

Santidrián Tomillo, P., V. S. Saba, R. Piedr, F. V. Paladino, and J. R. Spotila. 2008. Effects of illegal harvest of eggs on the population decline of leatherback turtles in Las Baulas Marine National Park, Costa Rica. Conservation Biology 22:1216–1224.

Santidrián Tomillo, P., E. Vélez, R. D. Reina, R. Piedra, F. V. Paladino, and J. R. Spotila. 2007. Reassessment of the leatherback turtle (*Dermochelys coriacea*) nesting population at Parque Nacional Marino Las Baulas, Costa Rica: effects of conservation efforts. Chelonian Conservation and Biology 6:54–62.

Sarti Martínez, L., A. R. Barragán, D. G. Muñoz, N. García, P. Huerta, and F. Vargas. 2007. Conservation and biology of the leatherback turtle in the Mexican Pacific. Chelonian Conservation and Biology 6:70–78.

Schwing, F. B., T. Murphree, and P. M. Green. 2002. The Northern Oscillation Index (NOI): A new climate index for the northeast Pacific. Progress in Oceanography 53: 115–139.

Seminoff, J. A., T. T. Jones, A. Resendiz, W. J. Nichols, and M. Y. Chaloupka. 2003. Monitoring green turtles (*Chelonia mydas*) at a coastal foraging area in Baja California, México: multiple indices describe population status. Journal of the Marine Biological Association of the United Kingdom 83:1355–1362.

Seminoff, J. A., A. Resendiz, and W. J. Nichols. 2002a. Diet of the East Pacific green turtle, *Chelonia mydas*, in the central Gulf of California, México. Journal of Herpetology 36:447–453.

Seminoff, J. A., A. Resendiz, and W. J. Nichols. 2002b. Home range of the green turtle (*Chelonia mydas*) at a coastal foraging ground in the Gulf of California, México. Marine Ecology Progress Series 242:253–265.

Seminoff, J. A., and P. Zárate. 2008. Satellite-tracked migrations by Galápagos green turtles and the need for multinational conservation efforts. Current Conservation 2:11–12.

Seminoff, J. A., P. Zárate, M. Coyne, D. Foley, D. Parker, B. N. Lyon, and P. H. Dutton. 2008. Post-nesting migrations of Galápagos green turtles, *Chelonia mydas,* in relation to oceanographic conditions: integrating satellite telemetry with remotely-sensed ocean data. Endangered Species Research 4:57–72.

Sharp, G. D., and D. R. McLain. 1993. Fisheries, El Niño–Southern Oscillation and upper-ocean temperature records: an eastern Pacific example. Oceanography 6:13–22.

Shillinger, G. L., D. M. Palacios, H. Bailey, S. J. Bograd, A. M. Swithenbank, P. Gaspar, B. P. Wallace, J. R. Spotila, F. V. Paladino, R. Piedra, S. A. Eckert, and B. A. Block. 2008. Persistent leatherback turtle migrations present opportunities for conservation. PLoS Biology 6:e171.

Spotila, J. R., R. D. Reina, A. C. Steyermark, P. T. Plotkin, and F. V. Paladino. 2000. Pacific leatherback turtles face extinction. Nature 405:529–530.

Strub, P. T., J. Mesias, and C. James. 1995. Satellite observations of the Peru-Chile countercurrent. Geophysical Research Letters 22:211–214.

Strub, P. T., J. M. Mesias, V. Montecino, J. Rutllant, and S. Salinas. 1998. Coastal ocean circulation off Western South America. In: Robinson A. R., and K. H. Brink (Eds.), The Sea, Vol. 11: The Global *Coastal Ocean*: Regional Interdisciplinary Studies and Synthesis. Wiley, New York. Pp. 273–313.

Suryan, R. M., V. S. Saba, B. P. Wallace, S. A. Hatch, M. Frederiksen, and S. Wanless. 2009. Environmental forcing on life history strategies: multi-trophic level response at ocean basin scales. Progress in Oceanography 81: 214–218.

Uchida, S., and H. Teruya. 1988. Transpacific migration of a tagged loggerhead, *Caretta caretta* and tag-return results of loggerheads released from Okinawa Island, Japan. In: Uchida, I. (Ed.), Proceedings of the International Sea Turtle Symposium, Hiwasa, Japan, 30 July–1 August 1988. Himeji City Aquarium and Hiwasa Chelonian Museum, Japan. Pp. 169–182.

Urteaga, J. R., and D. Chacón. 2007. Nesting activity and conservation of leatherback (*Dermochelys coriacea*) sea turtles, in the Rio Escalante-Chacocente Wildlife Refuge, Pacific coast of Nicaragua. In: Mast, R. B., B. J. Hutchinson, and A. H. Hutchinson (Comps.), Proceedings of the Twenty-Fourth Annual Symposium on Sea Turtle Biology and Conservation. NOAA Tech. Memo. NMFS-SEFSC-567. U.S. Department of Commerce, U.S. National Marine Fisheries Service, Miami, FL. Pp. 157–158.

U.S. National Marine Fisheries Service and U.S. Fish and Wildlife Service. 2007. Leatherback Sea Turtle (*Dermochelys coriacea*) 5-Year Review: Evaluation and Summary. U.S. National Marine Fisheries Service Office of Protected Resources, Silver Spring, MD, and U.S. Fish and Wildlife Service Southeast Region, and Jacksonville, FL.

van Dam, R. P., and C. E. Diez. 1996. Diving behavior of immature hawksbills (*Eretmochelys imbricata*) in a Caribbean cliff-wall habitat. Marine Biology 127:171–178.

Wallace, B. P., A. D. DiMatteo, A. B. Bolten, M. Y. Chaloupka, B. J. Hutchinson, F. A. Abreu-Grobois, J. A. Mortimer, J. A. Seminoff, D. Amorocho, K. A. Bjorndal, et al. 2011. Global conservation priorities for marine turtles. PLoS One 6(9):e24510; doi: 10.1371/journal.pone.0024510.

Wallace, B. P., and T. T. Jones, 2008. What makes marine turtles go: a review of metabolic rates and their consequences. Journal of Experimental Marine Biology and Ecology 356:8–24.

Wallace, B. P., S. S. Kilham, F. V. Paladino, and J. R. Spotila. 2006. Energy budget calculations indicate resource limitation in eastern Pacific leatherback turtles. Marine Ecology Progress Series 318:263–270.

Wallace, B. P., R. L. Lewison, S. L. McDonald, R. K. McDonald, C. Y. Kot, S. Kelez, R. K. Bjorkland, E. M. Finkbeiner, S. Helmbrecht, and L. B. Crowder. 2010. Global patterns of marine turtle bycatch. Conservation Letters 3(3):131–142.

Wallace, B. P., and V. S. Saba. 2009. Environmental and anthropogenic impacts on intraspecific variation in leatherback turtles: opportunities for targeted research and conservation. Endangered Species Research 7:1–11.

Wester, J. H., S. Kelez, and X. Velez-Zuazo. 2010. Nuevo limite sur de anidación de las tortuga verde *Chelonia mydas* y golfina *Lepidochelys olivacea* en el Pacifico Este. Paper presented at the II Congreso Nacional de Ciencias del Mar del Perú. Piura, Perú, 24–28 May 2010.

Worm, B., M. Sandow, A. Oschlies, H. K. Lotze, and R. A. Myers. 2005. Global patterns of predator diversity in the open oceans. Science 309:1365–1369.

2

Sea Turtles in the Tropical High Seas

Climate Variability, Oceanography, and
Ecosystem Responses

VINCENT S. SABA

Summary

The eastern tropical Pacific (ETP) is a unique marine ecosystem because it is subject to a high rate of interannual and multidecadal climate variability. Nutrient forcing in this part of the Pacific Ocean is dominated by equatorial and coastal upwelling, both of which are a function of the atmospheric climate. Within the ETP food web, climate variability has considerable effects on the population dynamics of a wide range of trophic levels, from phytoplankton to fishes to marine mammals. Sea turtles are by no means immune to these changes in the food web. Interannual climate change in the ETP is foremost evident by the El Niño–Southern Oscillation (ENSO), where sea surface temperatures fluctuate between warm and cold phases. Leatherback turtles nesting in Pacific Costa Rica, and likely those in Pacific México, respond strongly to warm El Niño and cool La Niña phase changes, as observed by their nesting numbers. During warm El Niño periods, food availability decreases throughout the ETP because of decreased cell size and biomass of primary producers inhabiting coastal and equatorial areas. Mature female leatherbacks, which feed primarily on gelatinous zooplankton, have a lower likelihood of nesting during El Niño periods because of the lower rate of food availability required to fuel the nesting process. The opposite holds true for La Niña periods. As a consequence of fluctuating food availability, females in the ETP are the smallest in body size and have the lowest reproductive output compared with other females worldwide. Although other species of sea turtle in the ETP have not yet been investigated in this context, it is likely that loggerhead, green, olive ridley, and hawksbill turtles are also affected by ENSO, given the strong response of leatherbacks.

Oceanography and Biological Hotspots of the Eastern Tropical Pacific

Phytoplankton

The eastern tropical Pacific (ETP) hosts a wide variety of marine life along the west coasts of the Americas and in the deeper pelagic waters. The biological hotspots in the ETP can be distinguished by their high phytoplankton biomass, which serves as the base of the marine food web driven by photosynthesis, also known as marine primary production. These primary producers serve as food resources for secondary producers (zooplankton), and so on, working up the food chain to fishes, marine mammals, and sea turtles.

Phytoplankton growth is dependent upon specific nutrients (i.e., nitrate, phosphate, silica, iron), in addition to the essential requirements of sunlight, carbon dioxide, and appropriate temperature, which varies by species (i.e., tropical vs. temperate). The nutrient supply is largely, but not entirely, controlled by physical forcing from both the atmosphere and ocean environment. Such physical forcing includes wind-driven vertical mixing, deep-ocean water upwelling, riverine input, and mesoscale eddies. To a lesser extent, biological forcing on nutrient supply includes nutrient recycling by bacteria, zooplankton, and upper-trophic organisms.

Physical Forcing, Nutrients, and Ecosystem Hotspots

In the ETP, the predominant physical forcing on nutrients derives from coastal and equatorial upwelling. Coastal upwelling is driven by the Coriolis effect, by which winds drive ocean currents to the right of the wind direction in the Northern Hemisphere and to the left in the Southern Hemisphere. Along eastern ocean boundaries such as coastal Perú and California, warm surface waters are driven away from the coastlines and are replaced, via Ekman transport, by cold, nutrient-rich deep water that can fuel phytoplankton growth. The western coastlines of South and North America have a narrow continental shelf that slopes abruptly into a deep trench, from which deep, nutrient-rich water is forced to the surface. On the other hand, equatorial upwelling is caused by the converging trade winds near the equator causing Ekman transport away from the equator, thus forcing the deeper, nutrient-rich equatorial water to the upper euphotic zone.

The result of upwelling on phytoplankton growth in the ETP can be discerned through satellite-based estimates of ocean net primary productivity (NPP). Figure 2.1A illustrates estimated NPP in the ETP that is largely driven by coastal and equatorial upwelling. The dark region along the equatorial Pacific is known as the equatorial cold tongue. The frontal boundaries separating warmer surface waters along the cold tongue are common foraging habitat for planktivorous seabirds and yellowfin tuna (Ballance et al. 2006). The coastal upwelling hotspots support major fisheries, especially the Peruvian coastal upwelling system (fig. 2.1A), which supports the largest-tonnage fishery in the world, producing 6 to 12 million metric tons of anchovy per year (Pennington et al. 2006). Off the coasts of California, Baja, and Chile, tuna and billfish fisheries are very productive.

In addition to the ecosystem hotspots driven by coastal and equatorial upwelling, the Gulf of Panamá, Gulf of Papagayo (eastern side of the Costa Rica Dome), and Gulf of Tehuantepec have high NPP driven by winds derived from the Atlantic Ocean (Gulf of México and Caribbean) through the mountains of Central America, causing wind-stress curl-induced upwelling (Pennington et al. 2006; see fig. 2.1A). The Costa Rica Dome (fig. 2.1A) is a tropical thermocline dome associated with equatorial circulation, cyclonic surface currents, large-scale seasonal wind patterns, and physical coupling with the Gulf of Papagayo. Mean biomass in the Costa Rica Dome can be as high as or higher than that in the equatorial upwelling zone. Within the dome, blue whales can be observed year-round feeding on krill (Fiedler 2002), while short-beaked common dolphins are in great abundance foraging further up the food chain on small fishes and squids (Ballance et al. 2006). The Galápagos Islands support a phytoplankton biomass that is at least double that of the surrounding equatorial upwelling system (fig. 2.1A), due to the intense mixing and upwelling where the equatorial undercurrent impinges on the western side (Pennington et al.

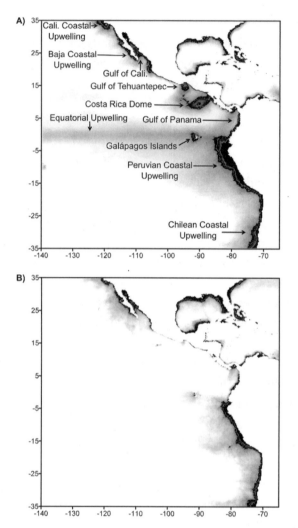

FIGURE 2.1. Ecosystem hotspots of the ETP characterized by net primary productivity (NPP) and the consequences of ENSO. (A) Mean daily NPP during the 1998–2000 La Niña and the ecosystem hotspots of the ETP. (B) Mean daily NPP during the 1997–1998 El Niño. NPP was calculated using the standard vertically generalized production model (VGPM) (Behrenfeld and Falkowski 1997). The average daily NPP for each ENSO period was calculated using Windows Image Manager (Kahru 2006), and monthly VGPM images were extracted from the Oregon State Ocean Productivity website (http://web.science .oregonstate.edu/ocean.productivity).

2006). Finally, the Gulf of California's high NPP (fig. 2.1A) is caused by northwest wind-induced mixing and upwelling, tidal mixing, and coastal-trapped waves that enhance nutrient supply (Douglas et al. 2007).

These biological hotspots provide a wide range of foraging habitat for the sea turtles of the ETP. Loggerhead turtles that nest in Japan migrate across the entire Pacific Ocean to the productive waters off the coast of Baja (Nichols et al. 2000) and the Gulf of California (Seminoff 2004) while those nesting in Australia migrate to the coast of South America (Boyle et al. 2009). The equatorial Pacific is a prime foraging area for leatherbacks (Saba et al. 2008a; see chapter 8 in this volume), green turtles (Seminoff et al. 2008), and olive ridleys (Kopitsky et al. 2001; Plotkin 2010), while hawksbill turtles can be found closer to the coastlines (see chapter 10). All five species can be found along the coastal upwelling areas stretching from Perú to Baja (plate 1). Highly productive foraging areas in tropical waters are essential for sea turtles because reptiles generally are sensitive to cooler temperatures. The tropical foraging areas are especially important for the smaller olive ridley and hawksbill turtles and for neonate and juvenile turtles that cannot tolerate the cooler waters of the temperate, highly productive North Pacific as larger adult leatherback and green turtles can (see chapter 4). However, there can be major consequences for sea turtles foraging in the ETP because this marine environment is subject to an extremely high rate of climate-induced ecosystem variability.

Climate Variability and Ecosystem Fluctuations

The El Niño–Southern Oscillation

Natural climate variability has major implications for the ecosystems of the ETP. The predominating natural phenomenon that affects the ETP most dramatically is the El Niño–Southern Oscillation (ENSO). Through ENSO, the ETP oscillates between warm and cool sea surface temperature (SST) phases forced by changes in atmospheric pressure and winds. The phase shift typically occurs every three to seven years. Although the effects of strong ENSO events can be observed throughout the global climate, the most prominent effects are in the ETP. When atmospheric pressure in the ETP drops below average, SST warms as the thermocline deepens. The warm phase of ENSO is termed an El Niño event. When atmospheric pressure rises above the norm in the ETP, the thermocline becomes shallower and SST cools. This cool phase of ENSO is referred to as a La Niña event. During El Niño events, precipitation patterns shift dramatically, and drought occurs throughout Australia and Indonesia while the western coast of South America experiences heavy rainfall (McPhaden et al. 2006).

The effects of ENSO on phytoplankton growth and ensuing ecosystem production in the ETP can be astounding. In the equatorial Pacific, the 1997–1998 El Niño event decreased biological production by 30 percent from the average, while the 1998–2000 La Niña increased production by 40 percent (fig. 2.1) (Turk et al. 2001). Warm El Niño events weaken the magnitude of equatorial and coastal upwelling in ETP, leading to a reduced nutrient supply to the surface waters. During La Niña, upwelling is enhanced, due to the shallow thermocline, allowing phytoplankton to thrive. In general, almost all of the ecosystem hotspots in the ETP become less biologically productive during El Niño. Figure 2.1 shows the change in NPP between La Niña and El Niño, and it is clear that the most dramatic difference occurs along the equatorial upwelling zone. However, some areas in the ETP are not affected in this manner. The waters just off the coast of Baja (100–700 km offshore) increased in NPP during the 1997–1998 El Niño (fig. 2.1B), and this has also been documented for previous El Niño events (Kahru and Mitchell 2000). The enrichment of nutrients in this area during El Niño may be caused by the periodic shoaling of a subsurface, subtropical water mass of high temperature, high salinity, and low oxygen (Ladah 2003). Furthermore, the northern and central Gulf of California may also increase in NPP during El Niño (Thunell 1998). Satellite-derived NPP data indicate that these two areas react differently to ENSO than do the other ecosystems of the ETP (fig. 2.1).

Multidecadal Climate Variability

In addition to climate shifts every three to seven years in the ETP, multidecadal fluctuations over the course of twenty to twenty-five years have also been observed throughout the entire Pacific Ocean. These fluctuations, or regimes, can be foremost characterized by the fisheries catch of

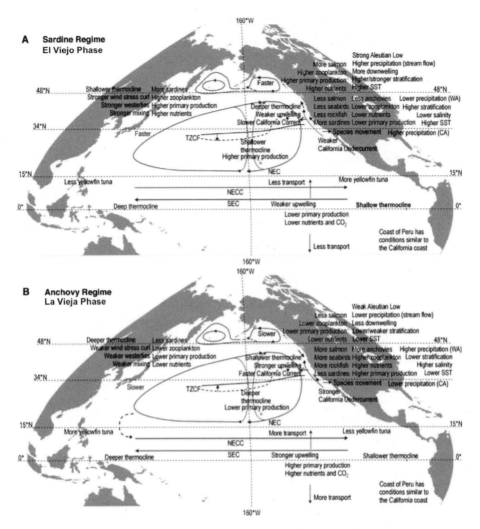

FIGURE 2.2. Biological and physical changes in the Pacific Ocean during multidecadal regimes. (A) Warm eastern tropical Pacific El Viejo phase, also characterized by a high fisheries catch of sardines. (B) Cool eastern tropical Pacific La Vieja phase, also characterized by a high fisheries catch of anchovies (Chavez et al. 2003).

anchovies and sardines. Changes in the catch rates of these species reflect the climate regime of the Pacific Ocean (fig. 2.2) (Chavez et al. 2003). When the ETP is cooler ("La Vieja") and has a shallow thermocline, high catch rates of anchovies occur off the coasts of Perú and California. During warm regimes in the ETP ("El Viejo"), sardines dominate the fisheries catch. Figure 2.2 demonstrates the physical and biological changes during each of the two multidecadal regimes in the Pacific Ocean (Chavez et al. 2003). Multidecadal regime shifts in the ETP likely have considerable effects on sea turtle populations in both an oceanographic (i.e., foraging area quality) and terrestrial (i.e., nesting beach temperature and precipitation) context.

ENSO and the Leatherback Nesting Response in Pacific Costa Rica

Energy Cost of Reproduction

Sea turtles in the ETP depend on a wide range of lower trophic levels for resources through both carnivory (e.g., leatherback, loggerhead, olive ridley, and hawksbill turtles) and omnivory (e.g., green turtles). Thus, these predators can be used as indicators of the biological conditions of their respective foraging areas as influenced by climate. Presently, the only ETP species studied in this context is the eastern Pacific (EP) leatherback turtle (Saba et al. 2007, 2008a, 2008b). Upon leaving nesting sites along the Pacific coasts of México (see chapter 9) and Costa Rica (see chapter 8), EP leatherbacks migrate south into the eastern equatorial and southeastern Pacific (Eckert and Sarti 1997; Shillinger et al. 2008), where they are subject to ecosystem variability forced by ENSO.

Leatherbacks are free-ranging reptiles that undertake vast migrations, some across an entire ocean basin (see chapter 4), to foraging areas that are rich in their primary food source, gelatinous zooplankton. Among the largest and fastest-growing reptiles in the world, leatherbacks require a high biomass of gelatinous prey to satisfy their energy demands for growth, migration, and reproduction (Wallace et al. 2006). The low energy content of gelatinous zooplankton (Doyle et al. 2007) renders leatherbacks dependent on either large organisms or large aggregations of prey. High densities of prey are even more important for mature females, which have to endure not only migration to nesting beaches and copulation with males but also the production of several clutches of nutrient-rich eggs over the course of a two-month time span. In addition, mature females must also go through the vigorous process of nesting on a high-energy beach, battling through strong currents and tides, sometimes as many as ten times in a single nesting season.

Leatherback Remigration Model

One of the most extensive sea turtle monitoring projects in the world, the time series of nesting leatherback data from Parque Nacional Marino Las Baulas (PNMB), Costa Rica, details not only their population status and overall trend (Spotila et al. 2000; Santidrián-Tomillo et al. 2007; see chapter 8) but also interannual variability in nesting numbers as a result of ENSO (Saba et al. 2007, 2008a). Nesting females at PNMB are comprehensively monitored, and almost every turtle is accounted for during each nesting season, which spans from October to February (Santidrián-Tomillo et al. 2007). Detailed data on both newly and previously tagged turtles provide an annual breakdown of each season's nesting cohort. Previously tagged turtles returning each year to nest are known as remigrants. The time between successive nesting seasons for a remigrant, the remigration interval, averages about four years for the leatherbacks at PNMB (Santidrián-Tomillo et al. 2007).

The length of the remigration interval for an individual turtle is a function of the quality of its foraging area between nesting years because of the high-energy demand of migration, reproduction, and the nesting process (Limpus and Nicholls 2000; Chaloupka 2001; Solow et al. 2002; Saba et al. 2007). When foraging areas are of high quality, the remigration interval should decrease, whereas the opposite should hold when foraging areas are of poor quality. A remigration probability model designed for the nesting population of EP leatherbacks at PNMB using ENSO indices demonstrates this relationship (fig. 2.3A) (Saba et al. 2007, 2008a). This model uses indices of ENSO in the eastern equatorial Pacific composed of SST anomaly data as measured by moored buoys (McPhaden et al. 2001) and sea surface chlorophyll data derived from an ocean color satellite sensor (Feldman and McClain 2006). The model predicts each year's number of remigrant leatherbacks returning to nest at Playa Grande, PNMB, using the number of previously tagged turtles and ENSO indices over specific time spans before the start of each nesting season. Anomalies of SST in the eastern equatorial Pacific are averaged 12 months before each nesting season, and sea surface chlorophyll is based on spring and summer blooms that can occur just after El Niño events.

Peaks in the number of remigrant leatherbacks at PNMB can be attributed to ENSO in two different ways: (1) the peak in remigrants in 1999 and 2000 were a result of the increase in NPP in the equatorial Pacific due to the 1998–2000 La Niña (figs. 2.1, 2.3A, and 2.4), and (2) the peak in remigrants in 2003 and 2005 was a result of an increase in NPP in the eastern equatorial Pacific that occurred at the end of the 2003 and 2005 El Niño events (figs. 2.3A and 2.4). It is important to note that the increase in NPP causing the remigrant peaks in 2003 and 2005 was not from La Niña (Saba et al. 2008a). Rather, the enhancement of biological production in the eastern equatorial Pacific was due to changes in the physical forcing of the western Pacific at the end of El Niño events, which caused an increase in the supply of iron to the euphotic zone in the east (Ryan et al. 2006). The dips in remigrant numbers can be attributed to El Niño events, which reduced NPP and the overall quality of EP leatherback foraging areas (figs. 2.1, 2.3A, and 2.4). If the ENSO indices are removed from the remigration probability model, the peaks and dips are no longer captured by the remigrant prediction (fig. 2.3B). This demonstrates the strong relationship between ENSO and the quality of leatherback foraging areas in the ETP. The effect of ENSO on the Pacific México nesting population (Sarti-Martínez et al. 2007; see chapter 9 in this volume) is probably similar to that of the nesting stock in Costa Rica, given their nearly identical post-nesting migration areas (Eckert and Sarti 1997).

Because of the strong response of EP leatherbacks to ENSO, it likely that this population also responds to the longer-term, multidecadal climate regimes in the ETP (Saba et al. 2008b). The EP nesting stock at Costa Rica had the strongest response toward the end and just after strong La Niña events (fig. 2.3A). This suggests that a cooler, more productive ETP enhances the reproductive output of females by increasing their probability of remigrating (Saba et al. 2007), thus decreasing their remigration interval (Reina et al. 2009). Therefore, La Vieja cool phases, or anchovy regimes, should enhance the overall reproductive output of EP leatherbacks because females will nest more frequently (fig. 2.4) (Saba et al. 2008b).

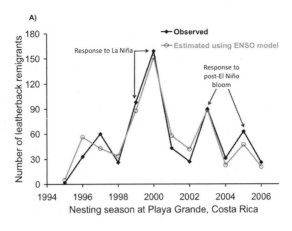

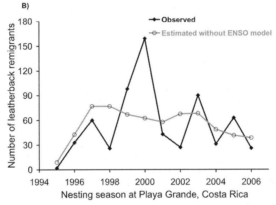

FIGURE 2.3. The effect of ENSO on the remigration probability of nesting leatherbacks in Pacific Costa Rica from 1995 to 2006: observed versus estimated number of remigrant leatherbacks nesting at Playa Grande, Parque Nacional Marino Las Baulas, with (A) and without (B) using the ENSO sea surface temperature remigration model (Saba et al. 2007) with the addition of a chlorophyll parameter (Saba et al. 2008a).

Eastern Pacific Leatherbacks: A Unique Population

Nesting Population Trends

The EP leatherback nesting stocks at México and Costa Rica have been declining precipitously over the past two decades (Sarti-Martínez et al. 2007; see chapters 8 and 9 in this volume). Although nesting beach pro-

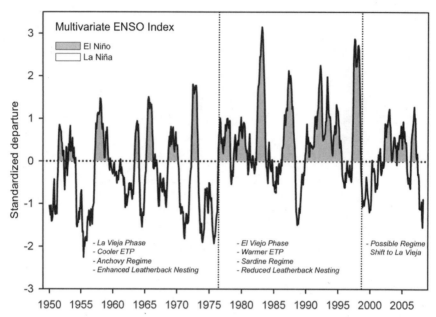

FIGURE 2.4. Time series of the Multivariate ENSO Index (MEI) from 1950 to early 2008 (Wolter and Timlin 1998), with regimes shifts in the EP (Chavez et al. 2003) and their likely influence on EP leatherback reproductive output. The influence is based on the results of the leatherback remigration model (Saba et al. 2007), which showed that cooler, more productive La Niña events increase the reproductive frequency of EP leatherbacks (see fig. 2.3). Negative departures of the MEI correspond to cool La Niña events, and positive departures correspond to warm El Niño events.

tection has been comprehensive at PNMB since the early 1990s, no signs of recovery have occurred there (Santidrián-Tomillo et al. 2007). The scientific observation of this population began in the early 1990s, when the decline in nesting numbers was first made evident. Anecdotal data suggest that the beaches of PNMB were robust in turtle numbers in the early 1970s, far outweighing what has been observed there in the 1990s through the present day (see chapter 8).

The drastic difference in nesting numbers in Costa Rica between the 1960s and 1970s and the 1990s to the present may be due to a combination of three factors: (1) The 1960s and 1970s were in the middle and toward the end of a La Vieja phase, when the ETP was cooler and possibly more biological productive than in the El Viejo phase of the 1980s and

1990s, thus enhancing nesting numbers and reproductive output of leatherbacks at PNMB (figs. 2.2 and 2.4) (Saba et al. 2008b; Wallace and Saba 2009). (2) The intense egg poaching that took place in the late 1970s and 1980s reduced the number of hatchlings, causing considerably less recruitment into future nesting cohorts (Santidrián-Tomillo et al. 2008). (3) Incidental capture among the coastal gill net fisheries along Perú and Chile in the late 1980s into the 1990s and possibly more recently may have caused a high leatherback mortality rate (de Paz et al. 2006; Alfaro-Shigueto et al. 2007, 2011; Eckert and Sarti 1997; Saba et al. 2008a). Ultimately, the shifting ecosystem states of the ETP have exacerbated the effects of anthropogenic mortality from egg poaching and fisheries interactions on the EP leatherback population, likely causing the precipitous decline (Saba et al. 2008b).

Life History Traits and Migration

The EP leatherback population is unique compared with populations that nest in the western Pacific, Atlantic, and Indian oceans. Nesting females in the EP are the smallest in body size, produce the smallest amount of eggs per clutch, and have the longest mean remigration interval, and the lowest nesting population size (Saba et al. 2008b). The high climatic variability in the ETP may be responsible for the dichotomies in life history traits, while the low population size and declining trend reflect the synergistic combination of ecosystem variability and anthropogenic mortality (Saba et al. 2008b).

Long-term satellite tracking of post-nesting females from both Costa Rica (Shillinger et al. 2008) and México (Eckert and Sarti 1997) have revealed another interesting dichotomy between EP leatherbacks and other populations: migration to coastal zones is uncommon among post-nesting EP females. The most extensive leatherback tracking study to date took place at PNMB, and although the sample size and average track durations exceeded those of any other leatherback tracking study (Shillinger et al. 2008), coastal migration was not nearly as common as it was with populations in the western Pacific (Benson et al. 2007a, 2007b), Atlantic (James et al. 2005, Eckert 2006), and Indian oceans (Lambardi et al. 2008).

In the context of ecosystem hotspots driven by high NPP fueling the food chain, coastal zones should be common foraging areas for leatherback turtles. Leatherbacks nesting in the western Pacific migrate across the entire basin to reach the productive coastal waters of northwest America (see chapter 4), while populations in the Atlantic and Indian oceans have been reported to target coastal zones at a much higher frequency than EP leatherbacks (James et al. 2005; Lambardi et al. 2008). In the ETP, not only are coastal zones the most productive, but they also remain high in NPP relative to pelagic areas during warm El Niño events (fig. 2.1). Coastal for- agers may be a minority within the EP leatherback population because of high mortality rates associated with coastal gill net fisheries (Saba et al. 2008a). This hypothesis does not suggest that post-nesting females are avoiding the coastal zones because of fishery activity; rather, it implies that turtles return to the same foraging areas year after year based on their previ- ous foraging success and survival in earlier years. Foraging area fidelity among individual turtles has been reported for leatherbacks (James et al. 2005), as well as for loggerheads and green turtles (Broderick et al. 2007). Therefore, if turtles foraging in coastal areas were subjected to a higher mortality rate than pelagic foragers, the probability of observing coastal for- agers would decrease over time.

Given the effects of ENSO on the ecosystems of the ETP, coastal areas should be common areas for post-nesting females, especially during El Niño events when the pelagic equatorial Pacific NPP is reduced (fig. 2.1B). But this is not the case for EP leatherbacks, suggesting that coastal fisheries may have drastically reduced the size of the coastal foraging population of adult female leatherbacks, thus increasing the entire population's sensitivity to ENSO (Saba et al. 2008a). In addition, recent reports have suggested that most leatherbacks caught as bycatch among the coastal gill net fisheries of Perú were juveniles and subadults (de Paz et al. 2006; Alfaro-Shigueto et al. 2007). This may be the result of a reduced number of adults in the popula- tion, or ontogenetic differences in habitat use.

Regime Shift

A regime shift may have occurred in the late 1990s, marking the end of El Viejo and the beginning of La Vieja (fig. 2.4) (Chavez et al. 2003). If La Vieja prevails over the next few decades in the ETP, the response of sea turtles could be evident through an increase in the nesting numbers and reproductive output of EP leatherbacks (Saba et al. 2008b). However, if coastal mortality of leatherbacks remains high, the increase in nesting numbers might not rise to expected levels.

Effects of Climate Variability on Sea Turtles Outside the ETP

Although the other four species of sea turtle in the ETP have not been studied in relation to ENSO and ecosystem variability, it is highly likely that their populations are also affected. However, studies at nesting beaches outside of the ETP have reported effects of climate variability on sea turtle nesting ecology. Loggerheads nesting in Australia and Japan responded to changes in SST at their regional foraging areas near nesting beaches (Chaloupka et al. 2008). As with EP leatherbacks, cooler SST periods resulted in higher nesting abundance, and vice versa during warmer SST periods. Chaloupka et al. (2008) used SST in the western Pacific and did not incorporate SST from the ETP, where juvenile and subadult loggerheads from these populations also commonly forage. Green turtle nesting numbers in the western Pacific were also a function of ENSO: cool SST periods increased nesting numbers (Limpus and Nicholls 2000; Chaloupka 2001). On the other hand, western Atlantic green turtles nesting at Tortuguero, Costa Rica, had a higher nesting probability just after warmer SST periods in the winter before each nesting season (Solow et al. 2002). That study, however, used SST from just one location in the Caribbean Sea (La Parguera, Puerto Rico) to represent the foraging area quality of the greens nesting at Tortuguero. If, in fact, green turtles have opposite responses in the western Pacific versus western Atlantic, this may be due to a difference in the trophic status of the populations, which could be the determining factor for how each population will respond to environmental changes in their foraging areas (Broderick et al. 2001).

Sea Turtle Trophic Status and Prey Responses to Climate Variability in the ETP

Olive Ridley and Loggerhead Turtles

It appears that gelatinous zooplankton in the ETP are affected by NPP transitions governed by ENSO, based on the strong response of leatherbacks nesting in Pacific Costa Rica. Gelatinous zooplankton are also dietary components of the other species of sea turtle in the ETP. Gelatinous organisms are short-lived and can respond quickly to environmental changes, and thus the time lag between the effects of ENSO and the response of nesting females should be short (one to two years before each nesting season [Saba et al. 2007]).

Olive ridley turtles foraging in the equatorial upwelling zone of the ETP have been reported to feed almost exclusively on gelatinous zooplankton (Kopitsky et al. 2001). This foraging behavior may represent a specific strategy while inhabiting the deep pelagic waters, and the turtles may change their diet when in shallower coastal waters. Nonetheless, if olive ridleys commonly forage on gelatinous zooplankton in the pelagic equatorial Pacific, as leatherbacks are, they should have similar responses to ENSO and multidecadal regime shifts. Although EP olive ridley nesting patterns and climate variability have not been analyzed in detail, limited data suggest the turtles are responding to ENSO. In the pelagic waters of the ETP, olive ridleys altered their migration pathways during warm El Niño events (Plotkin 2010). Unlike the other sea turtles of the ETP, many olive ridley nesting populations have increased over the past decade (see chapter 13), as observed through nesting trends (Chaloupka et al. 2005) and at-sea abundance estimates (Eguchi et al. 2007). The increase is attributed to protection of both turtles and nests (see chapter 13), and may have also been fueled partly by a regime shift to La Vieja in the late 1990s (fig. 2.4), especially given that these turtles are feeding on gelatinous zooplankton (Kopitsky et al. 2001) while in the pelagic ETP. Therefore, it is reasonable to surmise that EP olive ridleys should respond positively to cooler SST anomalies as leatherbacks do. The reason that EP leatherbacks have yet to respond to the late 1990s regime shift could be differences in life history traits (e.g., age at maturity), energy demands, and differential anthropogenic mortality between the two species.

Some loggerheads from nesting beaches in the western Pacific endure a trans-Pacific migration to the Baja California coast, where they typically consume red crabs (Peckham 2002). Along the Baja coastline in the California Current, aggregations of red crab have been linked to highly productive upwelling regions (Robinson et al. 2004). Most striking, one study reported that 90 percent of the total abundance of red crab along Baja was found in areas where upwelling and chlorophyll-a were intensified (Robinson et al. 2004). This trophic relationship suggests that periods of cool SST and more intensified upwelling are favorable for loggerhead turtles foraging off the Baja coast. Loggerheads foraging closer to nesting beaches in the western Pacific also respond positively to cool SST (Chaloupka et al. 2008). However, the effects of ENSO on loggerheads foraging off the Baja coast may differ from ENSO effects on leatherbacks and possibly olive ridleys. Figure 2.1 shows the increase in NPP along the offshore waters of Baja during El Niño, for reasons discussed earlier in this chapter. Red crab abundance in this area may increase during El Niño events, making these favorable foraging periods for loggerheads. However, this concept is purely speculative, and further research is required to better understand the effects of ENSO on the loggerheads of the ETP.

Green and Hawksbill Turtles

Although most green turtle populations are primarily herbivorous, those foraging in the Gulf of California have a mixed diet of seagrasses, algae, and some animal matter (Seminoff et al. 2002). Green turtles have also been reported to have an omnivorous diet along the Pacific coast of Colombia (Amorocho and Reina 2008) and off the west coast of Baja, occasionally foraging on red crabs (López-Mendilaharsu et al. 2005). The level of omnivory in green turtles (i.e., the amount of animal matter in the diet) is likely a function of the habitat where the turtles forage (Amorocho and Reina 2008). For example, green turtles foraging in the deep pelagic regions have been reported to consume mostly animal prey items (Seminoff et al. 2008), suggesting that the turtles may be more herbivorous in shallow coastal waters.

Satellite tracking of post-nesting green turtles from the Galápagos Islands revealed three distinct migration patterns: (1) to the coastal zones of the Galápagos Islands, (2) to the coastal zones of Central America, and (3) to the pelagic regions west and southwest of the Galápagos Islands (Seminoff et al. 2008). Although unresolved, diets of green turtles feeding in these three distinct foraging areas likely differ.

With the exception of the coast of Baja and the Gulf of California, NPP of green turtle migration areas decreases during El Niño events (fig.

2.1B). If the green turtle diet in the ETP is similar to that of green turtles foraging in the western Pacific, EP greens should benefit from cool SST periods. However, if the EP green turtle diet is more similar to those in the western Atlantic, warm SST periods may be favorable. This may be a function of the typical habitat used by the turtles. For example, if warm SST in the Caribbean increases the nesting probability of green turtles at Tortuguero, Costa Rica, as suggested by Solow et al. (2002), this might indicate that clear waters favor those turtles. Green turtles of the Caribbean are primarily herbivorous and target seagrasses, so warmer, clearer water that is low in NPP from microphytes should be favorable. Seagrasses are macrophytes that flourish in clear waters (Short et al. 2007) that are typically low in NPP derived from microphytes, and thus more light penetrates to the ocean floor to stimulate growth. Seagrasses have roots that can exploit sediment pore-water nutrient pools, and they have rhizomes that can convert otherwise unusable nitrogen into its nutrient form through nitrogen fixation (Welsh 2000). This allows seagrasses to thrive in waters that are nutrient poor and not favorable for most microphytic plankton that cannot undergo nitrogen fixation. Therefore, certain tropical clear waters do not show high NPP through ocean color measurements but may still be high in NPP in the benthic zones derived from macrophytes such as seagrasses. One way to examine possible effects of ENSO on EP green turtles would be to apply the ENSO remigration model (Saba et al. 2007) to nesting green turtles in the ETP. The results might give more insight to their habitat preference and diet.

Unfortunately, research focused on the migration and foraging of the EP hawksbill is extremely limited. Although anecdotal data suggest that hawksbills were more abundant in the past, these turtles are now rare in the ETP (see chapter 10). One study reported hawksbill turtles occurring off the coast of Baja and in the Gulf of California, where a few specimens had fragments of sponge in their stomachs (Seminoff et al. 2003). Hawksbills in other parts of the world are typically found in coastal waters, especially coral reef ecosystems, where they feed on benthic animals, including sponges and corals (León and Bjorndal 2002). Like seagrasses, coral reef ecosystems are associated with clear waters that are low in NPP from microphytes, so the response of EP hawksbills to ENSO might be the opposite to that of leatherbacks.

A Call for Future Research

Presently, only one of the five species of sea turtle occurring in the ETP has been studied extensively in relation to climate variability (Saba et al. 2007, 2008a, 2008b). Given the critical conservation status of EP sea turtles, with the exception of olive ridleys, future research in this area is crucial to better understand influences on their demography and recovery. Elucidating the influence of climate variability on sea turtles also helps distinguish the effects of anthropogenic mortality on their population dynamics. Human-induced climate change will inevitably affect the global ocean. The ETP is already subject to a high natural climate variability, and the added influence of global warming may disrupt these natural cycles. Whether this disruption will be positive or negative in the short and long term for sea turtles is unknown.

This chapter has focused only on the effects of climate variability on the foraging areas of sea turtles and has not discussed the effects on nesting beaches. A recent study at PNMB, Costa Rica, showed that La Niña events increased the hatching success and hatchling emergence rate of leatherback nests (Santidrián-Tomillo et al., unpublished data) and increased the proportion of male hatchlings (Sieg 2010), because higher rates of rainfall associated with La Niña keep the deep nest temperatures cooler. A recent modeling study that projected the impacts of climate change on EP leatherbacks nesting at PNMB, Costa Rica, suggested that this population will decline at a rate of 7 percent per decade over the twenty-first century (Saba et al., unpublished data). This decline was not due to changes in ENSO variability, NPP in the ETP, or hatchling sex ratio; rather, it was the dramatic decrease in hatching success and hatchling emergence rate due to the projected 3°C rise in air temperature in northwestern Costa Rica via continued greenhouse gas emissions globally (Saba et al., unpublished data). Therefore, the local climate of nesting beaches in the ETP and worldwide must also be considered when investigating the population dynamics of sea turtles.

ACKNOWLEDGMENTS

I am grateful to the local people of Playa Grande, Costa Rica, to countless Earthwatch volunteers, to the Ministerio de Ambiente y Energía, and to various staff and students for their help with the leatherback tagging

project at PNMB. I thank J. Spotila, F. Paladino, R. Reina, R. Piedra, P. Santidrián-Tomillo, G. Blanco, E. Vélez, C. Padilla, M. Boza, G. Shillinger, B. Wallace, and E. Chacón for their support to leatherback research and conservation at PNMB. Funding for the leatherback tagging study at PNMB was provided by Earthwatch Institute and The Leatherback Trust.

REFERENCES

Alfaro-Shigueto, J., Dutton, P. H., Van Bressem, M., Mangel, J. 2007. Interactions between leatherback turtles and Peruvian artisanal fisheries. Chelonian Conservation and Biology 6, 129–134.

Alfaro-Shigueto, J., J. C. Mangel, F. Bernedo, P. H. Dutton, J. A. Seminoff, and B. J. Godley. 2011. Small scale fisheries of Perú: a major sink for marine turtles in the Pacific? Journal of Applied Ecology 48:1432–1440.

Amorocho, D. F., Reina, R. D. 2008. Intake passage time, digesta composition and digestibility in East Pacific green turtles (*Chelonia mydas agassizii*) at Gorgona National Park, Colombian Pacific. Journal of Experimental Marine Biology and Ecology 360, 117–124.

Ballance, L. T., Pitman, R. L., Fiedler, P. C. 2006. Oceanographic influences on seabirds and cetaceans of the eastern tropical Pacific: a review. Progress in Oceanography 69(2–4), 360.

Behrenfeld, M. J., Falkowski, P. G. 1997. Photosynthetic rates derived from satellite-based chlorophyll concentration. Limnology and Oceanography 42, 1–20.

Benson, S. R., Dutton, P. H., Hitipeuw, C., Samber, B., Bakarbessy, J., Parker, D. 2007a. Post-nesting migrations of leatherback turtles (*Dermochelys coriacea*) from Jamursba-Medi, Bird's Head Peninsula, Indonesia. Chelonian Conservation and Biology 6, 150–154.

Benson, S. R., Kisokau, K. M., Ambio, L., Rei, V., Dutton, P. H., Parker, D. 2007b. Beach use, inter-nesting movement, and migration of leatherback turtles, *Dermochelys coriacea*, nesting on the north coast of Papua New Guinea. Chelonian Conservation and Biology 6, 7–14.

Boyle, M. C., FitzSimmons, N. N., Limpus, C. J., Kelez, S., Velez-Zuazo, X., Waycott, M. 2009. Evidence for transoceanic migrations by loggerhead sea turtles in the southern Pacific Ocean. Proceedings of the Royal Society of London, Series B 276, 1993–1999.

Broderick, A. C., Coyne, M. S., Fuller, W. J., Glen, F., Godley, B. J. 2007. Fidelity and over-wintering of sea turtles. Proceedings of the Royal Society of London, Series B 274, 1533–1538.

Broderick, A. C., Godley, B. J., Hays, G. C. 2001. Trophic status drives interannual variability in nesting numbers of marine turtles. Proceedings of the Royal Society of London, Series B 268, 1481–1487.

Chaloupka, M. 2001. Historical trends, seasonality and spatial synchrony in green sea turtle egg production. Biological Conservation 101, 263–279.

Chaloupka, M. L., Dutton, P. H., Nakano, H. 2005. Status of sea turtle stocks in the Pacific. In: Papers presented at the Expert Consultation on Interactions between Sea Turtles

and Fisheries within an Ecosystem Context: Rome, Italy, 9–12 March 2004. FAO Fisheries Rep. No. 738 Supplement. Food and Agriculture Organization, Fishery Resources Division, Rome. Pp. 135–164.

Chaloupka, M., Kamezaki, N., Limpus, C. 2008. Is climate change affecting the population dynamics of the endangered Pacific loggerhead sea turtle? Journal of Experimental Marine Biology and Ecology 356, 136–143.

Chavez, F. P., Ryan, J., Lluch-Cota, S. E., Ñiquen, M. C. 2003. From anchovies to sardines and back: multidecadal change in the Pacific Ocean. Science 299, 217–221.

De Paz, N., Reyes, J. C., Ormeño, M., Anchante, H. A., Altamirano, A. J. 2006. Immature leatherback mortality in coastal gillnet fisheries off San Andrés, southern Perú. In: Frick, M., Panagopoulou, A., Rees, A. F., Williams, K. (Eds.), Twenty-Sixth Annual Symposium on Sea Turtle Biology and Conservation. International Sea Turtle Society, Crete. Archelon, Athens, Greece. P. 376.

Douglas, R., Gonzalez-Yajimovich, O., Ledesma-Vazquez, J., Staines-Urias, F. 2007. Climate forcing, primary production and the distribution of Holocene biogenic sediments in the Gulf of California. Quaternary Science Reviews 26, 115–129.

Doyle, T. K., Houghton, D. R., McDevitt, R., Davenport, J., Hays, G. C. 2007. The energy density of jellyfish: estimates from bomb-calorimetry and proximate-composition. Journal of Experimental Marine Biology and Ecology 343, 239–252.

Eckert, S. A. 2006. High-use oceanic areas for Atlantic leatherback sea turtles (*Dermochelys coriacea*) as identified using satellite telemetered location and dive information. Marine Biology 149, 1257–1267.

Eckert, S. A., Sarti, L. 1997. Distant fisheries implicated in the loss of the world's largest leatherback nesting population. Marine Turtle Newsletter 78, 2–7.

Eguchi, T., Gerrodette, T., Pitman, R. L., Seminoff, J. A., Dutton, P. H. 2007. At-sea density and abundance estimates of the olive ridley turtle *Lepidochelys olivacea* in the eastern tropical Pacific. Endangered Species Research 3, 191–203.

Feldman, G. C., McClain, C. R. 2006. Ocean Color Web, SeaWiFS Reprocessing 5.0, NASA Goddard Space Flight Center. Available: http://oceancolor.gsfc.nasa.gov.

Fiedler, P. C. 2002. The annual cycle and biological effects of the Costa Rica Dome. Deep-Sea Research I 49, 321–338.

James, M. C., Myers, R. A., Ottensmeyer, C. A. 2005. Behaviour of leatherback sea turtles, *Dermochelys coriacea*, during the migratory cycle. Proceedings of the Royal Society of London, Series B 272, 1547–1555.

Kahru, M. 2006. Windows Image Manager (WIM) and WAM. Available: www.wimsoft .com.

Kahru, M., Mitchell, B. G. 2000. Influence of the 1997–98 El Niño on the surface chlorophyll in the California Current. Geophysical Research Letters 27, 2937–2940.

Kopitsky, K. L., Pitman, R. L., Dutton, P. H. 2001. Aspects of olive ridley feeding ecology in the eastern tropical Pacific. In: Coyne, M. S., Clark, R. D. (Eds.), Proceedings of the Twenty-First Annual Symposium on Sea Turtle Biology and Conservation. U.S. National Marine Fisheries Service, Philadelphia, PA. P. 368.

Ladah, L. B. 2003. The shoaling of nutrient-enriched subsurface waters as a mechanism to sustain primary productivity off central Baja California during El Niño winters. Journal of Marine Systems 42, 145–152.

Lambardi, P., Lutjeharms, J. R. E., Mencacci, R., Hays, G. C., Luschi, P. 2008. Influence of ocean currents on long-distance movement of leatherback sea turtles in the southwest Indian Ocean. Marine Ecology Progress Series 353, 289–301.

León, Y. M., Bjorndal, K. A. 2002. Selective feeding in the hawksbill turtle, an important predator in coral reef ecosystems. Marine Ecology Progress Series 245, 249–258.

Limpus, C. J., Nicholls, N. 2000. ENSO regulation of Indo-Pacific green turtle populations. In: Hammer, G. L., Nicholls, N., Mitchell, C. (Eds.), The Australian Experience. Kluwer, Dordrecht. Pp. 399–408.

López-Mendilaharsu, M., Gardner, S. C., Seminoff, J. A., Riosmena-Rodríguez, R. 2005. Identifying critical foraging habitats of the green turtle (*Chelonia mydas*) along the Pacific coast of the Baja California peninsula, México. Aquatic Conservation: Marine and Freshwater Ecosystems 15, 259–269.

McPhaden, M. J., Delcroix, T., Hanawa, K., Kuroda, Y., Meyers, G., Picaut, J., Swenson, M. 2001. The El Niño-Southern Oscillation (ENSO) Observing System. In: Koblinsky, C. J., Smith, N. R. (Eds.), Observing the Ocean in the 21st Century. Australian Bureau of Meteorology, Melbourne. Pp. 231–246.

McPhaden, M. J., Zebiak, S. E., Glantz, M. H. 2006. ENSO as an integrating concept in Earth science. Science 314, 1740–1745.

Nichols, W. J., Resendiz, A., Seminoff, J. A., Resendiz, B. 2000. Transpacific migration of a loggerhead turtle monitored by satellite telemetry. Bulletin of Marine Science 67, 937–947.

Peckham, H. 2002. Why did the turtle cross the ocean? Pelagic red crabs and loggerhead turtles along the Baja California coast. In: Seminoff, J. A. (Ed.), Proceedings of the Twenty-Second Annual Symposium on Sea Turtle Biology and Conservation. U.S. National Marine Fisheries Service, Miami, FL. P. 47.

Pennington, J. T., Mahoney, K. L., Kuwahara, V. S., Kolber, D. D., Calienes, R., Chavez, F. P. 2006. Primary production in the eastern tropical Pacific: a review. Progress in Oceanography 69, 285–317.

Plotkin, P. T. 2010. Nomadic behaviour of the highly migratory olive ridley sea turtle *Lepidochelys olivaea* in the eastern tropical Pacific Ocean. Endangered Species Research 13, 33–40.

Reina, R. D., Spotila, J. R., Paladino, F. V., Dunham, A. E. 2009. Changed reproductive schedule of leatherback turtles *Dermochelys coriacea* following the 1997–98 El Niño to La Niña transition. Endangered Species Research 7, 155–161.

Robinson, C. J., Anislado, V., Lopez, A. 2004. The pelagic red crab (*Pleuroncodes planipes*) related to active upwelling sites in the California Current off the west coast of Baja California. Deep-Sea Research II 51, 753–766.

Ryan, J. P., Ueki, I., Chao, Y., Zhang, H. C., Polito, P. S., Chavez, F. P. 2006. Western Pacific modulation of large phytoplankton blooms in the central and eastern equatorial Pacific. Journal of Geophysical Research, Biogeosciences 111, G02013.

Saba, V. S., Santidrián-Tomillo, P., Reina, R. D., Spotila, J. R., Musick, J. A., Evans, D. A., Paladino, F. V. 2007. The effect of the El Niño Southern Oscillation on the reproductive frequency of eastern Pacific leatherback turtles. Journal of Applied Ecology 44, 395–404.

Saba, V. S., Shillinger, G. L., Swithenbank, A. M., Block, B. A., Spotila, J. R., Musick, J. A., Paladino, F. V. 2008a. An oceanographic context for the foraging ecology of eastern Pacific leatherback turtles: consequences of ENSO. Deep-Sea Research I 55, 646–660.

Saba, V. S., Spotila, J. R., Chavez, F. P., Musick, J. A. 2008b. Bottom-up and climatic forcing on the worldwide population of leatherback turtles. Ecology 89, 1414–1427.

Santidrián-Tomillo, P., Saba, V. S., Piedra-Chacón, R., Paladino, F. V. 2008. Egg poaching: a major factor in the population decline of leatherback turtles, *Dermochelys coriacea*, at Parque Nacional Marino Las Baulas, Costa Rica. Conservation Biology 22, 1216–1224.

Santidrián-Tomillo, P., Vélez, E., Reina, R. D., Piedra, R., Paladino, F. V., Spotila, J. R. 2007. Reassessment of the leatherback turtle (*Dermochelys coriacea*) population nesting at Parque Nacional Marino Las Baulas: effects of conservation efforts. Chelonian Conservation and Biology 6, 54–62.

Sarti-Martínez, L., Barragán, A. R., Muñoz, D. G., García, N., Huerta, P., Vargas, F. 2007. Conservation and biology of the leatherback turtle in the Mexican Pacific. Chelonian Conservation and Biology 6, 70–78.

Seminoff, J. A. 2004. Occurrence of loggerhead sea turtles (*Caretta caretta*) in the Gulf of California, México: evidence of Life-History Variation in the Pacific Ocean. Herpetological Review 35, 24–27.

Seminoff, J. A., Nichols, W. J., Resendiz, A., Brooks, L. 2003. Occurrence of hawksbill turtles, *Eretmochelys imbricata* (Reptilia: Cheloniidae), near the Baja California peninsula, México. Pacific Science 57, 9–16.

Seminoff, J. A., Resendiz, A., Nichols, W. J. 2002. Diet of east Pacific green turtles (*Chelonia mydas*) in the central Gulf of California, México. Journal of Herpetology 36, 447–453.

Seminoff, J. A., Zárate, P., Coyne, M., Foley, D. G., Parker, D., Lyon, B. N., Dutton, P. H. 2008. Post-nesting migrations of Galápagos green turtles *Chelonia mydas* in relation to oceanographic conditions: integrating satellite telemetry with remotely sensed ocean data. Endangered Species Research 4, 57–72.

Shillinger, G. L., Palacios, D. M., Bailey, H., Bograd, S. J., Swithenbank, A. M., Gaspar, P., Wallace, B. P., Spotila, J. R., Paladino, F. V., Peidra, R., Eckert, S. A., Block, B. A. 2008. Persistent leatherback migration corridor presents opportunities for conservation. PLoS Biology 6, e171.

Short, F., Carruthers, T., Dennison, W., Waycott, M. 2007. Global seagrass distribution and diversity: a bioregional model. Journal of Experimental Marine Biology and Ecology 350, 3–20.

Sieg, A. E. 2010. Physiological constraints on the ecology of activity-limited ectotherms. Dissertation, Drexel University, Philadelphia, PA.

Solow, A. R., Bjorndal, K. A., Bolten, A. B. 2002. Annual variation in nesting numbers of marine turtles: the effect of sea surface temperature on re-migration intervals. Ecology Letters 5, 742–746.

Spotila, J. R., Reina, R. D., Steyermark, A. C., Plotkin, P. T., Paladino, F. V. 2000. Pacific leatherback turtles face extinction. Nature 405, 529–530.

Thunell, R. C. 1998. Seasonal and annual variability in particle fluxes in the Gulf of California: a response to climate forcing. Deep-Sea Research I 45, 2059–2083.

Turk, D., McPhaden, M. J., Busalacchi, A. J., Lewis, M. R. 2001. Remotely sensed biological production in the equatorial Pacific. Science 293, 471–474.

Wallace, B. P., Kilham, S. S., Paladino, F. V., Spotila, J. R. 2006. Energy budget calculations indicate resource limitation in eastern Pacific leatherback turtles. Marine Ecology Progress Series 318: 263–270.

Wallace, B. P., Saba, V. S. 2009. Environmental and anthropogenic impacts on intra-specific variation in leatherback turtles: opportunities for targeted research and conservation Endangered Species Research 7, 1–11.

Welsh, D. T. 2000. Nitrogen fixation in seagrass meadows: regulation, plant-bacteria interactions and significance to primary productivity. Ecology Letters 3, 58–71.

Wolter, K., Timlin, M. S. 1998. Measuring the strength of ENSO—how does 1997/98 rank? Weather 53, 315–324.

3

Offshore Oasis

Ecology of Sea Turtles at Oceanic Islands of the Eastern Pacific

PATRICIA ZÁRATE

Summary

Islands are oases for marine life and terrestrial biodiversity, and their isolation has resulted in unique flora and plant life with a more limited number of anthropogenic threats than found in mainland areas. Thus, it should be no surprise that sea turtles are attracted to these characteristics of marine areas around islands. The green turtle is the most common and abundant sea turtle species associated with islands of the eastern Pacific (EP). The Galápagos, Revillagigedos, and Tres Marías archipelagos, among others, host green turtle nesting and foraging, whereas Cocos Island, Easter Island, Malpelo Island, and Gorgona are suitable places just for feeding. Olive ridley turtles nest on Gorgona, and some leatherback nesting takes place on Coiba. Sea turtle nesting has disappeared from Clipperton Island. Most islands of the EP have relatively low human influence, and anthropogenic threats at nesting beaches are minimal, while introduced species have perhaps the greatest impact on nesting turtles, eggs, and hatchlings. In the marine realm, artisanal and industrial fishing fleets are ubiquitous in the region, and waters adjacent to these islands are no exception. Several important international conservation efforts are under way in the region, but further implementation and integration among these are necessary to provide the appropriate protection to nesting beaches and marine habitats of the EP.

Introduction

I came to the Galápagos Islands in July 2000 on a wonderful sunny day and found myself surrounded by a spectacular nature show: turquoise waters, a great diversity of seabirds flying overhead, marine iguanas sunning themselves along rocky shores, seemingly unaware of humans, and sea lions resting along the docks near the main street of Puerto Ayora, the arrival point for most of the archipelago's visitors on Santa Cruz Island. Visiting the place that provided inspiration for Darwin's theory of evolution by natural selection was a dream for me, as for any other biologist.

Fortunately, my dream continued when I received a position at Charles Darwin Foundation to study impacts of artisanal fisheries on marine resources in the Galápagos. However, before arriving in the Galápagos, I had attended a striking presentation by sea turtle luminary Dr. Peter Dutton that focused on the biology of the giant leatherback sea turtle and current threats for this and other sea turtle species. Dutton and I spoke after his presentation, and we agreed to establish a close collaboration between our institutions to study sea turtles in the Galápagos.

Two months later, Dutton and two other sea turtle specialists, Miguel Donoso from Chile and Gustavo Iturralde from Ecuador, came to Galápagos for our first sea turtle research trip—a two-day trip to foraging areas around Santa Cruz Island. We used the "turtle rodeo" technique to catch three long-tailed male green turtles (*Chelonia mydas*) at Caleta Tortuga Negra along the north coast of Santa Cruz Island. I learned how to apply flipper tags for identification, take morphological measurements, and collect blood and tissue samples. Although I was still officially studying artisanal fisheries, this short excursion cemented my desire to switch gears and pursue sea turtle research and conservation. A few weeks later, I was given the green light to dedicate full-time work to sea turtles, on the condition that I could raise sufficient funds to support myself and the research. With encouragement and assistance from Dutton and his team from the U.S. National Oceanic and Atmospheric Administration (NOAA), and my boss's blessing, I started the sea turtle program. From that point, I could think about nothing other than sea turtles in the Galápagos.

Oceanic Islands and Archipelagos in the Eastern Pacific

The Galápagos Archipelago is certainly the most famous island group in the eastern Pacific Ocean (hereafter EP), but other island formations in the region share similar traits. These small spits of land surrounded by vast blueness are oases for marine as well as terrestrial biodiversity. Ocean currents, carrying nutrients, oxygen, heat, and other ingredients necessary to create marine food webs, swirl around and run into these islands—like a roiling river washing against stones—creating concentrated areas of high productivity that support abundant marine life. Breathtaking aggregations of whales, hammerhead sharks, seabirds, and other migratory fauna bear testimony to the special conditions offered by these islands, many of which are designated World Heritage Sites by the United Nations Educational, Scientific and Cultural Organization (UNESCO) or enjoy some form of national or international recognition as protected areas. Anthropogenic threats to these marine ecosystems tend to be more limited than threats occurring closer to mainland areas, because of the islands' relative isolation. This isolation has resulted in not only unique animal and plant life but also unusual human encampments: several islands in the region are home to penal colonies or military installations. As Charles Darwin learned during his voyage on the HMS *Beagle*, oceanic islands act as "nature's laboratories," where animals and plants are separated from their mainland brethren, subject to natural processes in relatively isolated settings, but also can be highly susceptible to threats alien to their native environments.

Other oceanic islands in the EP include Revillagigedo Islands, Guadalupe Island, and Rocas Alijos in México; Clipperton Island, an overseas holding of France; Cocos Island in Costa Rica; and Easter Island in Chile (see plate 1). Several other islands, located not in the open ocean but on the continental shelf, function similarly to their oceanic counterparts as important habitats for unique flora and fauna; among these are the Tres Marías Islands in México, Coiba in Panamá, and Malpelo Island and Gorgona in Colombia. Like the Galápagos, all of these sites host sea turtles, and although information from these areas is relatively scarce, they provide other examples of ocean oases, where sea turtles thrive in relatively pristine marine habitats and nesting beaches. Given the preponderance of information available from the Galápagos compared with other EP island formations, this chapter focuses on current knowledge of sea turtle natural history and recent conservation efforts in the Galápagos Archipelago but also recognizes these other oceanic and continental islands and their local sea turtle populations.

Galápagos Archipelago, Ecuador

The oceanographic conditions surrounding the Galápagos Archipelago—located along the equator roughly 500 nautical miles west of mainland Ecuador—have created extreme biogeographic isolation and the greatest level of vertebrate endemism in the entire EP. The Galápagos Archipelago, volcanic in origin, comprises thirteen major islands (total area > 10 km²), five islands of medium size (area between 1–10 km²), and 215 islets (<1 km²) (Parque Nacional Galápagos 2006). Because of its uniqueness and incredible beauty, the Galápagos National Park was the first designated World Heritage Site in 1978 and was declared a Biosphere Reserve in 1984; in 2001 its World Heritage Site was expanded to include the breathtaking Galápagos Marine Reserve, considered one of the top sites in the world for diving (Sammon 1992; Scuba Diving Magazine 2010).

SEA TURTLE RESEARCH BEFORE 2000. The Galápagos Archipelago has been the primary island system for sea turtle research in the eastern tropical Pacific, as several "generations" of researchers have studied both nesting and foraging areas in the islands. The first scientific endeavor was the Joseph R. Slevin expedition (California Academy's scientific expedition in 1905–1906), during which biologists sacrificed numerous green turtles, taking flipper measurements and recording stomach contents (Fritz 1981). However, it was not until 1970 that research targeting green turtles began, initially with Peter Pritchard's exploratory surveys along the coasts of most islands to determine the extent of green turtle nesting in the archipelago (Pritchard 1971, 1972), and later with a flipper tagging program performed by Pritchard, Miguel Cifuentes, and Judy Webb to study green turtle nesting in the islands.

During the nearly seven decades of research inactivity between these expeditions, sea turtles were utilized as a food source by sailors and island inhabitants. In fact, some of the only accounts we have of sea turtles in the Galápagos during this period come from a canning operation on the islands that relied on local green turtles, and from sailors passing through the area that landed green turtles for food (fig. 3.1) (Hoff 1985). By the 1970s, commercial exploitation of sea turtles was common because they were abundant, large (>130 kg), and easy to obtain (Pritchard 1971). How-

FIGURE 3.1. (Top) Green turtles and marine iguanas provided food for visiting European sailors (Photo by Arne Eilertsen). (Bottom) Green turtles were butchered for the cannery in the Galápagos Archipelago (Photo courtesy Arne Falk-Rönne).

ever, Japanese longline fishermen had been operating in the islands for almost a decade, and their activity increased in the beginning of the 1970s, including exploitation of sea turtles. Turtles were caught by local fishermen and frozen onboard the Japanese vessels (Lundh 2004). The last official report of commercial exploitation of sea turtles at Galápagos occurred in 1971 and 1972, when the Japanese vessel *Chikuzen Marou* caught 2,000–3,000 turtles, most of which were adult female green turtles but also included juveniles and adult males (Green 1978; Lundh 2004).

Historically as well as today, green turtles are the only species that nests in the Galápagos and are by far the most common, occurring in great abundance in the marine habitats of virtually every island in the archipelago (plate 5). The Galápagos nesting population is one of two major rookeries for the species in the EP region, totaling more than 1,000 nesting females per year. The other is Playa Colola, Michoacán, México (Seminoff 2004; see chapter 11). Although green turtles in the EP are depleted relative to past levels due to the aforementioned human consumption of eggs and meat, numbers in recent years have been stable and perhaps increasing (Seminoff 2004).

The first exhaustive nesting beach and foraging habitat assessments in the Galápagos were performed by Derek Green from 1975 to 1980. During that time, Green tagged more than 2,300 green turtles and a few hawksbills at foraging areas and more than 4,000 turtles at nesting sites. This research resulted in a seminal paper describing the long-distance migrations of green turtles tagged on Galápagos beaches and recaptured in Central and South America, from Costa Rica to Perú, demonstrating an important linkage between the Galápagos and the rest of the region (Green 1984). Another influential paper described the growth of green turtles in foraging areas— growth so slow that Green suggested some turtles might take up to 50 years to reach maturity (Green 1993).

Green also reported an interesting dichotomy among green turtles at Galápagos foraging grounds that had been mentioned first by Joseph R. Slevin during his expedition (Slevin 1931) and later by Archie Carr (Carr 1967): the presence of a dark morph of turtle that nested and foraged locally, as well as a lighter morph known as the "yellow turtle" (plate 4), which was present in feeding areas but was never seen nesting. In addition to differences in coloration, these researchers described the relatively underdeveloped reproductive organs of yellow males and females compared with those of darker-morph turtles of a similar size. In addition, the yellow turtle was fattier, yielding six to eight times more oil than the dark turtle. Green suggested that only 1 percent of green turtles in the Galápagos were of the yellow morph, which recent findings have confirmed (see below).

For several years during Green's tenure on the islands, he often worked closely with Mario Hurtado, an Ecuadorian scientist who continued investigating the reproductive activity of green turtles (Hurtado 1984). These studies culminated in an estimated 1,400 nesting females nesting annually from 1970 to 1983 and a total of nearly 9,000 females tagged on Galápagos beaches during those years (Hurtado 1984). It was during these efforts, helped in part by Pritchard's earlier exploits, that beaches of Quinta Playa and Bahía Barahona on southern Isabela Island, Las Salinas on Baltra Island, Las Bachas on northern Santa Cruz Island, and Espumilla on Santiago Island were recognized as the most important nesting sites in the archipelago (Green 1994). However, 1983 marked the last year of this research, and not until 2000 did nesting beaches again become the focus of scientific investigation.

RECENT MONITORING WORK. Today, the conservation status of the green turtle nesting population in the Galápagos is positive, with adults afforded a high level of protection and the annual number of nesting females stable. Hatching and emerging success can vary from 40 percent to 70 percent among nesting sites, depending on the type of existing threats, and although feral animals such as cats, dogs, and pigs exist on some beaches, their numbers are controlled by the Galápagos National Park (GNP) guards' exotic species eradication program (see below). After successfully monitoring seven years of nesting activity at key sites beginning in 2001, more than 10,000 females have been tagged.

In addition to the nesting beach work, there also has been considerable research effort in foraging areas throughout the archipelago. In February 2003, an expedition was launched to revisit the same sites that Green had assessed during his five years of research to provide a comparison between past and present. A research team consisting of Dutton and Jeffrey Seminoff from NOAA, Nelly de Paz from Perú, the author and her teenage daughter, Mariantú, and several very enthusiastic local and international volunteers set off on an eight-day trip to the western side of the archipelago.

The team boarded a Charles Darwin Research Station research vessel boat *The Beagle* and traveled to Green's exact research sites, which probably had not been visited by humans for at least a decade. While approaching the first stop at Punta Espinoza, green turtles were in plain sight mating along the shoreline of the nesting beaches. At most of the sites visited, green turtles were in great abundance—some of them feeding on green algae (*Ulva* sp.), and others resting on the bottom of lagoons or shallow ponds or basking on the sea surface—making them very easy to capture by hand. However, the most exciting event was encountering the first "yellow turtle" documented since Green's time. A juvenile of 47.4 cm in length, it had a brightly colored carapace with white, orange, and brown streaks radiating from bright orange spots at the centers of each scute. The head was black with bright orange borders on facial scutes, the flippers were black with a narrow orange band along the outer and inner margins, and the plastron was uniformly light yellow (see plate 4; Zárate 2007). Of nearly 1,000 green turtles tagged at foraging grounds in Galápagos Archipelago from 2003 to 2007, fewer than 10 percent have been of the yellow morph. While it appears that they originate from green turtle rookeries in the Indo-Pacific (P. Zárate et al., unpublished data), their biology, behavior, and life cycle remain mysteries.

Green turtles nesting in the Galápagos Archipelago may represent a unique genetic stock (P. Zárate et al., unpublished data), but whether turtles within this stock overlap with habitats used by other nesting stocks was unknown until recently. Seminoff et al. (2008) tracked twelve adult female green turtles during their postnesting movements within and away from the Galápagos Archipelago. The researchers identified multiple postnesting behavior patterns, including utilization of oceanic habitats in the southeastern Pacific, long-distance migration to neritic habitats in Central America, and residential movements of some individuals that did not leave the archipelago. Thus, just as the Galápagos receives juvenile green turtles from multiple stocks, Galápagos nesting turtles appear to occupy multiple habitats used by other nesting stocks.

Besides green turtles, hawksbills (*Eretmochelys imbricata*), olive ridleys (*Lepidochelys olivacea*), and leatherbacks (*Dermochelys coriacea*) also have been recorded occasionally in the Galápagos. Six hawksbills were captured among the nearly 1,000 green (and yellow) turtles in marine habitats from 2003 to 2007. Olive ridleys were reported several times in the past and captured by Slevin's team in 1906 (Slevin 1931), and others have been reported as victims of boat strikes and bycatch in artisanal longline gear (Murillo et al. 2004). Leatherbacks were recorded three times from 1970 to 1983 (Green 1994) and once in 2003, when an individual was incidentally captured in an artisanal longline (Murillo et al. 2004). Satellite telemetry research has demonstrated that leatherbacks migrate through and around the Galápagos Archipelago en route between nesting beaches in México and

Costa Rica to feeding areas in the southeastern Pacific (Eckert and Sarti 1997; Shillinger et al. 2008).

The global notoriety and history of strong conservation programs make the Galápagos Archipelago a relative safe haven for abundant green turtles and other sea turtle species, as well as countless other flora and fauna species. Although anthropogenic threats to sea turtles persist and still require conservation action (see below), the Galápagos will continue to provide a home for sea turtles in the foreseeable future.

Revillagigedo Islands, México

The Revillagigedo Islands are located 400 nautical miles southwest of the southern tip of Baja California, México (Everett 1988). This oceanic archipelago encompasses a total area of 157.81 km^2, including four volcanic islands: San Benedicto, Socorro, Roca Partida, and Clarión (see plate 1). Socorro and Clarión both host stations of the Mexican Navy, with a population of 250 (staff and families) in the south of Socorro and a small garrison with approximately 10 men on Clarión (Brattstrom 1982; Awbrey et al. 1984; Everett 1988). However, other than this low-density military presence, the islands are uninhabited by humans. Thus, the islands host several endemic vertebrate and plant species, important seabird rookeries, and other marine megafauna. Furthering the protection of the wildlife in Revillagigedos, the Mexican government established the islands as a Biosphere Reserve in 1994.

As in the Galápagos, the green turtle is the only marine turtle species that nests in Revillagigedos, with relatively sparse nesting on Clarión and Socorro (Brattstrom 1982; Awbrey et al. 1984; Holroyd and Trefry 2010; plates 1 and 5). On Clarión, Sulphur Beach along the southern coast is the most important nesting site for green turtles, accounting for about 70 percent of nesting at Revillagigedos (Awbrey et al. 1984; Everett 1988). Nesting occurs year-round but increases seasonally from July to March, with peak nesting in October and November (Juárez et al. 2003). An average of 86 nests was deposited on Sulphur Beach during nesting seasons from 1999 to 2001.

Hawksbills, olive ridleys and leatherbacks are also reportedly found in coastal waters of the Revillagigedo Islands, but the only scientific survey information available is from research by Arturo Juárez, Laura Sarti, and colleagues (Juárez et al. 2003), who observed green turtles at sea around Socorro more frequently than around Clarión. Green turtles captured along the coasts of the islands were predominantly adults, ranging in curved carapace length from 80.0 to 108.8 cm.

As with Galápagos and other island systems, natural systems in the Revillagigedos have been affected by introduced animals, such as sheep, cats, and pigs, which have caused declines in marine and terrestrial bird populations. It is unclear whether these invasive animals have affected sea turtles, but known threats to sea turtles in waters around the Revillagigedos are predation by sharks and illegal fisheries.

Clipperton Island, France

Clipperton Island, also known as Île de la Passion, is located in the EP at 510 nautical miles southeast of Socorro in the Revillagigedo Islands (see plate 1). It is the only atoll of the northeastern Pacific, and its basement forms a seamount rising above the sea floor at 3,000 m (Glynn et al. 1996; Jost and Andrefouet 2006). Besides volcanic remnants composed of an isolated and conspicuous 29-m-high rock, the highest elevation above sea level is 4 m (Jost and Andrefouet 2006; Lorvelec et al. 2009).

The ring-shaped atoll is of approximately 9 km² and completely encloses a freshwater lagoon that was connected by two channels to the open ocean but because of hurricane effects was isolated from the ocean and became a brackish system sometime between 1839 and 1858 (Sachet 1960; Lorvelec and Pascal 2006; Lorvelec et al. 2009).

From the 1890s until 1910s human settlement took place on the island for mining activities, when the introduction of coconut trees and pigs occurred, with an evident impact on the island ecosystem. Reports from those days mentioned that the island was totally deserted with no vegetation cover, along with a huge abundance of crabs (Jost and Andrefouet 2006). The island is currently uninhabited but sporadically visited by crews of fishing boats and tourist. The flora and fauna in the Clipperton atoll has been characterized as low diversity (Jost and Andrefouet 2006).

The only record of sea turtles nesting on Clipperton Island comes from the notes of Benjamin Morrell (1832), where he mentions that green turtles come to the island to lay their eggs. However, there is no indication where he saw the turtles actually nesting or found evidence such as tracks from nesting activities or eggshells or eggs on the beach. The presence of nesting activity based on Morrell's comments, and the species identification as well, has been questioned because of a lack of description (Lorvelec and Pascal 2006). If nesting was effectively occurring at the time of Morrell's observation, it is possible that predation of eggs and adults by humans between 1893 and 1917 or the destruction of nests by the high abundance of pigs on the island during the first half of the twentieth century has caused the disappearance of the sea turtle population on Clipperton Island (Lorvelec and Pasacal 2006). Inventories carried out on the island register sea turtles (Cheloniidae) as native species that have disappeared from the island—they were first described in August 1825, with no further confirmation of reproduction on the island after that date (Lorvelec and Pascal 2009).

Lorvelec et al. (2009) visited the island in 2004 and recorded the presence of stranding turtles at different sites along the shoreline of the island. Of the nine carcasses they found, five corresponded to olive ridleys; the remaining four could not be identified because they were too decomposed or were reduced to just bones. The death of these strandings can almost certainly be attributed to longline fisheries, and to purse seine tuna fishermen to a lesser extent.

The origin of these strandings has been speculative; based on the distribution of nesting grounds in the region (Brown and Brown 1982; Fritts et al. 1982; Lopez Castro 1999; Alava et al. 2007; Eguchi et al. 2007) and on the current information of migration undertaken by olive ridleys in the region (Parker et al. 2003), the strandings found on Clipperton Island could come from any of the nesting ground populations within the region (Lorvelec et al. 2009).

Cocos Island, Costa Rica

Cocos Island is an oceanic island of both volcanic and tectonic origin located approximately 340 nautical miles from mainland Costa Rica (see plate 1). It is surrounded by deep waters and rich ocean currents that attract iconic aggregations of hammerhead sharks, rays, dolphins, and other large marine species—as well as scuba divers. It is the only oceanic island in the

EP region topped by dense tropical forests and their characteristic flora and fauna (Sinergia 69 2000). The island was never linked to a continent, so the island is home to a high proportion of endemic species.

No information in the peer-reviewed literature exists to date on sea turtles at Cocos Island, although a report by Montoya (1990) does mention their presence at the island. Sightings of sea turtles were first documented during the eighteenth and nineteenth centuries (M. Montoya, written communication), and over the past twenty years divers visiting the island have documented (in photographs and videos) the presence of hawksbills, greens turtles, olive ridleys, and leatherbacks.

Limited conditions suitable for sea turtle nesting can be found on Cocos Island. There are only two very small sandy beaches on the island, one located on Bahía Chatham and the other one on Bahía Wafer. A single record of the presence of sea turtles on the beach at Bahía Chatham corresponded to a sea turtle track found and photographed on 1989. More recent information regarding nesting on Cocos Island was obtained in 2008 from the beach at Bahía Wafer, where an olive ridley was observed laying eggs that successfully hatched and emerged (M. Montoya, written communication).

Sea turtle nesting could have been a more common situation at Cocos Island. However, its reduction could be explained as a consequence of the regional decline on the sea turtle population or by the depredation of sea turtle nets by feral pigs (*Sus scrofa*) introduced by whalers on the island at the end of the eighteenth century. In addition, the brown rat (*Rattus norvegicus*), whose presence on the island is dated at the end of the nineteenth century, could have also explained the decline of sea turtle nesting population at Cocos Island through predation of eggs (M. Montoya, written communication).

In recent years, the Programa Restauración de las Tortugas Marinas (PRETOMA) has led research expeditions to capture and track sea turtles and sharks in marine habitats around Cocos Island. To date, satellite transmitters have been deployed on numerous green turtles and at least one hawksbill turtle, and movement patterns have demonstrated individual variability, with some turtles staying close to the island and others making long-distance migrations toward mainland Central America (R. Arauz, pers. commun.).

Cocos Island was declared a national park by executive decree in 1978 and was designated a World Heritage Site by UNESCO in 1997. In addition, Cocos is home to a "Wetland of International Importance" as defined by the RAMSAR Convention. The only persons allowed to live on Cocos Island are Costa Rican park rangers. Tourists are allowed ashore only with permission of island rangers and are not permitted to camp, stay overnight, or collect any flora, fauna, or minerals from the island (Montoya 2007). A major threat to the Cocos marine ecosystem is illegal fishing.

Easter Island, Chile

Easter Island, or Rapa Nui in the island people's indigenous language, is a Polynesian island of volcanic origin and one of the world's most isolated inhabited islands, located more than 2,000 nautical miles from Chile (see plate 1). Famous for its monumental statues, or *moai*, Easter Island's palm forest was systematically deforested by native Easter Islanders in the process of erecting their statues (Hunt 2006). Chile first declared the island a national park in 1935, and UNESCO designated it a World Heritage Site in 1996.

Little information exists regarding marine turtles around Easter Island, but leatherbacks and green turtles have been documented (M. Donoso, pers. commun.); green turtles have even been satellite tracked within their foraging grounds on the Easter Island (P. Dutton, unpublished data). Leatherbacks have been satellite tracked from nesting beaches in Costa Rica to waters near Easter Island (Shillinger et al. 2008), and they have been caught around the islands by the Chilean artisanal swordfish fleet.

Island Systems on the Continental Shelf

Tres Marías Islands, México

The Tres Marías Islands are located just 65 nautical miles off the west coast of México (plate 1). These islands have been known since early in the history of the New World, and were first named as Las Islas de la Magdalena. The Tres Marías group comprises four islands: San Juanito, María Madre, María Magdalena, and María Cleofa. Since 1905, the Islas

Marías Federal Prison has been home to some of the most infamous and dangerous criminals in México.

References regarding sea turtles are scarce, but Stejneger (1899) reported the existence of nesting around May and June by green turtles and hawksbills, and Parsons (1962) reported a large number of hawksbills nesting on beaches of Tres Marías. Nesting activity of hawksbills on the islands has not yet been reconfirmed.

Coiba, Panamá

Coiba, the largest uninhabited tropical forested island in the Americas, is included in Coiba National Park, which comprises more than 2,700 km² of islands, forests, beaches, and mangroves (plate 1). The island is ringed by one of the largest coral reefs on the Pacific coast of the Americas (Cortés 1997). The remarkable preservation of Coiba is largely due to its use as a penal colony since 1920; the prisoners have served as a strong deterrent to further human colonization and to the extraction of the island's abundant resources (Castrellón 2008). Because of the pristine nature of the island and its surrounding oceans, it was declared a national park by the Panamanian government in 1992, and UNESCO declared the entire Coiba National Park a World Heritage Site in July 2005.

The Indo-Pacific current through the Gulf of Chiriquí provides a unique environment for marine life and, by extension, for recreational diving. The warm current carries tropical marine species from the other side of the Pacific, and larger animals such as humpback whales, sharks, whale sharks, and orcas, are also regular visitors (Aguilar et al. 1997). Regarding sea turtles, preliminary monitoring on the island reported the nesting activity of the green turtle in the beaches of Manila, Rio Amarillo, and Damas on Coiba and to a lesser extent the nesting activity of leatherbacks on Manila beach (Ruiz and Rodríguez 2011).

Malpelo Island, Colombia

Malpelo Island is an oceanic island of a volcanic origin emerging from the sea bottom at about 4 km of depth, roughly 270 nautical miles from the coast of Colombia (plate 1). It is uninhabited except for a small military post manned by the Colombian Army, which was established in 1986, and civilian visitors need a written permit from the Colombian Ministry of the Environment. It was declared a Flora and Fauna Sanctuary in 1995 and a World Heritage Site by UNESCO in 2006. Malpelo Island is a very popular diving location, as hundreds of hammerhead sharks and silky sharks are frequently seen by diving expeditions.

The marine environment is strongly influenced by the marine currents in the area, which create very productive habitats. These conditions make the island an important habitat for many migratory species, including marine mammals, schools of large pelagic fish and sharks, and sea turtles (Birkeland et al. 1975). All five EP species of marine turtles have been observed feeding around Malpelo Island: hawksbills, green turtles, olive ridleys, leatherbacks, and loggerheads. However, research surveys carried out in 2006 by the Fundación Malpelo and the Colombia's Centro de Investigación para el Manejo Ambiental y el Desarrollo (CIMAD) recorded only green turtles, all of them subadults in apparently good health conditions and associated with coral reefs habitats. This first survey represented the first step in the implementation of the Sea Turtle Sighting Program at Malpelo Island (Pavía and Amorocho 2006).

Gorgona, Colombia

Home to a now defunct but once notoriously harrowing prison, this island (and its smaller sister island, Gorgonilla) was christened in the sixteenth century by Francisco Pizarro, who, after losing dozens of his men to venomous snake bites, likened the place to the mythical Gorgon sisters. Since the prison closed in 1984 and the island was named a national park in 1985, the only humans on the island are temporarily stationed park rangers and visiting tourists.

Like other islands in the region, Gorgona is a refuge for marine and terrestrial biodiversity. The island is a popular tourist destination for whale watching, as female humpback whales pass close to shore with their newborn calves in tow every year, and its rich coral reef habitats draw divers year-round. Among the inhabitants of Gorgona's fringing reefs, relatively abundant juvenile green turtles, as well as less abundant hawksbills, have been observed during an extensive mark-recapture study conducted since

2003 by CIMAD. During this period, the research group has hand captured nearly 500 free-swimming green turtles (and fewer than ten hawksbills and olive ridleys) during night dives around Gorgona (Amorocho 2009). In addition, low-density olive ridley nesting occurs on sandy beaches on the island (Payan 2010). Like other known foraging areas for juvenile green turtles in the EP, Gorgona appears to host a mixed stock of juvenile turtles, comprising individuals reflecting genetic haplotypes from different rookeries from the region, as well as individuals of the yellow morph observed in Galápagos that exhibit haplotypes from western Pacific green turtle stocks (Amorocho 2009, Amorocho et al. 2012).

As consistent monitoring efforts have shown in Gorgona (and other sites), islands in the EP often represent important feeding areas for individuals from distinct rookeries in the region, which highlights the importance of protecting these areas to ensure persistence of multiple breeding populations of green turtles in the region.

Challenges and Advances in Island Conservation: A Case Study of the Galápagos Archipelago

Despite the isolation of islands from mainland areas and their associated anthropogenic effects, human-induced threats to island ecosystems certainly exist, as mentioned previously. These remote natural systems have evolved a unique and delicate balance that often differ greatly from those on the closest mainland and thus are particularly sensitive to impacts from other environments. In particular, a common theme for island systems in the EP and elsewhere are introduced species, including plants, but also vertebrates, such as pigs, dogs, livestock, and rodents.

Among all the islands in the region, Galápagos represents the most complex system of critical biophysical, socioeconomic, and cultural resources, which have a profound impact on the archipelago's natural resources and biodiversity. The number of visitors has increased 9 percent annually over the last twenty-five years, and the resident population in Galápagos more than doubled from 1990 to 2006, now topping 20,000 inhabitants. Population growth is increasing pressure on natural resources and the demand for improved public services, motor vehicles, commercial flights, fuel consumption, and electricity, among others. Increased food demand for the growing Galápagos human population is reflected in the decline of marine resources such as lobster, sea cucumber, and cod.

Major causes of concern regarding sea turtles at Galápagos Archipelago are related to the increasing tourism activity, artisanal fisheries within the Galápagos Marine Reserve (GMR) and the national park, and introduced mammal predators. Intensive marine traffic and illegal fishery practices in the GMR have been linked to observations of dead and injured sea turtles. Additionally, feral pigs are voracious predators of sea turtle eggs and hatchlings.

Despite the challenges and problems the archipelago is facing, Galápagos is the only oceanic archipelago with 95 percent of its original biodiversity still intact, owing to a strong legal framework for conservation and the achievements of the conservation institutions in Galápagos (CDF, GNP and INGALA. 2008). However, a strong commitment among all Galápagos stakeholders is necessary to ensure sustainable cohabitation between humans and nature. In recent years, management entities of the Galápagos have implemented several measures to address these threats, which might serve as models for other island systems in the region facing similar threats.

Coastal Zonation Scheme Implementation

In 2000, a coastal zonation scheme was established to regulate the human exploitation of natural resources within the GMR, to avoid conflicts between stakeholders, and to protect high-biodiversity sites. In some areas, fishing and other activities are permitted; in other areas, fishing is prohibited but tourism is allowed; and in others, only research and management activities are permitted. The Charles Darwin Foundation's marine research team is carrying out systematic biological surveys to provide a baseline for future evaluation (Watkins et al. 2008). Zonation is an extremely important management tool but has been very difficult to implement due to stakeholder resistance; for example, fishermen were very reluctant to accept the idea of "no-take areas." The zonation scheme is to be reevaluated in terms of both socioeconomic impacts and preliminary ecological impacts, but many of the benefits of no-take areas will become apparent only with time (Novy 2000; Heylings et al. 2002; WWF-USAID 2006).

Longlining Banned Within the Galápagos Marine Reserve

Cold, hot, and warm marine currents come together in the waters of the GMR, generating a wide diversity of animal life, especially around submarine volcanoes whose peaks nearly reach the surface. These areas are important for both fishermen and tourism because they host a wide variety of commercial fishes, as well as sharks, sea lions, sea turtles, and dolphins, among others (Oviedo 1999; Banks 2002). Although industrial fishing is forbidden within the GMR, local fishermen have used these areas for targeting tunas and swordfish using longline gear (Zárate 2002; Murillo et al. 2004; Galápagos National Park, unpublished data). Considering the high bycatch associated with this fishing gear, the Charles Darwin Foundation did a study in 2003 to determine if small-scale longlining of yellowfin tuna and swordfish should be permitted in the reserve. The researchers found that most of the catch comprised nontarget species, mostly sharks but also the four sea turtle species recorded in the archipelago (Murillo et al. 2004). As a result, longlining was banned in Galápagos waters in 2005 (Registro Oficial 2006).

Quarantine and Inspection System and Eradication Programs

Since the New World discovery of the Galápagos Archipelago by Bishop Tomás de Berlanga, humans have introduced exotic (i.e., nonnative) species to the islands, some intentionally, including goats, pigs, cats, and both ornamental and food plants (vegetables and fruits), and some accidentally, including rodents, insects, and weedy plants (Watkins et al. 2008). Herbivores, such as goats, compete with native creatures for the little available food, so there is not enough to support the tortoises and land iguanas, while introduced plants compete with the native plants for scarce nutrients in depauperate Galápagos soil. Pigs and goats destroy nests and eat bird and reptile hatchlings and eggs.

Immigration and tourism in recent decades to the Galápagos have increased the risk of introduced species entry through various pathways, such as cargo boats and airplanes. To prevent the entry and spread of potentially threatening exotic species, the Galápagos inspection and quarantine system (SICGAL) was established in 2000 (Zapata 2008). Trained SICGAL inspectors now search incoming cargo shipments from boats and planes, as well as luggage carried by tourists and residents (Zapata 2008). It is much more cost-effective to prevent the arrival of introduced species, because the costs of implementing mitigating activities after their arrival can be high and continuous.

Although the quarantine and inspection programs can prevent new biological invasions, eradication programs have been established by the Galápagos National Park to address alien species already present in the archipelago. Successful eradication programs include a program to eradicate feral cats (*Felis catus*) from Baltra Island (Phillips et al. 2005); the largest removal of an insular goat population, using ground-based methods, from Pinta island (Campbell et al. 2004); and the multiyear, multimillion dollar Isabela Project, which resulted in complete eradication of goats and donkeys from Santiago Island and most of Isabela Island (Carrión et al. 2006).

Most important for sea turtles, feral pigs (*Sus scrofa*) on Santiago Island have been responsible for a near zero recruitment rate of giant tortoises and green turtles (Green 1979; Calvopiña 1985). Like the goat eradication efforts, the eradication of pigs from Baltra Island is considered the largest insular pig removal to date: more than 18,000 pigs were removed during a thirty-year eradication campaign (Cruz et al. 2005). As of 2007, several islands and islets in the archipelago are now free of cats, goats, pigeons, donkeys, and pigs. The inhabited islands have more abundances and greater incidences of introduced species, all of which are considered high priority for control and eradication efforts in the coming years.

International Conservation Efforts for Island Systems in the EP

Oceanic and continental shelf islands play an important role as developmental refugia for sea turtles—especially green turtles—in the EP region. They act as juvenile nursery and foraging areas where individuals from multiple genetic stocks mix, and provide reproduction sites for adult turtles. Because these island systems not only share functions for sea turtle life cycles and natural history but also share individual turtles and turtle populations, international conservation strategies have begun to address islands as a network of important habitats. An important framework for sea turtle conservation in the EP is the Eastern Tropical Pacific Marine Corridor (CMAR) Initiative, which is a voluntary multilateral agreement among the governments of Costa Rica, Panamá, Colombia, and Ecuador to work toward integrated, sustainable use and conservation of marine resources in these countries' waters. A related program, the Eastern Tropical Pacific Seascape (ETPS) Initiative, managed by Conservation International, supports interinstitutional, cooperative scientific research and marine management among the same four countries. The Comisión Permanente del Pacífico del Sur (or the Lima Convention) has developed an Action Plan for Sea Turtles in the Southeast Pacific among signatory countries Panamá, Colombia, Ecuador, Perú, and Chile (Seminoff and Zárate 2008). The Inter-American Tropical Tuna Commission (IATTC) and its bycatch reduction efforts are globally recognized to be among the world's finest for regional fisheries management organizations. The Inter-American Convention for the Protection and Conservation of Sea Turtles (IAC) is another policy instrument designed to decrease impacts on sea turtles from fisheries and other human impacts. Clearly, the region benefits from having several strong, complementary conservation instruments and organizational structures in place to promote and enhance sustainable resource use and biodiversity protection.

Nonetheless, sea turtle conservation in the EP requires successful implementation and greater integration among the region's international instruments and accords. New legislation and enforcement of existing laws that curb unsustainable exploitation of sea turtle products in the region's coastal communities are also necessary. It is hoped that coordinated management efforts will provide habitat protection that extends from nesting beaches to marine habitats for sea turtles and other species within and among the island oases of the EP.

REFERENCES

Aguilar, A., J. Forcada, M. Gazo, and E. Badosa. 1997. Los cetáceos del Parque Nacional Coiba (Panamá). In: Castroviejo, S. (Ed.), *Flora y Fauna del Parque Nacional de Coiba (Panamá)*. Spanish Agency for International Cooperation, Madrid. Pp. 75–106.

Alava, J. J., P. C. H. Pritchard, J. Wyneken, and H. Valverde. 2007. First documented record of nesting by the olive ridley turtle (*Lepidochelys olivacea*) in Ecuador. Chelonian Conservation Biology 6:282–285.

Amorocho, D. F. 2009. Green sea turtle (*Chelonia mydas*) in the eastern Pacific island of Gorgona National Park, Colombia. Ph.D. dissertation, Monash University, Melbourne, Australia.

Amorocho, D. F., F. A. Abreu-Grobois, P. H. Dutton, and R. D. Reina. 2012. Multiple distant origins for green sea turtles aggregating off Gorgona Island in the Colombian eastern Pacific. PLoS One 7(2):e31486.

Awbrey, F. T., S. Leatherwood, E. D. Mitchell, and W. Rogers. 1984. Nesting green sea turtles (*Chelonia mydas*) on Isla Clarión, Islas Revillagigedos, México. Bulletin of the Southern California Academy of Sciences 83:69–15.

Banks, S. 2002. Ambiente físico. In: Danulat, E., and G. J. Edgar (Eds.), Reserva Marina de Galápagos. Línea Base de la Biodiversidad. Fundación Charles Darwin/Servicio Parque Nacional Galápagos, Santa Cruz, Galápagos, Ecuador. Pp. 22–27.

Birkeland, C., D. Meyer, J. Stames, and C. Buford. 1975. Subtidal communities of Malpelo Island. In: Graham, J. (Ed.), The Biological Investigation of Malpelo Island, Colombia. Smithsonian Contrib. Zool. No. 176. Smithsonian Institution Press, Washington, DC. Pp. 55–68.

Brattstrom, B. H. 1982. Breeding of the green sea turtle, *Chelonia mydas* on the Revillagigedo Islands, México. Herpetological Review 13:71.

Brown, C. H., and W. M. Brown. 1982. Status of sea turtles in the southeastern Pacific: emphasis on Peru. In: K. A. Bjorndal (Ed.), Biology and Conservation of Sea Turtles. Smithsonian Institution Press, Washington, DC. Pp. 235–240.

Calvopiña, L. 1985. Annual Report of the Charles Darwin Research Station 1984–1985. Department of Introduced Mammals, Quito, Ecuador.

Campbell, K., C. J. Donlan, F. Cruz, and V. Carrión. 2004. Eradication of feral goats *Capra hircus* from Pinta Island, Galápagos, Ecuador. Oryx 38:328–333.

Carr, A. 1967. So Excellent a Fishe. A Natural History of Sea Turtles. Natural History Press, New York.

Carrión, V., J. Donlan, K. Campbell, C. Lavoie, and F. Cruz. 2006. Feral donkey (*Equus asinus*) eradications in the Galápagos. Biodiversity and Conservation 16(2):437–445.

Castrellón, Z. 2008. Parque Nacional Coiba. Available: www.mapa.es/rmarinas/jornada_rrmm/coiba.pdf.

CDF, GNP, and INGALA. 2008. Galápagos Report 2006–2007. Charles Darwin Foundation, Galápagos National Park, and Galapagos National Institute, Puerto Ayora, Galápagos, Ecuador.

Cortés, J. 1997. Biology and geology of eastern Pacific coral reefs. In: *8th International Coral Reef Symposium*. Smithsonian Tropical Research Institute, Balboa, Republic of Panama. Pp. 57–64.

Cruz, F., C. J. Donlan, K. Campbell, and V. Carrión. 2005. Conservation action in the Galápagos: feral pig (*Sus scrofa*) eradication from Santiago Island. Biological Conservation 121:473–478.

Eckert, S. A., and A. L. Sarti. 1997. Distant fisheries implicated in the loss of the world's largest leatherback population. Marine Turtle Newsletter 78:2–7.

Eguchi, T., T. Gerrodette, R. L. Pitman, J. A. Seminoff, and P. H. Dutton. 2007. At-sea density and abundance estimates of the olive ridley turtle *Lepidochelys olivacea* in the eastern tropical Pacific. Endangered Species Research 3:191–203.

Everett, W. T. 1988. Notes from Clarión Island. Condor 90:512–513.

Fritts, T. H., M. L. Stinson, and R. Márquez. 1982. Status of sea turtle nesting in southern Baja California, México. Bulletin of the Southern California Academy of Sciences 81:51–60.

Fritz, T. 1981. Marine turtles of the Galápagos Islands and adjacent areas of the eastern Pacific on the basis of observations made by SR Slevin, 1905–1906. Journal of Herpetology 15:293–301.

Glynn, P. W., J. E. N. Veron, and G. M. Wellington. 1996. Clipperton Atoll (eastern Pacific): oceanography, geomorphology, reef-building coral ecology and biogeography. Coral Reefs 15:71–99.

Green, D. 1978. The East Pacific green sea turtle in Galápagos. Noticias de Galápagos 28:9–12.

Green, D. 1979. Double tagging of green turtles in the Galápagos Islands. Marine Turtle Newsletter 13:4–9.

Green D 1984. Long-distance movements of Galápagos green turtles. Journal of Herpetology 18:121–130.

Green, D. 1993. Growth rates of wild immature green turtles in the Galápagos Islands, Ecuador. Journal of Herpetology 27:338–341.

Green, D. 1994. Galápagos sea turtles: an overview. In: Schroeder, B. A., and B. S. Witherington (Comps.), Proceedings of the Thirteenth Symposium on Sea Turtle Biology and Conservation. NOAA Technical Memorandum NMFS-SEFSC-314. U.S. National Marine Fisheries Service, Miami, FL. Pp. 65–68.

Heylings, P., R. Bensted-Smith, and M. Altamirano. 2002. Zonificación e historia de la Reserva Marina de Galápagos. In: Danulat, E., and G. J. Edgar (Eds.), Reserva Marina de Galápagos. Línea Base de la Biodiversidad. Fundación Charles Darwin/Servicio Parque Nacional Galápagos, Santa Cruz, Galápagos, Ecuador. Pp. 10–21.

Hoff, S. 1985. Drømmen om Galapagos. Grøndahl und Sønn, Oslo. Unpublished manuscript. Available: www.galapagos.to/TEXTS/HOFF-2.HTM.

Holroyd, G. L., and H. E. Trefry. 2010. The importance of Isla Clarión, Archipelago Revillagigedo, Mexico, for green turtle (*Chelonia mydas*) nesting. Chelonian Conservation and Biology 9(2):305–309.

Hunt, T. L. 2006. Rethinking the fall of Easter Island. American Scientist 94:412.

Hurtado, M. 1984. Registro de la anidación de la tortuga negra, *Chelonia mydas* en las Islas Galápagos. Boletín Científico y Técnico 4:77–106.

Jost, C. H., and S. Andrefouet. 2006, Long term natural and human perturbations and current status of Clipperton Atoll, a remote island of the eastern Pacific. Pacific Conservation Biology 12(3):207–218.

Juárez, J. A., L. Sarti, and P. H. Dutton. 2003. First results of the green/black turtles of the Revillagigedos Archipelago: a unique stock in the eastern Pacific. In: Seminoff, J. A. (Comp.), Proceedings of the Twenty-Second Annual Symposium on Sea Turtle Biology and Conservation. NOAA Technical Memorandum NMFS-SEFSC-503. U.S. National Marine Fisheries Service, Miami, FL. P. 70.

López Castro, M. C. 1999. Nesting of sea turtles in south Baja California. Proceedings of the

First Annual Meeting of the Baja California Sea Turtle Group, 23 January 1999, Loreto, Baja California Sur, México. Oceanic Resource Foundation, San Francisco, CA. Pp. 6–7.

Lorvelec, O., and M. Pascal. 2006. Vertebrates of Clipperton Island after one and a half century of ecological disruptions. Revue d'Écologie (La Terre et la Vie) 61:135–158.

Lorvelec, O., and M. Pascal. 2009. Les vertébrés de Clipperton soumis à un siècle et demi de bouleversements écologiques. In: Charpy, L. (Ed.), Clipperton: Environnement et biodiversité d'un microcosme oceanique. Muséum national d'Histoire naturelle, IRD, Paris. Pp. 393–420.

Lorvelec, O., M. Pascal, and J. Fretey. 2009. Sea turtles on Clipperton Island (eastern tropical Pacific). Marine Turtle Newsletter 124:10–13.

Lundh, J. P. 2004. Galápagos: a brief history. Oslo, Norway. Unpublished manuscript. Available: www.galapagos.to/TEXTS/LUNDH-0.HTM.

Montoya, M. 1990. Plan de manejo. Parque Nacional Isla del Coco. Comisión Técnica de Ambientes Marinos, Ministerio de Recursos Naturales, Energía y Minas, San José, Costa Rica.

Montoya, M. 2007. Conozca la Isla del Coco: una guía para su visitación. In: Biocursos para amantes de la naturaleza: Conozca el Parque Nacional Isla del Coco, la isla del tesoro (26 abril al 6 de mayo 2007). Organization for Tropical Studies, San José, Costa Rica. Pp. 35–176.

Morrell, B. 1832. A Narrative of Four Voyages to the South Sea, North and South, Pacific Ocean, Chinese Sea, Ethiopic and Southern Atlantic Ocean, Indian and Antarctic Ocean from the Years 1822 to 1831. Harper, New York.

Murillo, J. C., H. Reyes, P. Zárate, S. Banks, and E. Danulat. 2004. Evaluación de la captura incidental durante el Plan Piloto de Pesca de Altura con Palangre en la Reserva Marina de Galápagos. Fundación Charles Darwin y Dirección Parque Nacional Galápagos, Santa Cruz, Galápagos, Ecuador.

Novy, J. W. 2000. Incentive measures for conservation of biodiversity and sustainability: a case study of the Galápagos Islands. World Wildlife Fund and United Nations Environment Programme. Available: www.cbd.int/doc/case-studies/inc/cs-inc-ec-galapagos-en.pdf.

Oviedo, P. 1999. The Galápagos Islands: conflict management in conservation and sustainable resource management. In: Buckles, D. (Ed.), Cultivating Peace: Conflict and Collaboration in Natural Resource Management. International Development Research Centre/World Bank, Singapore. Pp. 163–182.

Parker, D. M., P. H. Dutton, K. Kopitsky, and R. L. Pitman. 2003. Movement and dive behavior determined by satellite telemetry for male and female olive ridley turtles in the eastern tropical Pacific. In: J. A. Seminoff (Ed.), Proceedings of the 22nd Annual Symposium on Sea Turtle Biology and Conservation, Miami, Florida, USA. NOAA Tech. Memo. NMFS-SEFSC-503. U.S. National Marine Fisheries Service, Miami, FL. Pp. 48–49.

Parque Nacional Galápagos. 2006. Plan de Manejo Parque Nacional Galápagos. Parque Nacional Galápagos, Islas Galápagos, Ecuador.

Parsons, J. 1962. The Green Turtle and Man. University of Florida Press, Gainesville.

Pavía, A., and D. F. Amorocho. 2006. Las tortugas marinas del Santuario de Fauna y Flora

Malpelo. Primer Acercamiento. Centro de Investigación para el Manejo del Ambiente y el Desarrollo, Cali, Colombia.

Payan, L. 2010. Fortalecimiento del Programa de monitoreo de tortugas marinas CIMAD–UAESPNN en el Parque Nacional Natural Gorgona. Informe final Septiembre 21–Febrero 14 de 2010. Informe Final de Consultoria WWF. Cali, Colombia.

Phillips, R. B., B. D. Cooke, K. Campbell, V. Carrión, C. Márquez, and H. L. Snell. 2005. Eradicating feral cats to protect Galápagos land iguanas: methods and strategies. Pacific Conservation Biology 11:57–66.

Pritchard, P. C. H. 1971. Galápagos Sea turtles—preliminary findings. Journal of Herpetology 5:1–9.

Pritchard, P. C. H. 1972. Sea turtles in the Galápagos Islands. IUCN Publications, New Series, Supplementary Papers 31:34–37.

Registro Oficial. 2006. R.O. No. 354. Resolución No. 009-2005, Autoridad Interinstitucional de Manejo de la Reserva Marina de Galápagos (AIM). Tribunal Constitucional del Ecuador. Revista Judicial. Available: www.derechoecuador.com.

Ruíz, A., and J. Rodríguez. 2011. Caracterización de las playas de anidación de tortugas marinas en el Parque Nacional Coiba, provincia de Veraguas, Panamá. Conservación Internacional, Coiba, Panamá.

Sachet, M. H. 1960. Histoire de l'île Clipperton. Cahiers du Pacifique 2:3–32.

Sammon, R. 1992. The Galápagos Archipelago. In: Seven Underwater Wonders of the World. Thomasson-Grant, Charlottesville, VA. Pp. 93–117.

Scuba Diving Magazine. 2010. Top 100 reader's choice survey—overall rating of the destination. January/February.

Seminoff, J. A. 2004. 2004 Global Status Assessment: Green Turtle (*Chelonia mydas*). IUCN Species Survival Commission. Available: http://mtsg.files.wordpress.com/2010/07/mtsg_chelonia_mydas_assessment_expanded-format.pdf.

Seminoff, J. A., and P. Zárate. 2008. Satellite tracked migrations by Galápagos green turtles and the need for multinational conservation efforts. Current Conservation 2:1–12.

Seminoff, J. A., P. Zárate, M. S. Coyne, D. G. Foley, D. Parker, B. Lyon, and P. H. Dutton. 2008. Post-nesting migrations of Galápagos green turtles, *Chelonia mydas*, in relation to oceanographic conditions of the eastern tropical Pacific Ocean: integrating satellite telemetry with remotely-sensed ocean data. Endangered Species Research 4:57–72.

Shillinger, G. L., D. M. Palacios, H. Bailey, S. J. Bograd, A. M. Swithenbank, P. Gaspar, B. P. Wallace, J. R. Spotila, F. V. Paladino, R. Piedra, S. A. Eckert, and B. A. Block. 2008. Persistent leatherback turtle migrations present opportunities for conservation. PLoS Biol 6(7):e171.

Sinergia 69. 2000. Aspectos meteorológicos y climatológico del ACMIC y su área de influencia. Proyecto GEF/PNUD Conocimiento y uso de la biodiversidad del ACMIC, Vol. 2. Sinergia, San José, Costa Rica.

Slevin, J. R. 1931. Log of the schooner "Academy" on the voyage of scientific research to the Galápagos Islands, 1905–1906. Occasional Papers of the California Academy of Sciences 17:1–162.

Stejneger, L. 1899. Reptiles of the Tres Marias. In: Merriam, C. H. (Ed.), Natural History of the Tres Marias, Mexico. North American Fauna No. 14. Department of Agriculture, Division of Biological Survey, Washington, DC. P. 64.

Watkins, G., S. Cardenas, and W. Tapia. 2008. Introduction. In: Galápagos Report 2006–2007. Charles Darwin Foundation, Galápagos National Park, and Galápagos National Institute, Puerto Ayora, Galápagos, Ecuador.

WWF-USAID. 2006. Pasos hacia la sustentabilidad de la Reserva Marina de Galápagos. Proyecto Conservacion de la Reserva Marina de Galápagos. World Wildlife Fund–U.S. Agency for International Development.

Zapata, C. E. 2008. Evaluation of the quarantine and inspection system for Galápagos (SICGAL) after seven years. In: Galápagos Report 2006–2007. Charles Darwin Foundation, Galápagos National Park, and Galápagos National Institute, Puerto Ayora, Galápagos, Ecuador.

Zárate, P. 2002. Tiburones. In: Danulat, E., and G. J. Edgar (Eds.), Reserva Marina de Galápagos. Línea Base de la Biodiversidad Fundación Charles Darwin y Servicio Parque Nacional Galápagos, Santa Cruz, Galápagos, Ecuador. Pp. 373–388.

Zárate, P. 2007. Assessment of the foraging areas of marine turtles in the Galápagos Islands: 2000–2006. Final report to NOAA—U.S. National Marine Fisheries Service. Charles Darwin Foundation, Puerto Ayora, Galápagos.

4

Sea Turtles of the U.S. West Coast

Life in the Higher Latitudes

SCOTT R. BENSON AND PETER H. DUTTON

Summary

Sea turtles are typically associated with tropical sandy beaches or coral reef habitat, not with the cool and dynamic waters off the U.S. West Coast. Although some sea turtle species are indeed rare, wayward visitors during warm water events in temperate high latitudes of the eastern North Pacific, leatherbacks (*Dermochelys coriacea*) and green turtles (*Chelonia mydas*) are regularly found in this region. The California Current is known to contain some of the most productive marine habitats in the world, but little has been known about the movements and ecology of green and leatherback turtles within this ecosystem. Using a variety of research techniques, including genetic studies, satellite telemetry, aerial surveys, boat-based capture operations, and analysis of blood and tissue samples, scientists have recently uncovered some of the mysteries of these ancient mariners. Endangered leatherbacks perform trans-Pacific movements from tropical western Pacific nesting beaches to forage in offshore and neritic waters off the North American west coast during summer and fall. Drawn by dense aggregations of brown sea nettles (*Chrysaora fuscescens*) and other sea jellies, leatherbacks are rarely seen at sea, and their cryptic behavior beneath the dense fog that often blankets the west coast adds to their intrigue. In contrast with the seasonal presence of leatherbacks in the cool open sea, green turtles occur year-round in estuarine and coastal marine ecosystems within the warmer Southern California Bight. Although evidence of the occurrence of green turtles has been available since the late 1800s, scientists have only recently learned that their long-term residence in coastal embayments, most notably San Diego Bay, is a natural behavior and part of their normal life history. Aided by a warm-water effluent from a nearby power plant, the thriving population in San Diego Bay includes the largest eastern Pacific green turtle on record and exhibits the fastest growth rates among Pacific green turtles. Large populations of whales and seabirds and important fisheries are also well-known components of the California Current ecosystem. Recent revelations about leatherback and green turtle populations in this high-latitude, temperate region have enhanced our knowledge of the diverse assemblage of marine vertebrates.

Introduction

When we think about the amazing variety of marine creatures off the U.S. West Coast, we might imagine gray whales, sea otters, or white sharks, but sea turtles usually do not come to mind. The waters of the California Current are thought to be too cold and inhospitable for these marine reptiles, which are typically associated with coral reefs, seagrass pastures, or sandy beaches in tropical climes. Until recently, occurrence of sea turtles in California was characterized as a rare, exotic event—the wayward wandering of a lost animal entrained in a warm-water pool sliding toward the coast, or perhaps in the warm-water effluent of a power plant. However, this view has changed dramatically over the last ten years as research has uncovered new information about these creatures that no doubt discovered California long before the Gold Rush. This chapter is a tale of two turtles (species) and how we now know that they are an integral part of two contrasting local marine ecosystems: leatherback turtles in the productive and dynamic upwelling regions off the central coast of California, and green turtles in the coastal lagoons in southern California (plate 1). In this chapter we describe the ecology of marine turtles at high latitudes in the eastern Pacific, providing an overview of recent discoveries as well as the stories behind the research.

Leatherbacks in the California Current Ecosystem

Habitat Overview

The California Current is among the most productive marine ecosystems in the world, supporting a diverse assemblage of marine species and large commercial and recreational fisheries. Dominated by wind-driven up-welling, these cool, nutrient-rich waters support abundant year-round resident species and attract far-ranging migratory animals that forage here seasonally, including seabirds, baleen whales, sharks, and large predatory fishes, such as swordfish and tunas. Marine turtles, a conspicuous icon of tropical latitudes, historically were considered rare visitors to these waters during warm-water periods, such as El Niño events. Recent studies, however, have brought increasing recognition of the importance of this temperate habitat as a major foraging destination for leatherback turtles, *Dermochelys coriacea*.

Leatherback turtles can tolerate extreme temperature variations and exhibit the most extensive geographic range of any reptile, including temperate waters of the northeastern Pacific. Originating from nesting beaches in the tropical western Pacific, leatherbacks perform trans-Pacific migrations to foraging grounds along the continental margin between British Columbia, Canada, and Point Conception, California, USA. They tend to arrive in coastal waters between May and August and reach peak densities during late summer and early fall, when upwelling winds begin to subside, sea surface temperatures rise, and large blooms of sea jellies (Scyphomedusae) become conspicuous. The sequence and exact timing of these events vary from year to year, influencing the development of sea jelly prey fields and the occurrence of leatherbacks.

Developing Awareness of Leatherbacks

Historically, records of leatherback occurrence in the California Current ecosystem were rare, most likely because of the cryptic nature of leatherbacks at sea. The first compilation of leatherback occurrence in the northeastern Pacific Ocean was done by a San Diego State University graduate student, based on anecdotal records, opportunistic sightings, and stranding reports going back as far as 1887 and ranging from Baja California to Alaska (Stinson 1984). Subsequent examination of stranding data and sightings made by recreational boat skippers revealed that coastal areas near Monterey Bay, California, were a common region of leatherback occurrence when sea surface temperatures reached 15–16°C during the summer months (Starbird et al. 1993). Systematic aerial surveys off the coast of Oregon and

Washington during 1989–1992 also revealed the presence of leatherback turtles there during summer months, when sea surface temperatures were warmest (Bowlby et al. 1994). Collectively, these data revealed a summer/ fall occurrence pattern, but nothing was known about the significance of individual foraging areas or the population origin of leatherbacks seen off the U.S. West Coast. As recently as 1998, the Pacific Leatherback Recovery Plan (NMFS and USFWS 1998a) presumed that most of these turtles originated from eastern Pacific nesting beaches in nearby México.

The first evidence of a more complex migratory pattern came from genetic studies of leatherback turtles caught incidentally in fisheries in the central and eastern North Pacific and stranded specimens along the U.S. West Coast (Dutton et al. 2000). To the surprise of everyone, the genetic signatures (i.e., DNA sequences) of leatherbacks sampled in the central and eastern North Pacific did not match those found in nesting females in México, but rather pointed to the tropical western Pacific, thousands of miles away. The precise origin, however, was still a mystery because the western Pacific leatherback population was known to nest at multiple beaches scattered throughout Papua New Guinea, Papua Barat (Indonesia), Solomon Islands, and possibly other undocumented sites. No information on postnesting movement patterns of leatherback turtles was available for any of these beaches, but identifying links between foraging areas and nesting sites was considered critical for implementing successful multinational management and conservation actions to protect and recover this Critically Endangered species (International Union for Conservation of Nature 2010).

Planes, Boats, and Satellites

The first step in linking foraging areas and nesting sites was to locate a foraging area where leatherbacks could be studied and outfitted with satellite-linked transmitters. Although Monterey Bay, California, was known to have a seasonal occurrence of leatherbacks (Starbird et al. 1993) (plate 2), finding and capturing free-swimming turtles proved challenging. In the early 1990s Scott Eckert, a leatherback expert who had pioneered satellite tracking of sea turtles, had raised funds to charter a purse-seiner to find and catch a free-swimming leatherback in Monterey Bay. After all, how hard could it be? He set out in the dense fog that typically shrouds the central coast of California during August and September—paradoxically, also when most turtle sightings occur—crisscrossing the bay in search of leatherbacks. After several days of poor visibility, he realized this was harder than he first thought and abandoned the search.

The story picks up again in 1999, when the U.S. National Marine Fisheries Service (NMFS) was focusing resources on implementing the newly rolled-out Pacific sea turtle recovery plan, and one of the priorities at the Southwest Fisheries Science Center (SWFSC) was to address leatherback research needs. The eastern Pacific nesting populations in México and Costa Rica had collapsed in the late 1990s, underscoring the desperate need to learn more about movement and habitat requirements of these animals in the marine environment. Karin Forney, a marine mammal biologist at the SWFSC, had regularly seen leatherbacks in California while flying aerial surveys along the coast to count harbor porpoise (Forney et al. 1991). Forney had noticed, among other things, that leatherback turtles associated with dense sea jelly patches in the waters close to shore. She had the foresight to record these sightings beginning in 1990 and recorded nearly 100 observations of leatherbacks off the central California coast. Based on these sightings, the authors, together with Eckert and Forney conceived a plan to study leatherbacks using the combined efforts of an aerial spotting team to locate turtles and a boat-based team that could safely capture and bring aboard these large animals to outfit them with satellite-linked transmitters.

In September 2000, which turned out to be a particularly good year for leatherbacks in Monterey Bay, the plan went into motion. With cooperative weather, the observers in the chartered plane quickly reported several leatherbacks less than a mile off the beach. The boat-based team set off in Moss Landing Marine Laboratory's (MLML) R/V *John H. Martin* with a crew of staff and volunteers, who were astounded at their first sight of an enormous leatherback. Using a sturdy, oversized hoop net, we managed to catch the turtle and safely hauled her onto the deck of the boat despite the objections of our groaning A-frame (fig. 4.1, top). After releasing her with a satellite-linked transmitter attached, she became the first turtle to reveal the movement patterns of leatherbacks foraging in temperate waters of the northeastern Pacific.

In the following years, the capture methods were further refined, especially with the involvement of Professor Jim Harvey (MLML), who

FIGURE 4.1. (Top) First leatherback caught and sampled aboard R/V *John H. Martin* in Monterey Bay, California, September 2000 (© Scott Eckert). (Bottom) A 607-kg leatherback aboard R/V *Sheila B.*, September 2007 (© Heather Harris).

procured a custom-made boat that was better suited for leatherback in-water work. The R/V *Sheila B.* is a 35-ft Munson featuring a retractable bow that can be lowered into the water, allowing researchers to quickly slide the captured leatherback straight onto the foredeck instead of having to lift the huge creatures out of the water and onto the deck with a cargo net (fig. 4.1, bottom). During the pilot study in 2000, two free-swimming leatherbacks were safely and successfully brought aboard, which not only demonstrated that such research was possible but also provided previously unavailable access to foraging leatherbacks to study their ecology and movements.

Crossing Borders and Connecting Countries

The first four tagged leatherbacks moved southwestward from Monterey Bay, into equatorial waters and toward the western Pacific, thus confirming the original genetics results and highlighting the importance of coordinating international conservation efforts in these two regions (Dutton and Squires 2008). However, track lengths were not sufficiently long to determine nesting beach origin, reaching only as far as the Marianas Trench, north of New Guinea. Therefore, an additional tactic was adopted, and transmitters were deployed on leatherbacks at known western Pacific nesting beaches to identify postnesting movement patterns and potential links to the northeastern temperate Pacific.

The first deployments from western Pacific nesting beaches took place along the north coast of Papua New Guinea during the December–February 2001/2002 and 2002/2003 nesting seasons. All nineteen tagged leatherbacks moved southeastward toward higher latitudes of the western South Pacific near New Caledonia and Vanuatu (Benson et al. 2007c). Although these results presented important new information about the western Pacific leatherback population, they did not provide the desired link to the temperate eastern North Pacific.

The existence of a sizable leatherback nesting population on the remote northern coast of Bird's Head Peninsula (Papua Barat, Indonesia) subsequently provided a basis for additional telemetry studies in the western Pacific. Deployments conducted during the primary July nesting peak in 2003, 2005, 2006, and 2007, and during a secondary nesting peak in January–February 2005 and 2007, linked a few individuals to the U.S. West Coast (fig. 4.2), including the first recorded trans-Pacific migration of a leatherback: the turtle traveled more than 10,000 km from Papua Barat to Oregon, USA (Benson et al. 2007a). These deployments also documented additional diverse postnesting movements into tropical and temperate waters throughout the Pacific Ocean and adjacent tropical seas, including the South China Sea, pelagic waters of the central North Pacific, tropical Indonesian Seas, and the East Australia Current system (Benson et al. 2011).

Subsequent telemetry deployments on free-swimming leatherbacks captured off central California during 2002–2007 further confirmed Papua Barat beaches as nesting destinations and established an additional

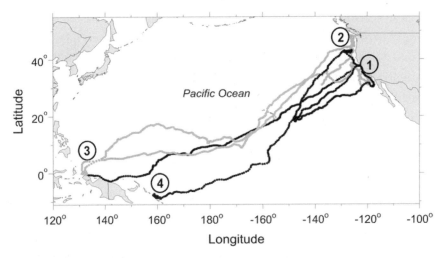

FIGURE 4.2. Western Pacific leatherback movements. Two black tracks show trans-Pacific migrations from central California foraging grounds (1) to nesting beaches in Jamursba-Medi, Papua Barat, Indonesia (3), and Santa Isabel Island, Solomon Islands (4). The third black track is a round-trip between central California and equatorial north Pacific. Gray tracks show migrations from Jamursba-Medi (3) to central California (1) and Oregon/Washington (2).

link to a previously undocumented nesting beach in the Solomon Islands (fig. 4.2). Because trans-Pacific migrations of this magnitude are energetically costly, the California Current ecosystem must be a reliable and productive foraging area for the western Pacific nesting population.

Leatherback Ecology: Foraging in the Shadows

Although there have been myriad shipboard surveys and oceanographic sampling programs in the California Current spanning many decades, the cryptic behavior and low surfacing profile of leatherback turtles make them difficult to detect and have hindered efforts to quantify their abundance and distribution. The most comprehensive information on leatherback abundance, distribution, and habitat at northeastern Pacific foraging grounds has been obtained from aerial surveys, which are the most effective method for locating and quantifying leatherbacks at sea.

Based on the sightings recorded by Forney during harbor porpoise aerial surveys, Benson et al. (2007b) estimated the abundance of leatherbacks at nearshore California foraging areas to be highly variable, ranging from 12 to nearly 400 individuals per year during the summer/fall foraging season. No multiyear trend was evident between 1990 and 2003, but a positive correlation was identified between leatherback abundance and the Northern Oscillation Index (NOI; Schwing et al. 2002), an indicator of upwelling strength. When the NOI was positive, upwelling conditions were favorable and leatherback abundance was high. Conversely, during years with negative NOI values, upwelling was poor and leatherback abundance was low. This pattern was likely linked to patterns of sea jelly abundance and distribution in nearshore waters of central California, particularly the brown sea nettle, *Chrysaora fuscescens*, a large scyphomedusa. Our observations of leatherbacks during sampling and tagging operations in 2000–2007, and deployments of a suction-cup–mounted video camera during 2008, have confirmed that leatherbacks feed almost exclusively on brown sea nettles off central California (fig. 4.3). Although sea jellies are often considered an indicator of poor ocean health (Mills 2001), brown sea nettle blooms in

FIGURE 4.3. Leatherback with suction-cup–mounted video camera approaches a brown sea nettle off the central California coast (© Bill Watson).

coastal waters of central California have been linked to productive upwelling conditions. During years with delayed or interrupted upwelling, when productivity is reduced, brown sea nettles are rare or absent (Graham 1994; Benson et al. 2011). Sea jellies have a low nutritional value, and leatherback turtles must locate dense concentrations (about 20–30 percent of their body weigh per day) to meet their energetic needs (Davenport and Balazs 1991; Wallace et al. 2006).

The 1990–2003 aerial surveys also indicated that leatherbacks were most abundant along the central California coast between Monterey Bay (36.5° N) and Point Arena (39° N). This part of the coast is characterized by multiple headlands, bays, and submarine canyons, which interact with local hydrographic features to create localized retention areas and upwelling shadows (Graham 1994). In these regions, nutrient-rich, upwelling-modified water is entrained nearshore, particularly during wind relaxation, supporting the growth and retention of zooplankton, larval fish, crabs, and gelatinous organisms, leading to dense aggregations of the predatory brown sea nettle (Graham 1994; Wing et al. 1995; Graham and Largier 1997; Graham et al. 2001). This sea jelly has relatively high carbon content among the scyphomedusae and can sustain extraordinarily high growth rates in productive environments, making it the ideal prey for leatherbacks (Shenker 1984; Graham 2009). Not surprisingly, the greatest densities of leatherbacks were found in or near these oceanographic retention areas, where brown sea nettle abundance was greatest, including portions of the Gulf of the Farallones, northern Monterey Bay, and continental shelf habitat between Point Reyes and Bodega Bay (Benson et al. 2007b).

Similar processes have been reported off Oregon, where scyphomedusae become denser and larger in size during summer, when the movement of surface and near-surface waters concentrates plankton in nearshore retention areas (Shenker 1984; Suchman and Brodeur 2005). Telemetry data (Benson et al. 2011) and aerial surveys (Bowlby et al. 1994) show that leatherbacks also forage seasonally in such areas off Oregon and Washington, and in some years off Vancouver Island, British Columbia (Spaven et al. 2009).

The timing of arrival of leatherbacks in foraging areas throughout the California Current during May–August appears directly linked to the phenology of upwelling and sea jelly aggregation in each area. Telemetry data indicate that the earliest arrivals are turtles that foraged along the U.S. West Coast during the previous summer/fall. These individuals departed the California Current during October–December to spend the colder winter months in warmer waters of the central equatorial Pacific. Beginning in March, they approached the central California foraging areas via the Southern California Bight, where coastal water temperatures were warmest, and followed a relatively nearshore path (within about 50 nautical miles) northward toward central California. Postnesting leatherbacks that performed trans-Pacific movements from their western Pacific beaches tend to arrive later, during July–September. The approach of these individuals tends to follow a more direct, northern path toward either central California or Oregon/Washington waters. Leatherbacks appear to be mystic masters of time and space, arriving in the right place at the right time, wherever that may be.

Looking Forward

In the first decade of the twenty-first century, we have been in a period of perpetual discovery, transitioning from virtual ignorance about the origin of leatherbacks off the U.S. West Coast, to a recognition and appreciation of the complexity of the ecological seascape experienced by this population. The diversity of movements has broad implications for stock structure, foraging ecology, and conservation. As our research continues, each question answered creates two new puzzles. Why leatherbacks migrate across the Pacific to distant temperate waters of the eastern North Pacific, when other available foraging areas are closer to the nesting beaches, is a question central to leatherback ecology, evolution, and conservation. The answer may remain elusive during our lifetimes, but we hope that our research will provide some clues and tools for future inquisitive minds seeking to understand the role of leatherbacks in the California Current and other marine ecosystems of the Pacific.

Green Turtles in Southern California

In contrast to widely distributed, gigantothermic leatherbacks, green turtles (*Chelonia mydas*) spend most of their lives not in the cold waters of the California Current but instead in neritic habitats of coastal lagoons of the North American West Coast, where they feed primarily on algae and seagrasses (Dutton and Eckert 1994). Green turtles have been reported seasonally in waters along the California coast and as far north as Alaska, although in cold winter months they become moribund and strand at northern latitudes (Hodge and Wing 2000). Adult female green turtles nest on tropical beaches and then migrate, sometimes thousands of kilometers, back to their feeding grounds. The nearest nesting sites to southern California are found in the tropics along the coast of México, Central America, the Galápagos, and the Hawaiian Archipelago (plate 5). Juvenile and adult green turtles are resident year-round at feeding grounds extending along the coast of Baja California, México, and southern California, USA (plate 1).

Green turtles generally had become regarded as a rare exotic species in southern California. However, newspaper accounts indicate that sea turtles were not uncommon in the late 1800s in San Diego Bay and Mission Bay and were routinely caught by local fishermen. In addition, thousands of green turtles moved through the San Diego port, mostly brought in by whaling ships that caught them to feed their crews and then began selling them. Schooners brimming with green turtles caught on feeding grounds in lagoons at the southern end of the Baja California peninsula would come up to San Diego Bay to offload their catch (Stinson 1984). Turtle soup and meat were featured menu items at many San Diego restaurants, and in 1919 a sea turtle processing plant, Blackman Cannery, opened on the Bayfront at National City. By the 1930s, however, the sea turtles seemed to have disappeared from the public eye; the cannery had closed, and newspapers had stopped reporting sea turtles in San Diego Bay.

Turtles in the Jacuzzi

In 1984, Margie Stinson presented work for her Master's thesis about a small group of green turtles she had discovered in the south part of San Diego Bay. According to Stinson, the turtles had taken up winter residence in the effluent channel of the San Diego Gas and Electric power plant in the 1960s when it was built. The turtles had apparently been attracted to the warm water discharge from the power plant that had inadvertently cre-

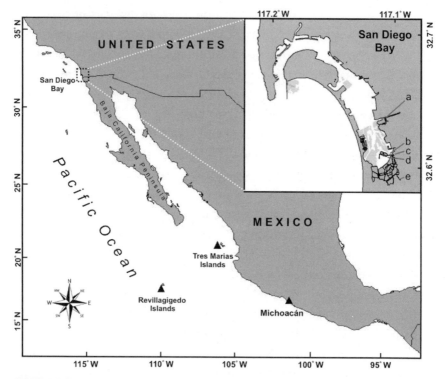

FIGURE 4.4. Map of San Diego Bay and known nesting areas for the green turtles in San Diego Bay. In the inset, gray shaded areas within San Diego Bay indicate known eelgrass habitat: (a) Sweetwater River; (b) intake channel; (c) power plant; (d) effluent channel; (e) Otay River. Known nesting areas are Tres Marías Islands, Revillagigedo Islands, and Michoacán on the mainland of México.

ated a tropical refuge in the southern part of the bay (fig. 4.4). With advice from Mexican fishermen and logistical support from U.S. Navy SEALs, Stinson captured and tagged seven green turtles, including a huge one she fondly named Wrinklebutt because of the distinctively upturned posterior end of its deformed carapace. This unmistakable turtle had been sighted by power plant workers in the 1960s, which suggested that it was a San Diego Bay regular. Wrinklebutt was recaptured in 2007 and weighed in at 241 kg (530 lbs), the largest eastern Pacific green turtle on record. Stinson used some early tracking devices to study the turtles and concluded that at least seven green turtles had moved into the "Jacuzzi"—the swirling, warm water discharged by the power plant—in November to stay warm during the cold winter months, and that they left the bay in the late spring, perhaps to breed in México (Stinson 1984).

Following Stinson's work, Donna McDonald, then at the Hubbs-Sea World Research Institute, and one of the authors (P.H.D.) conducted regular visits to the Jacuzzi, where they systematically recorded turtle sightings and water temperatures. Surprisingly, turtles were present throughout the year, even during the height of summer, when the water temperatures reached 100°F/38°C (Dutton and McDonald 1990). In 1991, McDonald and Dutton began catching and tagging the turtles, and the consistency with which turtles were captured indicated that more than just a few peculiar turtles had been lured in from the cold California Current by the warm water from the power plant. McDonald and Dutton attached ultrasonic tracking devices and confirmed that they were year-round residents, as had been observed in the lagoons in the Gulf of California, where green turtles were known to overwinter in a state of quasi hibernation when water temperatures dropped below 14.5°C (Felger et al. 1976). Perhaps green turtles, that undoubtedly had used San Diego Bay as foraging pastures long before humans arrived, were now using the warm water of the power plant to stay warm and thrive during the winter (McDonald et al. 1994).

Since then, a multitude of volunteers, researchers, and graduate students have become involved in monitoring and research of the San Diego Bay green turtles, which have now become recognized as part of the local ecosystem. The U.S. Fish and Wildlife Service, the U.S. Navy, and the Port of San Diego have partnered with NMFS to address various management mandates that concern this endangered species. A project is under way with a new generation of San Diego State University students using ultrasonic telemetry to map precise habitat use patterns within the bay.

By 2009, a total of ninety-six green turtles had been tagged, ranging in size from 44.0 to 110.4 cm straight carapace length (Eguchi et al. 2010). The smaller of these turtles would have moved recently into the bay from the oceanic waters where young juveniles live during the first years of their lives before seeking out coastal neritic feeding grounds. It is now clear that this is a viable natural population, with new turtles periodically settling into the bay and taking up long-term residence. Recaptures of some turtles first tagged in the early 1990s have revealed growth rates among the fastest reported for Pacific green turtles (Dutton and Dutton 1999); at least one ani-

FIGURE 4.5. Donna Dutton with a small juvenile green turtle tagged in 1991 in San Diego Bay (left) and the same turtle as a mature male in 1996 (right; photos by P. Dutton). The history of multiple captures of this turtle is typical of many turtles captured in San Diego Bay: he was first captured in 1991, then again in 1996, 2004 (three times!), 2007, and 2011.

mal tagged as a small juvenile in 1991 was recaptured five years later as a large mature male; at least one animal tagged as a small juvenile in 1991 was recaptured five years later as a large mature male (fig. 4.5). Tracking revealed that the turtles were swimming over to the eelgrass pastures around the bay during the day and returning periodically to sit in the Jacuzzi and stay warm (Lyon et al. 2006). This optimal foraging strategy allows the turtles to remain active through the cold winter months and apparently reach the upper limits of their growth rate potential. New research led by Jeffrey Seminoff using stable isotope tools indicates that the San Diego Bay green turtles have a unique trophic feeding ecology compared with their conspecifics at foraging grounds farther south (J. A. Seminoff, unpublished data). This finding suggests that San Diego Bay green turtles may be growing faster and reaching maturity sooner on average than elsewhere in the Pacific, where growth rates are highly variable and generally slower (Seminoff et al. 2002; Chaloupka et al. 2004).

Origin of California Green Turtles

The green turtles in San Diego Bay and elsewhere off the coast of California were generally assumed to have originated from the nesting beaches in Michaocán, México, where thousands of eastern Pacific green turtles were known to nest (Seminoff 2004; see chapter 11). Dutton and McDonald began to discover that not all the San Diego Bay turtles had the distinct dark gray plastron and black carapace that characterized the eastern Pacific morphotype typical of the Michoacán breeding population (a.k.a. the "black" turtle). They found a variety of colors and shapes, many resembling those typical of the Hawaiian green turtles (fig. 4.6). Clearly, shape and color are not reliable diagnostics to identify the source population of green turtles found at foraging areas, so Dutton set out on a quest to solve the mystery of their stock origin using newly developed genetic tools.

Knowing which stock these turtles belong to has implications for

FIGURE 4.6. Color variation for green turtles in San Diego Bay (photo by J. Seminoff).

their conservation status under the U.S. Endangered Species Act, because the Mexican breeding populations are considered Endangered, whereas the Hawaiian population is listed as Threatened (NMFS and USFWS 1998b), with the main nesting population at French Frigate Shoals in the Hawaiian Islands steadily increasing over the last thirty-five years (Balazs and Chaloupka 2004). The answer was elusive for many years, because the genetic signatures of most California green turtles did not match those of the Hawaiian or the Michaocán breeding populations. Finally, Laura Sarti, with México's National Commission for Natural Protected Areas, and her team collected samples from a little-known nesting population in the Revillagigedo Islands off the coast of México that may have solved the mystery (Dutton et al. 2008). Preliminary genetic results along with satellite tracking indicate that most of the San Diego Bay green turtles are likely from Revillagigedos, although analysis is still ongoing to determine whether any are part of the Hawaiian and Michaocán breeding populations (Dutton 2003).

"Mud, Leeches, and Turtle Tumors"

Catching turtles in the shallow estuarine environment of south San Diego Bay is usually a visceral experience since it requires wallowing, sometimes chest deep, in mud to retrieve nets and observe the turtles close-up. Starting in 1991, Dutton and McDonald noticed many of the San Diego Bay green turtles had lesions and "white flecks" in the corners of their eyes. They sent photographs to George Balazs of the U.S. National Marine Fisheries Service, Pacific Islands Science Center, who had been documenting the spread of a devastating tumor disease, fibropapillomatosis (FP) among the Hawaiian green turtles, and he confirmed these eye lesions as early stages of FP (McDonald and Dutton 1990). They also noticed that these turtles were covered in small marine leeches in the axial or "armpit" areas, a condition commonly associated with FP, and genetic analysis of skin samples revealed that the herpes virus associated with FP in green turtles was present (Greenblatt et al. 2005; S. Roden and P. H. Dutton, unpublished data). However, the FP did not progress beyond these early stages in the two decades of observation, except in one case, where a turtle was found in 2001 with more than eighty small tumors. This particular turtle was captured recently and was tumor free. It is possible that the herpes virus infecting the San Diego Bay turtles is a less virulent than other strains, although the environmental cofactors associated with FP are poorly understood (Greenblatt et al. 2005). The San Diego Bay turtles appear robust and generally healthy, despite their existence in a polluted urbanized environment. A recent study detected trace metals and persistent organic pollutants in the turtles, although further work is needed to understand the long-term health consequences of accumulating these pollutants (Komoroske et al. 2011).

An Uncertain Future?

In sum, green turtles are more common than previously thought in southern California, and as research continues, we are coming to better understand that they are natural members of the estuarine and coastal marine ecosystem communities. In the summer, green turtles are increasingly sighted in the kelp beds off San Diego and in the coastal lagoons up the coast of southern California. A resident green turtle population has been discovered in the San Gabriel River in Los Angeles, associated with the warm-water discharge of another power plant. Although green turtles are associated with the warm-water effluent from power plants along the coast, their long-term residence in coastal embayments, most notably San Diego Bay, is a natural behavior that is part of their normal life history. The San Diego Bay turtles face continued challenges as they try to share the bay with a growing human population and burgeoning coastal development. Ironically, their most immediate environmental challenge will come from the removal of the power plant that has kept the southern portion of the bay comfortably warm for them. After several years of decreasing power generation, the aging power plant was finally decommissioned in late 2010 and is scheduled to be completely dismantled by 2014. What will the turtles do without the plume of warm water? We are fortunate to have two decades of data on the biology, habitat use, and behavior of this population that will serve as a baseline to monitor changes resulting from the shutdown.

Currently, the greatest threat to the San Diego Bay green turtles has been from boat strikes, which have been associated with most of the dead turtles that have been found. The relative seclusion of the high-use area within a southern portion of the bay, which was established as a wildlife refuge, has provided a haven away from the heavy boat traffic throughout the central and northern part of the bay, but if the turtles disperse to other areas, they will have to increasingly dodge this boat traffic to avoid injury. Perhaps they will remain mostly in the southern portion and dig into the mud to sleep during the winter if the water cools enough, just like their ancestors did before humans built the city and changed their environment. Whatever the turtles do, local, regional, and federal agencies now recognize that the southern California green turtle residents need to be accounted for in management and development decisions that affect the coastal ecosystems they depend on for their survival and species recovery.

ACKNOWLEDGMENTS

Investigations of leatherback turtles in the challenging coastal environment of central California would not be possible without the support of many colleagues and volunteers. We thank the staff of the Marine Operations Department at Moss Landing Marine Laboratory, particularly J. Douglas, L. Bradford, and S. Hansen, for their efforts to keep us safe and close to the turtles. Finding leatherbacks would have been impossible without the support of the pilots and crew of Aspen Helicopters, Inc., and the many aerial observers that have participated in the surveys, especially K. Forney, E. LaCasella, and K. Whitaker. We thank J. Harvey, B. Watson, and multiple cohorts of MLML graduate students for their enthusiasm, input, and participation in the foraging ecology studies. S. Eckert originally noted the significance of leatherback turtle sighting data collected during the aerial surveys and provided the initial inspiration and expertise for in-water sampling and tagging studies. Financial support and personnel were provided by the U.S. National Marine Fisheries Service (Southwest Fisheries Science Center, Southwest Region, and Office of Protected Resources). Leatherback research was conducted under NMFS scientific research permits 1227 and 1596. Aerial surveys were conducted under NMFS permits 748 and 773-1437, and National Marine Sanctuary permits GFNMS/MBNMS-03-93, MBNMS-11-93, GFNMS/MBNMS/CINMS-09-94, GFNMS/MBNMS/CINMS-04-98-A1, MULTI-2002-003, and MULTI-2008-003. Green turtle field studies in San Diego Bay were inspired by Margie Stinson, encouraged by George Balazs, and initially led by Donna McDonald-Dutton with help from numerous volunteers over the years, notably Stephen Johnson, Marilyn Dudley, Elyse Bixby, Amy Frey, Luana Galver, Lauren Hansen, Lauren Hess, Amy Jue, Erin LaCasella, Vicki Pease, Dan Prosperi, Suzanne Roden, Manjula Tiwari, Boyd Lyon, and Lisa Komoroske. Tomo Eguchi, Robin LeRoux, and Jeff Seminoff were co-principal investigators with Peter Dutton in recent years. Don Waller, Kent Miles, and Chris Hawkins of San Diego Gas and Electric Co., Tom Liebst of Dynergy, and Eileen Mahler of the Unified Port of San Diego shared information and helped with logistics. Funding and other resources were provided by the San Diego Fish and Wildlife Commission, the U.S. Fish and Wildlife Service, Hubbs-Sea World Research Institute, San Diego State University, and NOAA—U.S. National Marine Fisheries Service. Research in San Diego Bay was conducted under NMFS scientific research permits 697, 988, 1297, and 1591 and California Department of Fish and Game permit 0411.

REFERENCES

Balazs, G. H., and Chaloupka, M. 2004. Thirty-year recovery trend in the once depleted Hawaiian green sea turtle stock. Biological Conservation 117:491–498.

Benson, S. R., Dutton, P. H., Hitipeuw, C., Samber, B., Bakarbessy, J., and Parker, D. 2007a. Post-nesting migrations of leatherback turtles (*Dermochelys coriacea*) from Jamursba-Medi, Bird's Head Peninsula, Indonesia. Chelonian Conservation and Biology 6:150–154.

Benson, S. R., Eguchi, T., Foley, D. G., Forney, K. A., Bailey, H., Hitipeuw, C., Samber, B. P., Tapilatu, R. F., Rei, V., Ramohia, P., Pita, J., and Dutton, P. H. 2011. Large-scale movements and high-use areas of western pacific leatherback turtles, *Dermochelys coriacea*. Ecosphere 2:art84; doi:10.1890/ES11-00053.1.

Benson, S. R., Forney, K. A., Harvey, J. T., Caretta, J. V., and Dutton, P. H. 2007b. Abundance, distribution, and habitat of leatherback turtles (*Dermochelys coriacea*) off California 1990–2003. Fisheries Bulletin 105:337–347.

Benson, S. R., Kisokau, K. M., Ambio, L., Rei, V., Dutton, P. H., and Parker, D. 2007c. Beach use, internesting movement, and migration of leatherback turtles, Dermochelys coriacea, nesting on the north coast of Papua New Guinea. Chelonian Conservation and Biology 6:7–14.

Bowlby, C. E., Green, G. A., and Bonnell, M. L. 1994. Observations of leatherback turtles offshore of Washington and Oregon. Northwestern Naturalist 75:33–35.

Chaloupka, M., Limpus, C., and Miller, J. 2004. Green turtle somatic growth dynamics in a spatially disjunct Great Barrier Reef metapopulation. Coral Reefs 23:325–335.

Davenport, J., and Balazs, G. H. 1991. Fiery bodies: are pryosomas an important component of the diet of leatherback turtles? British Herpetological Society Bulletin 37:33–38.

Dutton, P. H. 2003. Molecular ecology of the eastern Pacific green turtle. In: Seminoff, J. A.

(Comp.), Proceedings of the Twenty-Second Annual Symposium on Sea Turtle Biology and Conservation. NOAA Tech. Memo. NMFS-SEFSC-503. U.S. National Marine Fisheries Service, Miami, FL. P. 69.

Dutton, P. H., Balazs, G. H., LeRoux, R. A., Murakawa, S. K. K., Zárate, P., and Sarti-Martínez, L. 2008. Genetic composition of Hawaiian green turtle foraging aggregations: mtDNA evidence for a distinct regional population. Endangered Species Research 5:37–44.

Dutton, D. L., and Dutton, P. H. 1999. Accelerated growth in San Diego Bay green turtles? In: Epperly, S. P., and J. Braun (Comps.), Proceedings of the Seventeenth Annual Symposium on Sea Turtle Biology and Conservation. NOAA Tech. Memo. NMFS-SEFSC-415. U.S. National Marine Fisheries Service, Miami, FL. Pp. 164–165.

Dutton, P. H., and Eckert, S. K. 1994. The endangered marine environment; the leatherback sea turtle; the green sea turtle; the loggerhead sea turtle; the olive ridley sea turtle. In: Thelander, C. G. (Ed.), Life on the Edge: A Guide to California's Endangered Natural Resources: Wildlife. Heyday Books, Berkeley, CA. Pp. 454–462.

Dutton, P. H., Frey, A., LeRoux, R., and Balazs, G. 2000. Molecular ecology of leatherbacks in the Pacific. In: Pilcher, N., and G. Ismael (Eds.), Sea Turtles of the Indo-Pacific. Research, Management and Conservation. ASEAN Academic Press, London. Pp. 248–253.

Dutton, P. H., and McDonald, D. L. 1990. Sea turtles present in San Diego Bay. In: Richardson, T. H., J. I. Richardson, and M. Donnelly (Comps.), Proceedings of the Tenth Annual Workshop on Sea Turtle Biology and Conservation. NOAA Tech. Memo. NMFS-SEFC-278. U.S. National Marine Fisheries Service, Miami, FL. Pp. 139–141.

Dutton, P. H., and Squires, D. 2008. Reconciling biodiversity with fishing: a holistic strategy for Pacific sea turtle recovery. Ocean Development and International Law 39:200–222.

Eguchi, T., J. A. Seminoff, R. LeRoux, P. H. Dutton, and Dutton, D. M. 2010. Abundance and survival rates of green turtles in an urban environment—coexistence of humans and an endangered species. Marine Biology 157:1869–1877.

Felger R. S., Clifton, S. K., and Regal, P. S. 1976. Winter dormancy in sea turtles: independent discovery and exploitation in the Gulf of California by two local cultures. Science 191:283.

Forney, K. A., Hanan, D. A., and Barlow, J. 1991. Detecting trends in harbor porpoise abundance from aerial surveys using analysis of covariance. Fishery Bulletin 89:367–377.

Graham, T. 2009. Scyphozoan jellies as prey for leatherback turtles off central California. Master's Thesis, San Jose State University, San Jose, CA.

Graham, W. M. 1994. The physical oceanography and ecology of upwelling shadows. Ph.D. dissertation, University of California, Santa Cruz.

Graham, W. M., and Largier, J. L. 1997. Upwelling shadows as nearshore retention sites: the example of northern Monterey Bay. Continental Shelf Research 17:509–532.

Graham, W. M., Pagès, F., and Hamner, W. M. 2001, A physical context for gelatinous zooplankton aggregations: a review. Hydrobiologia 451:199–212.

Greenblatt, R. J., Work, T. M., Dutton, P. H., Sutton, C. A., Spraker, T. R., Casey, R. N., Diez, C. E., Parker, D., St. Leger, J., Balazs, G. H., and Casey, J. W. 2005. Geographic variation in marine turtle fibropapillomatosis. Journal of Zoo and Wildlife Medicine 36:527–530.

Hodge, R., and Wing, B. 2000. Occurrences of marine turtles in Alaska waters, 1960–1998. Herpetological Review 31:148–151.

International Union for Conservation of Nature. 2010. IUCN Red List of Threatened Species. Version 2010.1. Available: www.iucnredlist.org (downloaded on 1 June 2010).

Komoroske, L. M., Lewison, R. L., Seminoff, J. A., Deheyn, D. D., and Dutton, P. H. 2011. Pollutants and the health of green sea turtles resident to an urbanized estuary in San Diego, CA. Chemosphere 84(5):544–552.

Lyon, B., Seminoff, J. A., Eguchi, T., and Dutton, P. H. 2006. *Chelonia* in and out of the Jacuzzi: diel movements of east Pacific green turtles in San Diego Bay, USA. In: Frick, M., A. Panagopoulou, A. F. Rees, and K. Williams (Comps.), 26th Symposium on Sea Turtle Biology and Conservation, Island of Crete, Greece, 3–8 April 2006. Available: www.seaturtlesociety.org/docs/26turtle.pdf.

McDonald, D. L., and Dutton, P. H. 1990. Fibropapillomas on sea turtles in San Diego Bay, California. Marine Turtle Newsletter 51:9–10.

McDonald, D., Dutton, P. H., Mayer, D., and Merkel, K. 1994. Review of the Green Turtles of South San Diego Bay in Relation to the Operations of the SDG&E South Bay Power Plant. Doc. 94-045-01, C941210311. San Diego Gas and Electric Co., San Diego, CA.

Mills, C. E. 2001. Jellyfish blooms: are populations increasing globally in response to changing ocean conditions? Hydrobiologia 451:55–68.

NMFS and USFWS. 1998a. Recovery Plan for U.S. Pacific Populations of the Leatherback Turtle (*Dermochelys coriacea*). U.S. National Marine Fisheries Service, Silver Spring, MD.

NMFS and USFWS. 1998b. Recovery Plan for U.S. Pacific Populations of the East Pacific Green Turtle (*Chelonia mydas*). U.S. National Marine Fisheries Service, Silver Spring, MD.

Schwing, F. B., Murphree, T., and Green, P. M. 2002. The Northern Oscillation Index (NOI): a new climate index for the northeast Pacific. Progress in Oceanography 53:111–139.

Seminoff, J. A. 2004. *Chelonia mydas*. In: IUCN 2011. IUCN Red List of Threatened Species. Version 2011.2. Available: www.iucnredlist.org/apps/redlist/details/4615/0.

Seminoff, J. A., Resendiz, A., Nichols, W. J., and Jones, T. T. 2002. Growth rates of wild green turtles (*Chelonia mydas*) at a temperate foraging area in the Gulf of California, México. Copeia 2002:610–617.

Shenker, J. M. 1984. Scyphomedusae in surface waters near the Oregon coast, May–August 1981. Estuarine Coastal Shelf Science 19:619–632.

Spaven, L. D., Ford, J. K. B., and Sbrocchi, C. 2009. Occurrence of Leatherback Sea Turtles (*Dermochelys coriacea*) off the Pacific coast of Canada, 1931–2009. Can. Tech. Rep. Fish. Aquat. Sci. No. 2858. Fisheries and Oceans Canada, Nanaimo, British Columbia.

Starbird, C. H., Baldridge, A., and Harvey, J. T. 1993. Seasonal occurrence of leatherback sea turtles (*Dermochelys coriacea*) in the Monterey Bay region, with notes on other sea turtles 1986–1991. California Fish and Game 79:54–62.

Stinson, M. L. 1984. Biology of sea turtles in San Diego Bay, California, and in the northeastern Pacific Ocean. Master's Thesis, San Diego State University, San Diego, CA.

Suchman, C. L., and Brodeur, R. D. 2005. Abundance and distribution of large medusae in surface waters of the northern California Current. Deep-Sea Research II 52:51–72.

Wallace, B. P., Kilham, S. S., Paladino, F. V., and Spotila, J. R. 2006. Energy budget calculations indicate resource limitation in eastern Pacific leatherback turtles. Marine Ecology Progress Series 318:263–270.

Wing, S. R., Largier, J. L., Botsford, L. W., and Quinn, J. F. 1995. Settlement and transport of benthic invertebrates in an intermittent upwelling system. Limnology and Oceanography 40:316–329.

PART TWO

International Management and Policy Frameworks

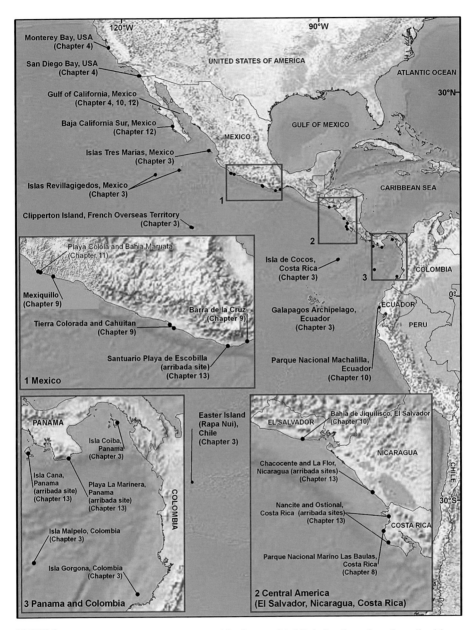

PLATE 1: Locator map of the eastern Pacific Ocean, highlighting key sites described in the text.

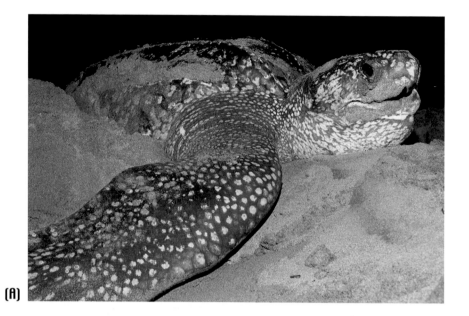

PLATE 2: Leatherback turtle (*Dermochelys coriacea*). (A) Nesting female. (B) Hatchlings, Parque Nacional Marino Las Baulas, Costa Rica (© Jason Bradley; © Sam Friederichs).

(C)

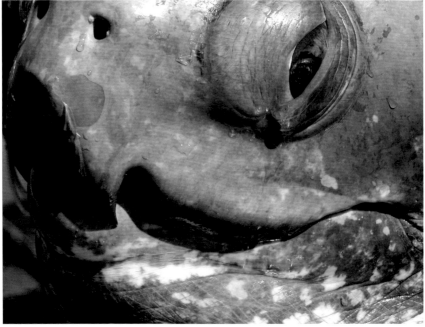

(D)

PLATE 2: (C) A leatherback foraging near Monterey Bay, California, USA (© Alice A. "Alex" Eilers). (D) Close-up of leatherback's characteristically hooked upper jaw (© Scott Benson).

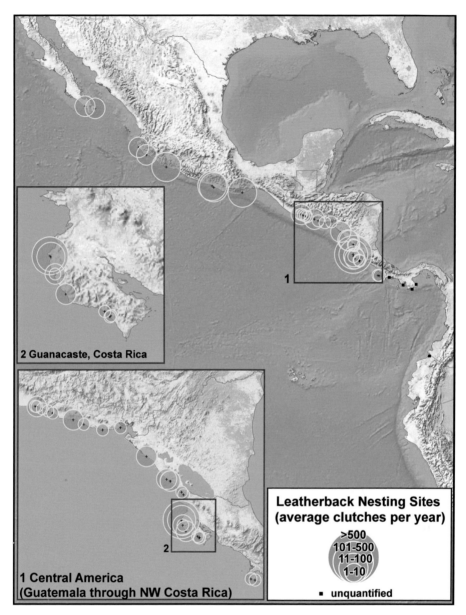

PLATE 3: Map of nesting distribution of the leatherback turtle (*Dermochelys coriacea*) in the eastern Pacific Ocean. Circle sizes indicate relative abundances. Data from the SWOT database and related references (see appendix).

FACING PAGE:
PLATE 4: Green (or black) turtle (*Chelonia mydas*). (A) Nesting green (a.k.a. "black") turtle, Parque Nacional Marino Las Baulas, Costa Rica (© Sam Friederichs). (B) A "yellow" turtle captured in Gorgona National Park, Colombia (© Bryan Wallace). Note the marked color differences between plates A and B.

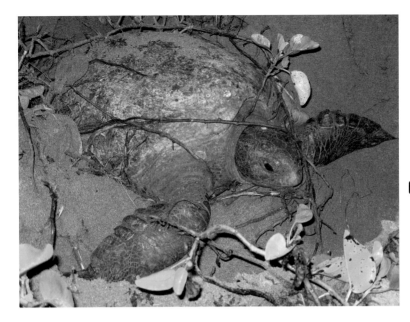

(A)

(B)

(C)

(D)

PLATE 4: (C) Hatchling, Galápagos Islands, Ecuador (© Boyd Lyon). (D) Juvenile, tangled in a research net, Isla La Plata, Ecuador (© Alexander Gaos).

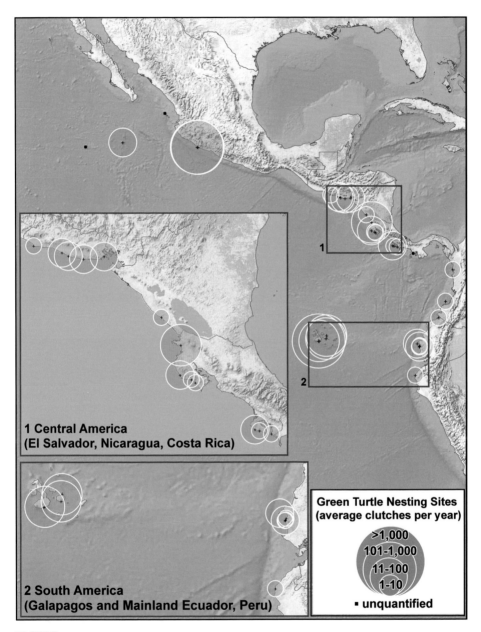

PLATE 5: Map of nesting distribution of the green (or black) turtle (*Chelonia mydas*) in the eastern Pacific Ocean. Circle sizes indicate relative abundances. Data from the SWOT database (see appendix).

PLATE 6: Hawksbill turtle (*Eretmochelys imbricata*). (A) Nesting female. (B) Hatchling, Bahía de Jiquilisco, El Salvador (© Michael Liles).

(C)

(D)

PLATE 6: (C) Juvenile caught by researchers, Gulf of California, México (© Ingrid Yañez). (D) Close-up showing this species' eponymous mouth shape (© Michael Liles).

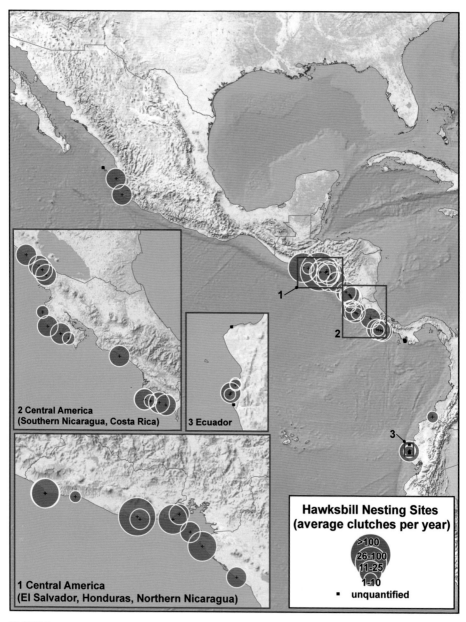

PLATE 7: Map of nesting distribution of the hawksbill turtle (*Eretmochelys imbricata*) in the eastern Pacific Ocean. Circle sizes indicate relative abundances. Data from the SWOT database (see appendix).

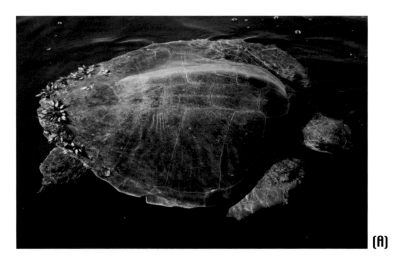

PLATE 8: Olive ridley turtle (*Lepidochelys olivacea*). (A) Adult turtle in the open ocean (© Lindsey Peavey). (B) Hatchling, Parque Nacional Marino Las Baulas, Costa Rica (© Sam Friederichs). (C) An arribada (mass nesting), Refugio de Vida Silvestre, Ostional, Costa Rica (© David Sherwood).

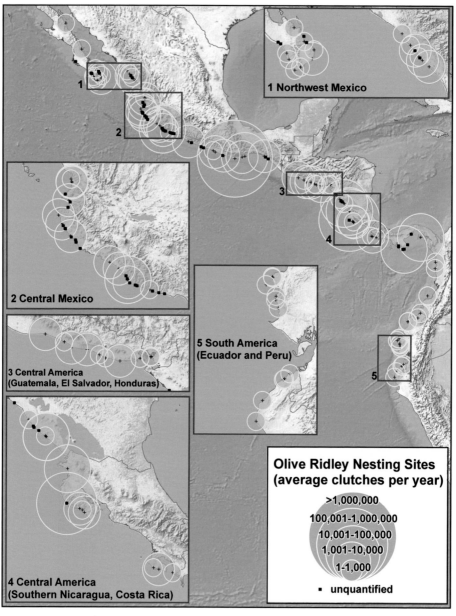

PLATE 9: Map of nesting distribution of the olive ridley turtle (*Lepidochelys olivacea*) in the eastern Pacific Ocean. Circle sizes indicate relative abundances. Data from the SWOT database (see appendix).

(A)

(B)

PLATE 10: Loggerhead turtle (*Caretta caretta*). (A) Juvenile loggerheads caught in Peruvian longlines (© Pro Delphinus). (B) Close-up view of loggerhead's large head and powerful jaws (© Jeffrey Seminoff). (A map of the loggerhead nesting distribution is not included because no loggerhead nesting occurs in the eastern Pacific Ocean region.)

PLATE 11: Threats to sea turtles in the eastern Pacific Ocean: fisheries bycatch. (A) An olive ridley turtle hooked by a longline, Costa Rica (© Sam Friederichs). (B) Loggerhead carcasses due to mortality in nearshore bottom longlines and gillnets, Baja California Sur, México (© S. Hoyt Peckham). (C) Juvenile green turtle entangled in a fishing net, Perú (© Pro Delphinus).

(A)

PLATE 12: Threats to sea turtles in the eastern Pacific Ocean: overconsumption by humans. (A) Egg harvest by a local community member, Oaxaca, México (© J. A. Seminoff).

(B)

(C)

PLATE 12: (B) Harvest of nesting turtles, Mazatlán, México, ca. 1967 (woman inspecting turtles is coeditor J. A. Seminoff's grandmother) (© Richard H. Jones, courtesy of J. A. Seminoff). (C) Jewelry fashioned from hawksbill carapace material in a market, San Salvador, El Salvador (© Bryan Wallace).

PLATES 13 AND 14: Research initiatives directed toward sea turtles in the eastern Pacific Ocean. (A) A leatherback carrying a transmitter is tracked by researchers aboard the R/V *David Starr Jordan* (© Jim Harvey). (B) A researcher scans a leatherback caught in fishing nets in Perú for a microchip tag used to identify individual turtles (© Pro Delphinus). (C) A researcher captures a free-swimming loggerhead turtle using the "rodeo" technique off Baja California Sur, México (© S. Hoyt Peckham).

(D)

PLATES 13 AND 14: (D) A male green turtle carries a new satellite transmitter while being carried to the sea for release (© Bryan Wallace).

PLATES 13 AND 14: (E) A field research team, including indigenous Comcáac community members with coeditor J. A. Seminoff (second from left), measures green turtles in Sonora, Gulf of California, México (© Timothy Dykman).

(F)

(G)

PLATES 13 AND 14: (F) A researcher buries equipment to monitor changes in oxygen levels in sea turtle nests due to fluctuations in ocean tides (© Bryan Wallace). Plates 13 and 14: (G) A female leatherback carries a newly attached satellite transmitter while being measured, Mexiquillo, México (© Ana Barragán).

PLATES 15 AND 16: Conservation efforts directed toward sea turtles in the eastern Pacific Ocean. (A) A team of researchers and fishermen work to free a green turtle from a net, Perú (© ecOceánica).

PLATES 15 AND 16: (B) Local community member supervising a beach hatchery in Astillero, Nicaragua (© Brian Hutchinson). (C) Ecuador's minister of the environment assists Equilibrio Azul in a release of a juvenile hawksbill equipped with a satellite transmitter (© Felipe Vallejo).

(D)

PLATES 15 AND 16: (D) Public release of hawksbill hatchlings, Reserva Natural Estero Padre Ramos, Nicaragua (© Bryan Wallace).

(E)

(F)

PLATES 15 AND 16: (E) Environmental education program featuring conservationists and Mr. Leatherback and schoolchildren near Chacocente, Nicaragua (© Brian Hutchinson). (F) A biologist reviews the contents of a sea turtle nest to determine hatching success, Mexiquillo, México (© Ana Barragán).

5

Fisheries Management off the U.S. West Coast

A Progressive Model for Sea Turtle Conservation

MARK HELVEY AND CHRISTINA FAHY

Summary

Federal fisheries in the United States, managed under the Magnuson-Stevens Fishery Conservation and Management Act, require that fishing activities reduce or minimize the capture and subsequent discard of nontarget species lacking commercial value or prohibited from being landed. In addition, these fisheries may be further managed under other federal laws, such as the Endangered Species Act (ESA) and the Marine Mammal Protection Act (MMPA), to reduce incidental interactions or mortalities of protected or listed marine species. The U.S. National Marine Fisheries Service (NMFS) has used several management approaches to minimize interactions and the subsequent discard of marine mammals and sea turtles protected and/or listed under the MMPA and the ESA, respectively, in the swordfish drift gillnet fishery off the U.S. West Coast. To reduce jeopardy to leatherback sea turtles, NMFS implemented a 213,500 mi^2 time/area closure in 2001 to reduce potential entanglement. While time/area closures may be the most effective strategy for minimizing ecosystem impacts with this gear type, this regulatory action may explain the downward trend in fishing effort since the management action was implemented. Because consumer demand for swordfish remains high, the continuing decline in swordfish production from U.S. West Coast fishermen suggests that reliance on foreign imports will persist unless fishery managers seek new fishing opportunities. A potential alternative is longline gear, which over the last two decades has seen improved technological modifications that have significantly reduced interactions with and associated mortality rates for not only sea turtles but also fish, seabirds, and marine mammals. Holistic approaches for restoring Pacific sea turtle populations could include greater promotion of a U.S. West Coast swordfish fishery using selective longline gear technologies now in use by the Atlantic and Hawaii U.S. swordfish fisheries. Such a strategy could result in reduced reliance on swordfish caught by foreign fleets operating with less effective management measures and provide for the continued transfer of U.S. bycatch reduction measures to these fleets.

Introduction

Fisheries managed by the U.S. federal government are conserved and managed under the authority of the Magnuson-Stevens Fishery Conservation and Management Act (MSA). The primary intent of the MSA is to contribute sustainable fish resources to the national food supply, economy, and health of the United States, and to provide recreational opportunities. Specific provisions in the MSA, however, require these objectives be met while protecting the environment through reducing or minimizing bycatch, that is, the capture and subsequent discard of nontarget species that lack commercial value or are under management measures requiring they not be landed (Benaka and Dobrzynski 2004). In addition, commercial fisheries that interact with marine species protected or listed under U.S. federal laws, such as the Endangered Species Act (ESA) and the Marine Mammal Protection Act (MMPA), may be further managed to reduce interactions or mortalities.

Since these three acts have been in existence, U.S. fishery managers have used various strategies to reduce interactions with protected species, and their solutions have varied depending on the information available at the time, the practicality of the approach, and the latest research, as well as adherence to relevant laws. A common approach providing definitive and often immediate protection for sensitive species involves time/area closures. While such an avoidance strategy seems simple (Kennelly and Broadhurst 2002), the measure may only displace effort to other areas to meet demand, thereby shifting the problem elsewhere without really achieving intended conservation (Hall et al. 2000; Santora 2003; Gilman et al. 2006; Sarmiento 2006; Rausser et al. 2009; Bartram et al. 2010). Time/area closures also affect fishermen by limiting, or even eliminating, opportunities to fish. Consequently, fishery managers have also relied on gear selectivity and deter-

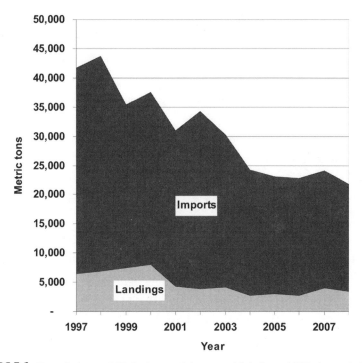

FIGURE 5.1. Trends in total U.S. demand for swordfish from 1997 through 2008, defined as the sum of U.S. landings, including Pacific landings, and imports. (The United States does not export swordfish.)

rence techniques to reduce encounters with protected species (Hall and Mainprize 2005; Werner et al. 2006). Compliance with these solutions usually places less of an economic burden on fishermen.

The broadbill swordfish (*Xiphias gladius*) drift gillnet fishery off the U.S. West Coast is an example of where the U.S. National Marine Fisheries Service (NMFS) has used several management approaches to minimize interactions and the subsequent discard of marine mammals and sea turtles protected and/or listed under the MMPA and the ESA, respectively. Swordfish is a popular seafood in the United States, but demand has progressively been met through foreign imports (fig. 5.1). As U.S. fishermen have attempted to meet this demand with more efficient fishing methods, coincident concerns about ecosystem sustainability have substantially reduced effort in the fishery with this particular gear. Because the North Pacific swordfish stock is healthy (ISC 2009), U.S. swordfish fishermen could

benefit from gear alternatives that sustain fishery production while minimizing environmental impacts. In this chapter, we review the history of the
U.S. West Coast swordfish fishery and the current legislative landscape that
has shaped swordfish management. We discuss the evolution of conservation strategies to reduce incidental take of protected species, primarily sea
turtles, while contributing sustainable swordfish harvest to U.S. markets,
and comment on the direction of this fishery in the broader context of Pacific sea turtle recovery.

The Legal Framework

During the late 1960s through the mid-1970s, the conservation
movement in the United States shifted toward a maturing environmental
sentiment that used legislative channels to achieve objectives. The U.S.
Congress passed three particular statutes that shaped how federal fisheries
are now conducted: the MSA, the MMPA, and the ESA.

The Magnuson-Stevens Fishery Conservation
and Management Act

Congress passed the MSA in 1976 in response to U.S. concerns
that foreign fleets were overexploiting U.S. fishery resources and that international agreements were not conserving fish stocks. The purpose of this
Act was to extend U.S. jurisdiction 200 nautical miles offshore (defined as
the exclusive economic zone, or EEZ), develop American fisheries in that
zone, and manage for the conservation and maximum yield of fish stocks.
Amended in 1996 by the Sustainable Fisheries Act and again by the Magnuson-Stevens Fishery Conservation and Management Reauthorization
Act of 2006, the MSA is implemented through eight regional fishery management councils throughout the country, which have the responsibility of
developing fishery management plans (FMPs). Ten national standards are
incorporated into the Act, with measures for fishery conservation and management. National Standard 9 of the MSA requires that conservation and
management measures minimize bycatch and, when bycatch cannot be
avoided, minimize the mortality of such bycatch. The MSA also requires
that FMPs establish a standardized reporting methodology to assess the

amount and type of discards occurring in the fishery, and that they include conservation and management measures that, to the extent practicable, effectively minimize discards and their mortality when fishing interactions cannot be avoided (section 303(A)(11)).

The Endangered Species Act

The ESA (U.S. Code, Title 16, sections 1531 et seq.) was created by the 93rd U.S. Congress in 1973 to respond to concerns about inadequate conservation efforts on behalf of species at risk of extinction. The purpose of the ESA is to protect plant, vertebrate, and invertebrate species, and "the ecosystems upon which they depend." The ESA is administered by two federal agencies, the U.S. Fish and Wildlife Service (USFWS) and NMFS. While NMFS has jurisdiction for most marine species (including anadromous fish), the USFWS has jurisdiction over freshwater fish and all other species. Species that occur in both habitats may be jointly managed by both agencies. For example, NMFS manages activities that affect sea turtles in their marine habitat, and the USFWS manages activities that may affect sea turtles on land, typically on nesting beaches. The ESA provides protection only for species that have been officially "listed" and, in general, prohibits the "take"[1] of these protected species without permission from the responsible agency. Critical habitat may also be designated for a listed species if specific areas contain important physical and biological features that are found to be essential to the conservation of the species and that can be managed or protected.

Through periodic status assessments or when petitioned by the public, NMFS and USFWS will review and determine whether listing determinations should be changed (e.g., delisted or classified differently than the current listing). In general, most sea turtle species are globally listed, with "species" defined by its formal taxonomy. The 1978 amendments to the ESA further defined "species" as "any subspecies of fish or wildlife or plants, and any distinct population segment of any species of vertebrate fish and wildlife which interbreeds when mature." In 1996, NMFS and USFWS published a policy to define as "distinct population segments" (DPSs) populations that are both "discrete" and "significant" relative to their taxa. "Discrete" and "significant" are further defined through satisfaction of several conditions, including but not limited to physical separation of populations, ecological or behavioral differences, and distinct genetic characteristics. The two agencies published status reviews for the six sea turtle species (green turtle, hawksbill turtle, Kemp's ridley turtle [exclusive to the Atlantic Ocean], leatherback turtle, loggerhead turtle, and olive ridley turtle) under their jurisdiction in 2007, and while NMFS and USFWS did not recommend changes to the listings, they did recommend that analyses be conducted to determine whether to further separate the listed entities into DPSs. In 2008, NMFS and USFWS initiated an assessment of the status of loggerhead sea turtles, which were globally listed as a threatened species in 1978. Following the assessment, and given the criteria contained within the DPS policy, in 2011 the agencies determined that the loggerhead sea turtle is composed of nine DPSs, with four DPSs listed as threatened and five DPSs listed as endangered, including North Pacific loggerheads (the population that has been documented off southern California).

Section 7(a)(2) of the ESA requires that *all* federal agencies (U.S. Navy, U.S. Army Corps of Engineers, etc.) shall ensure that none of their actions are likely to jeopardize the continued existence of the "listed" species or destroy or adversely modify critical habitat, if designated. When the action of a federal agency may affect an ESA-listed species, that agency is required to consult with either NMFS or USFWS, depending on jurisdiction. "Formal consultation" results if it is determined that the action will "adversely affect" listed species found within the area. Following a biological assessment by the action agency, the consulting agency will write a biological opinion. This opinion ultimately determines whether the action, taken together with the cumulative effects, is likely to "jeopardize the continued existence" of the listed species and/or "result in the destruction or adverse modification" of designated critical habitat of the listed species.

Most sea turtles protected under the ESA are listed globally or by ocean basin, so the determination of jeopardy is based on the effects of the action on the continued existence of the entire population of that listed species, not on individuals or a particular nesting population, even though these entities are initially considered in the analysis. Given the aggregate effects (current status, baseline of threats throughout its life history, and additive effects of the federal action), NMFS will ask whether the species can be expected to both survive and recover, based on its reproduction, numbers, and distribution in the wild.

If a "jeopardy" conclusion is reached by the consulting agency, then the agencies will evaluate the availability of reasonable and prudent alternatives that an applicant (e.g., a fisherman/collective fishery) could take to avoid violation (i.e., the prohibition of take of a listed species) under the ESA. Reasonable and prudent alternatives are actions, identified during formal consultation, that (1) can be implemented in a manner consistent with the intended purpose of the action, (2) can be implemented consistent with the scope of the action agency's legal authority and jurisdiction, (3) are economically and technologically feasible, and (4) would avoid the likelihood of jeopardizing the continued existence of listed species, as determined by NMFS or USFWS.

The Marine Mammal Protection Act

In its original form, the MMPA, passed by the U.S. legislature in 1972, placed a general moratorium on the taking[2] of marine mammals in U.S. waters, with the exception of some incidental takes during commercial fishing activity. The 1994 amendments established section 118 of the MMPA, which requires that NMFS classify each U.S. fishery according to the likelihood of incidental mortality and serious injury to marine mammals. It also establishes a process to develop take reduction plans for fisheries causing frequent or occasional incidental mortality or serious injury of "strategic" marine mammal stocks, that is, that the removal of individuals from a stock exceeds the "potential biological removal" (PBR) level. The PBR level is the maximum number of animals, not including natural mortalities, that may be removed annually from a marine mammal stock while still allowing that stock to reach or maintain its optimum population. Thus, a low PBR level indicates that the size of the stock is likely small and cannot withstand additional mortalities if it is to recover to healthy and sustainable levels. While PBR levels have not been established for sea turtles, they would likely benefit from such a management tool; however, the uncertainties in population size estimates, particularly the early life stages, as well as complexities in the dynamics of sea turtle populations continue to challenge the agencies in managing anthropogenic threats. The MMPA also grants NMFS the authority to place observers on fishing vessels that are known to have higher levels of marine mammal bycatch. Lastly, section 101(a)(2) of the MMPA allows the U.S. Secretary of the Treasury to ban the importation of commercial fish that have been caught with fishing technology that results in the serious injury or mortality of marine mammals in excess of U.S. standards. NMFS is currently in the process of drafting criteria that will help to define "U.S. standards." While regulations implementing the MMPA apply solely to marine mammals and their products, other marine species such as sea turtles may benefit from a regulation, especially one that may reduce interactions or mortalities in fisheries.

The History of the Swordfish Fishery off the U.S. West Coast

The broadbill swordfish is a high-value, highly migratory species harvested worldwide in tropical, subtropical, temperate, and sometimes cold waters (FAO 1985; Ward and Elscot 2000). The quality of swordfish flesh is excellent and marketed in both fresh and frozen form (Holts 2001). In the Pacific Ocean, swordfish occur between about 50° N and 50° S latitudes and routinely move between surface waters and great depths (Ward and Elscot 2000). These swordfish are caught mostly by the longline fisheries of Far East and Western Hemisphere nations, with lesser amounts caught by gill net and harpoon fisheries. Since 2005, the largest catches in the eastern Pacific Ocean have been taken by Spain, Chile, and Japan, which together harvest about 70 percent of the total swordfish catch taken in the region, followed by México and the United States (IATTC 2008).

Off the U.S. West Coast, the swordfish fishery predates European settlement. From at least the first century AD, the Chumash tribe in California's Santa Barbara region caught swordfish with harpoons thrown from plank canoes (Davenport et al. 1993). This method depended on a behavioral trait called "finning" where swordfish periodically surface, exposing their dorsal fin, which can easily be sighted on clear, relatively windless days. The harpoon fishery was revived in the early 1900s by southern California fishermen (Coan et al. 1998) and grew in response to consumer demand for swordfish (Sakagawa 1989).

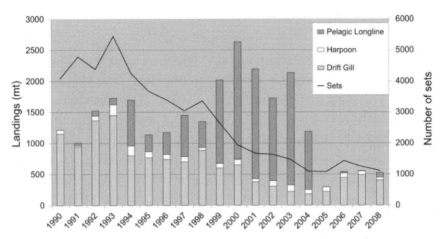

FIGURE 5.2. Estimated annual U.S. West Coast commercial swordfish landings by fishery (in metric tons), and estimated sets fished for drift gillnets, 1990–2008. (Adapted from the PacFIN database, Pacific States Marine Fisheries Commission.)

Harpoon fishing remained the only means of harvesting swordfish until the late 1970s, when a few vessels began targeting common thresher sharks (*Alopias vulpinus*) using drift gillnets (DGN). Almost immediately, swordfish and shortfin mako sharks (*Isurus oxyrinchus*) became important components of the catch (Hanan and Coan 1993). It was soon discovered that the nets were more cost-effective than harpoons in terms of fuel economy and yielded greater catches (fig. 5.2). Swordfish was also worth nearly four times the dockside value of sharks (Bedford 1987; Holts 1988), and by the early 1980s swordfish became the primary target species for the DGN fleet. Because of the greater efficiency of the DGN fishery and the harpoon fishery's dependency on relatively calm seas (Sakagawa 1989; Coan et al. 1998), the DGN fishery became the primary means of harvesting swordfish off the U.S. West Coast EEZ by the early 1990s (PFMC 2009).

Another gear type used to harvest highly migratory pelagic species such as tuna and swordfish is longlines, now primarily used worldwide for commercially harvesting this species complex (Watson and Kerstetter 2006). When targeting swordfish, the gear is set at night in waters shallower than 100 m (i.e., "shallow-set longline" or SSLL), to coincide with their nocturnal movements into surface waters to feed. First attempts at exploring the commercial use of longlines in California occurred in the late 1960s but did not expand due to a high blue shark bycatch (Kato 1969). During the 1991–1992 fishing season, three high-seas[3] longline vessels relocated from the U.S. Gulf of México to California, and by 1994, the number of vessels had grown to thirty-one (Vojkovich and Barsky 1998).

Beginning in 1995, most of these vessels began following swordfish movements by operating out of Hawaii in the spring and summer and out of California during the fall and winter (PFMC 2003). These vessels fished seaward of the 200-mile EEZ off the West Coast because the gear was not authorized by the state of California. The West Coast fishery continued until 2004, when the Pacific Fishery Management Council's (PFMC) Highly Migratory Species Fishery Management Plan (HMS FMP) was implemented, effectively prohibiting SSLL between the mainland and the 150° W longitude, because of inadequate sea turtle protections.[4]

Regulatory Controls for Reducing Sea Turtle Bycatch off the U.S. West Coast

The harpoon swordfish fishery operating in the Southern California Bight has remained free of bycatch reduction regulations because of negligible interactions with nontarget species. The DGN fishery, however, has been subjected to a variety of regulatory controls, under the MMPA, the ESA, and the MSA.

In accordance with section 118 of the MMPA, a Pacific Offshore Cetacean Take Reduction Team (POCTRT) was formed in 1996 and prepared a Take Reduction Plan (TRP), finalized through the passage of regulations in October 1997 to address the incidental take (serious injury or mortality) of several species of cetaceans in the DGN fishery (NMFS 1997). The TRP required the top of the gill nets be set at a minimum depth of 36 feet below the surface, that pingers be attached to those nets to alert marine mammals, and that vessel operators attend educational workshops to learn about provisions of the ESA and MMPA, TRP regulations, and gear, fishing practices, and potential strategies for reducing interactions with protected species, including sea turtles.

While the TRP was developed to reduce the take of marine mammals, some of its measures may have unintentionally benefited sea turtles. For example, during skipper workshops, NMFS has incorporated the dissemination of information regarding proper sea turtle handling requirements as well as resuscitation techniques already mandated by the ESA for all fishermen (when practicable and safe). This knowledge may have improved the survivability of sea turtles entangled in fishing gear, especially those brought up on deck and allowed time to recover from forced submergence. Information on the hearing capabilities of sea turtles is limited, but the available data suggest the auditory capabilities of sea turtles are in the low-frequency range (<1 kHz) (Ridgway et al. 1969; Bartol et al. 1999; Ketten and Bartol 2005; Southwood et al. 2008), below the 10 kHz frequency broadcasted by the pingers, so the turtles do not hear them. Concern for the incidental take of sea turtles in the DGN fishery has prompted the POCTRT to advise NMFS to explore the potential use of lower frequency pingers to alert sea turtles to the presence of nets.

In 2000, NMFS conducted an ESA section 7 consultation on the issuance of a federal permit to incidentally take limited numbers of endangered and threatened marine mammals in commercial fisheries. The resulting biological opinion provided a comprehensive summary of threats facing four species of sea turtles (green, leatherback, loggerhead, and olive ridley) throughout their life history, from the time they begin as incubating eggs on beaches thousands of miles from the U.S. West Coast, to the time they return as females to lay their own eggs. Given the individual threats each sea turtle species faced throughout the Pacific Ocean, coupled with the added threat of the fishery off the U.S. West Coast, NMFS concluded that the current operations of the DGN fishery, as it operated from 1990 through 1999, were likely to jeopardize the continued existence of loggerhead and leatherback sea turtles (NMFS 2000).

In analyzing impacts of the DGN fishery on sea turtles, NMFS found that loggerhead sea turtles had been inadvertently caught in the Southern California Bight during El Niño events (i.e., 1992–1993, and 1997–1998). During these oceanographic episodes, unusually warm sea surface temperatures, due in part to a weakened California Current and northward-flowing equatorial currents, carry hundreds of thousands of pelagic red crabs from Baja California, México, up the coast to southern and central California (Aurioles-Gamboa 1992). Loggerheads taken by the DGN fishery, normally not found with regularity in that area, may have followed their primary food source north from Baja, where a foraging hotspot has been

identified (Pitman 1990; Peckham et al. 2007, 2008). To avoid the likelihood
of the DGN fishery jeopardizing the continued existence of loggerheads,
NMFS designed a unique time/area closure regulation based on forecasted
El Niño conditions for southern California. When anomalously warm sea
surface conditions are forecasted or present or if an El Niño has been declared during the months of June, July, and August, NMFS is authorized to
close the DGN fishery during any or all of those summer months south of
Point Conception, California, and east of 120° W longitude (fig. 5.3).

Since this 2003 regulation was put in place, NMFS has never had
to implement this time/area closure, and fishermen have reported productive and successful catches in this area. In addition, only two loggerheads
have been observed taken by the fishery, both released alive (L. Enriquez,
NMFS Southwest Region Observer Program, pers. commun. 2010).

In analyzing DGN impacts on leatherbacks, NMFS found a stronger correlation between the season the turtles were encountered off central

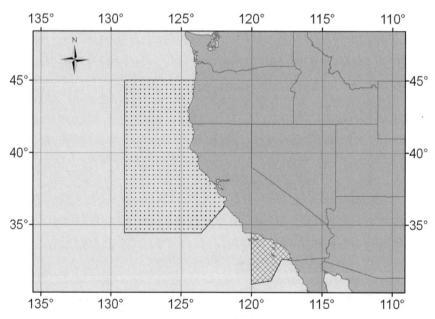

FIGURE 5.3. Time/area closures for the U.S. West Coast drift gillnet fishery to
minimize impacts with Pacific leatherback sea turtles annually from 15 August to
15 November (large area) and with Pacific loggerheads during El Niño years only
from 1 June to 31 August (small area).

and northern California by the fishery than with anomalous oceanographic conditions. All leatherbacks observed taken by the DGN fishery, except for one, were located north of Point Conception, and all were observed taken in the late summer to early winter, where earlier researchers had noted peak sightings off Monterey Bay (Stinson 1984; Starbird et al. 1993). Researchers have since confirmed that the peak time for leatherbacks occurring in the area, late summer through fall, coincides with a relaxation of upwelling and dense aggregations of brown sea nettles (*Chrysaora fuscescens*) that occur off central California and the Pacific Northwest during this same time period (Benson et al. 2007).

Although NMFS had limited information on the genetic origin of the leatherbacks observed taken in the fishery or the importance of the area offshore of the U.S. West Coast to the animals, the agency took a conservative approach and assumed the leatherbacks originated from both the eastern and western Pacific nesting beaches. Two of the animals were genetically tested and found to originate from western Pacific nesting beaches (i.e., Indonesia or Solomon Islands), which at least confirmed that a fraction of leatherbacks from these populations traveled thousands of miles across the Pacific Ocean to forage on their primary prey species of jellyfish in nearshore waters off California and Oregon. Since 2000, extensive research on leatherbacks foraging off central California, including in-water capture, satellite tracking, and genetic analysis, has revealed that the vast majority of the animals originate from nesting beaches in the western Pacific (Benson et al. 2007).

To reduce the likelihood of jeopardy to leatherbacks, NMFS implemented a 213,500 mi^2 time/area closure to reduce by 78 percent the number of leatherback turtles that would potentially be entangled in the fishery (NMFS 2000). The 2001 regulations prohibit DGN fishing in the area north of Point Conception, California, to mid-Oregon and west to 129° W longitude from 15 August to 15 November (fig. 5.3). Through the 2008 fishing season, no observed interactions with leatherbacks turtles have been recorded since the time/area closure has been in effect (L. Enriquez, NMFS Southwest Region Observer Program, pers. commun. 2010). However, with historical DGN fishing grounds now substantially reduced for 93 days of the 170-day primary fishing season, it is not surprising the number of vessels making landings in the swordfish fishery has decreased from 78 in 2000 to 37 in 2008. In addition, the number of net sets has been almost halved, with 1,936 sets made in 2000 down to 1,103 sets in 2008 (fig. 5.2). This decrease in fishing effort comes at a time when U.S. demand for swordfish remains high (fig. 5.1), for example, just under 22,000 metric tons in 2008, with 84 percent being met by imports. Assuming U.S. demand remains constant, the continuing decline in swordfish production from U.S. West Coast fishermen suggests that reliance on foreign imports will persist unless fishery managers seek fishing opportunities to reverse this trend.

In the case of entangling gear, like DGN, time/area closures may be the most effective approach for minimizing sea turtle interactions (Valdemarsen and Suuronen 2003). This is especially true given the limited information on the dynamic nature of the spatial and temporal distribution of sea turtles and their association with various oceanographic features. While new scientific information may eventually allow for some reconfiguration of the time/area closure, including consideration of the development of an ecosystem-based decision support tool (e.g., TurtleWatch; Howell et al. 2008), opportunities for technological fixes are limited, leaving substitution for more selective fishing gears as a key means to reduce bycatch (Harrington et al. 2005; McShane et al. 2007). In 1996, the POCTRT explored harpoons and SSLL as substitutes for use in the DGN fishery. Harpoons were considered operationally unreasonable because the area fished was limited and could not provide sufficient swordfish to make it profitable for the existing DGN fleet to switch (NMFS 1997). Longlines were considered administratively and logistically unreasonable because the DGN fishery at the time was under state of California jurisdiction, which prohibited longlines, and the expense of conversion and unknown resource impacts would create a long-term constraint on the fleet (NMFS 1997). Since then, the HMS FMP was implemented, listing longlines as authorized gear.

Over the past decades, there have been successful campaigns against the use of longline gear in California. While such arguments undoubtedly had merit based on past practices, longlines have become a more selective fishing gear: since 1996, more technological fixes have been applied to longlines than to all other fishing methods combined (Werner et al. 2006). Recent research has demonstrated that selective modifications in deployment and hook design can significantly reduce bycatch rates of and mortality for fish, seabirds, sea turtles, and marine mammals (Watson et al. 2005; Gilman et al. 2006; Kerstetter and Graves 2006; Watson and Kerstetter 2006; Werner et al. 2006; Kaplan et al. 2007; Diaz 2008; Serafy et al. 2008; FAO Fisheries Department 2009). Kerstetter and Graves (2006) noted that circle hooks are more likely to hook animals externally rather than internally (i.e., "gut-hooking"). In addition, fish caught on circle hooks have longer survival times on the line, which not only translates into a higher percentage of discards being returned alive but also benefits fishermen by providing a fresher product for the retained catch.

Sea Turtle Conservation Strategies

With the diminishing U.S. West Coast DGN swordfish fleet, fishery managers have few options for ensuring a continued delivery of domestically caught swordfish to the U.S. market while attaining the goals of ecosystem sustainability. Questions remain regarding whether the use of selective SSLL can revitalize the West Coast swordfish fishery while minimizing impacts to sea turtles. On the surface, this may not seem necessary—why contribute to the Pacific-wide longline fishing juggernaut when valued species such as leatherbacks could find refuge in California's waters during the fall months? That is the position taken by well-organized campaigns concerned about sea turtle preservation, among other conservation issues.[5] This position was clearly evident when the California Coastal Commission voted against NMFS's issuance of an exempted fishing permit that was proposed to evaluate selective SSLL off California in the summer of 2007. The commission's decision seemed to ignore scientific analyses,[6] which determined that sea turtles would rarely interact with selective SSLL in California waters, and those that did had high probability of surviving the interaction/encounter (NMFS 2007).

Efforts to stop selective SSLL from occurring off rich fishing grounds such as California represent bold actions to protect Pacific sea turtles. But are these initiatives really providing intended conservation benefits, or are they misdirected? It is possible that more sea turtles would annually forage in California waters if they were able to escape the gauntlet of foreign fishing fleets using unregulated and uncontrolled SSLL gear across the Pacific. Such a supposition requires implementing a key fishing strategy: increasing domestic and international sources of sustainably caught

Pacific Ocean swordfish. California swordfish fishermen, and Hawaii-based ones as well, can assist with such an endeavor in a couple of ways.

U.S. consumers place a high demand on swordfish (fig. 5.1). Because this demand surpasses U.S. supply, consumers rely heavily on swordfish caught by foreign fleets, most of which are considered to be governed by less stringent environmental laws than are U.S. fleets. However, if consumer demand for swordfish were based on sustainable fishery management, they would rely less on imports and more on U.S. caught swordfish. While such a shift in market sources may seem counter to more intuitive conservation approaches such as a complete prohibition, the reality is that there will always be a global demand for swordfish. Therefore, reliance on sustainably caught sources can actually contribute to Pacific sea turtle recovery. Pacific sea turtles cross international boundaries during their migrations, and any unilateral action in one area will only outsource potential sources of mortality to other areas that may look at ecosystem sustainability differently (Gilman et al. 2006; Sarmiento 2006; Gjertsen 2011). For example, some assert that during the four-year closure of the Hawaii longline swordfish fishery, the benefit of protecting sea turtles may have been lost by shifting fishing effort to other fishing nations with possibly less effective management measures (Gilman et al. 2006; Sarmiento 2006; Rausser et al. 2009). In other words, sources of swordfish provided by regulated U.S. fishermen were replaced by less-regulated fisheries, resulting in little or no net conservation gain for sea turtles (Dutton and Squires 2008).

Another opportunity revolves around capacity building or the promotion of sustainable fishing technologies and practices to other fleets. Hall and Mainprize (2005) discuss the importance of widely disseminating successful technologies to foreign fleets and encouraging their use to reduce bycatch. Such efforts are frequently hampered by fishermen more concerned about the risk of losing marketable catch than any incentive to reduce bycatch (Jennings and Revill 2007). Because of variable differences in fishing gear, target species, location of fishing grounds, and sea turtle species and age classes incidentally taken by a fishery, methods found suitable for one fishery may not be as effective or commercially viable in others, calling for fleet-specific assessments (Gilman et al. 2006). Therefore, a "lessons learned" approach needs to be pursued at every chance. More opportunities to evaluate selective gears in different ecosystems could advance global knowledge of bycatch reduction approaches while maintaining target species catch rates. The colder, fresher waters carried by the California Current, compared to the outer or inshore waters of this ecosystem (PFMC 2003), provide an excellent opportunity to evaluate how selective gears potentially could be shared with other fishermen operating in a similar environment.

Besides employing selective gear technologies to reduce sea turtle bycatch, there have also been calls for taking a more broad-based or holistic approach to restore Pacific sea turtle populations (Bellagio Steering Committee 2004; Kaplan 2005; Seminoff et al. 2007; Dutton and Squires 2008; Benson et al. 2009). This is an essential strategy, considering that coastal mortality (e.g., harvest of eggs and foraging/nesting turtles; entanglement in and ingestion of marine debris; unregulated coastal commercial or artisanal fisheries) may present a greater threat to sea turtle populations, at least for leatherbacks (Kaplan 2005). These other strategies include the protection of nesting beaches, encouraging the sustainability of the traditional use of sea turtles, and undertaking multilateral policy actions (Bellagio Steering Committee 2004). Gjertsen (2011) found that nesting beach conservation may be the most cost-effective means to achieve increases in leatherback populations relative to time/area closures or other restrictions on fishing effort.

Though seemingly paradoxical, reliance on swordfish harvested by U.S. fishermen may be an important contribution to the recovery of Pacific sea turtles. The United States imposes some of the most stringent and measurable restrictions on its fishing industry, regardless of gear types, compared with other fishing nations. In the case of SSLL, NMFS manages these fishermen not only by technological and operational requirements necessitated by mandated environmental protections, but also limits on interactions with protected species imposed by ESA and MMPA requirements to further ensure ecosystem sustainability.

Because of the global scale and interconnections of fish production and marketing systems, unintended fishery interactions do not disappear as a result of regulating one area or one nation's fishery (Crespo and Hall 2002)—the impacts are simply transferred to other places. As the global population and corresponding demand for seafood continue to grow, sustainable fishing practices need to be part of the holistic approach for meeting food production needs while simultaneously assisting in sea turtle recovery along the full geographic scale of their range. Consequently, a more pragmatic approach to satisfy U.S. demand for swordfish might be for the United States to rely less on foreign fleets and demonstrate by example that U.S. technological innovations can produce comparable and sustainable catches while minimizing ecosystem impacts.

Abbreviations

DGN	drift gillnet
DPS	distinct population segment
EEZ	exclusive economic zone
ESA	Endangered Species Act
FAO	Food and Agriculture Organization
FMP	fishery management plan
HMS FMP	Highly Migratory Species Fishery Management Plan
IATTC	Inter-American Tropical Tuna Commission
ISC	International Scientific Committee for Tuna and Tuna-like Species in the North Pacific Ocean
MMPA	Marine Mammal Protection Act
MSA	Magnuson-Stevens Fishery Conservation and Management Act
NOAA	National Oceanic and Atmospheric Administration
NMFS	U.S. National Marine Fisheries Service
PBR	potential biological removal
PFMC	Pacific Fishery Management Council
POCTRT	Pacific Offshore Cetacean Take Reduction Team
SSLL	shallow-set longline
TRP	take reduction plan
USFWS	U.S. Fish and Wildlife Service

ACKNOWLEDGMENTS

We thank the editors and N. Fahy for their review of earlier drafts of this chapter. We also appreciate the technical assistance of Craig D'Angelo, Lyle Enriquez, Sunee Sonu, and Charles Villafana on various aspects of this chapter. The views expressed herein are those of the authors and do not necessarily reflect the views of their agency.

1. "Take," as defined under the ESA, includes harassment, harm, pursuit, hunting, shooting, wounding, killing, trapping, capturing or collecting, or attempting to engage in any such conduct.

2. "Take," as defined under the MMPA, means to harass, hunt, capture, collect, or kill a marine mammal, or attempt any of these actions (Code of Federal Regulations, Title 50, section 216.3).

3. "High seas" refers to beyond the 200-nautical-mile U.S. EEZ.

4. An ESA section 7 consultation was conducted by NMFS on the SSLL fishery operating out of California as part of the HMS FMP, and a jeopardy conclusion was reached for loggerhead sea turtles. Regulations written under the authority of the ESA essentially prohibited this component of the fishery in order to protect loggerhead sea turtles.

5. Some sport fishermen are also well organized to prevent any efforts for longlines, claiming that longlines will deplete target fisheries for tuna and striped marlin.

6. NMFS conducted an ESA section 7 consultation on the action and, based on the best available scientific information available, came to a no jeopardy conclusion for leatherback sea turtles (NMFS 2007) and a "finding of no significant impact" on the human environment, another federal requirement under the National Environmental Policy Act (NMFS and PFMC 2007).

REFERENCES

Aurioles-Gamboa, D. 1992. Inshore-offshore movements of pelagic red crabs *Pleuroncodes planipes* (Decapoda, Anomura, Galatheidae) off the Pacific coast of Baja California Sur, Mexico. Crustaceana 62:71–84.

Bartol, S. M., J. A. Musick, and M. L. Lenhardt. 1999. Auditory evoked potentials of the loggerhead sea turtle (*Caretta caretta*). Copeia 1999:186–840.

Bartram, P. K., J. J. Kaneko, and K. Kucey-Nakamura. 2010. Sea turtle bycatch to fish catch ratios for differentiating Hawaii longline-caught seafood products. Marine Policy 34:145–149.

Bedford, D. W. 1987. Shark management: a case history—the California pelagic shark and swordfish fishery. In: Cook, S. (Ed.), Sharks: An Inquiry into Biology, Behavior, Fisheries, and Use. Oregon State University Extension Service, Corvalis. Pp. 161–171.

Bellagio Steering Committee. 2004. What can be done to restore Pacific turtle populations? The Bellagio Blueprint for Action Pacific Sea Turtles. Bellagio Conference on Sea Turtles. World Fish Center, Penang, Malaysia.

Benaka, L. R., and T. J. Dobrzynski. 2004. The U.S. National Marine Fisheries Service's national bycatch strategy. Marine Fisheries Review 66:1–8.

Benson, S., H. Dewar, P. Dutton, C. Fahy, C. Heberer, D. Squires, and S. Stohs. 2009.

Swordfish and Leatherback Use of Temperate Habitat (SLUTH). Administrative Report LJ-09-06. U.S. National Marine Fisheries Service, La Jolla, CA.

Benson, S. R., K. A. Forney, J. T. Harvey, J. V. Carretta, and P. H. Dutton. 2007. Abundance, distribution, and habitat of leatherback turtles (*Dermochelys coriacea*) off California, 1990–2003. Fisheries Bulletin 105:337–347 (2007).

Coan, A. L., M. Vojkovich, and D. Prescott. 1998. The California harpoon fishery for swordfish, *Xiphias gladius*. In: Barrett, I., O. Sosa-Nishizaki, and N. Bartoo (Eds.), International Symposium of Pacific swordfish, Ensenada, México, 11–14 December 1994. NOAA Tech. Rep. NMFS 142. U.S. Department of Commerce, Seattle, WA. Pp. 37–49.

Crespo, E. A., and M. A. Hall. 2002. Interactions between aquatic mammals and humans in the context of ecosystem management. In: Evans, P. G. H., and J. A. Raga (Eds.), Marine Mammals: Biology and Conservation. Kluwer Academic/Plenum, New York. Pp. 463–490

Davenport, D., J. R. Johnson, and J. Timbrook. 1993. The Chumash and the swordfish. Antiquity 67:257–272.

Diaz, G. A. 2008. The effect of circle hooks and straight (J) hooks on the catch rates and numbers of white marlin and blue marlin released alive by the U.S. pelagic longline fleet in the Gulf of Mexico. North American Journal of Fisheries Management 28:500–5006.

Dutton, P. H., and D. Squires. 2008. Reconciling biodiversity with fishing: a holistic strategy for Pacific sea turtle recovery. Ocean Development and International Law 39:200–222.

FAO. 1985. FAO Species Catalogue. Vol. 5. Billfishes of the World. An Annotated and Illustrated Catalogue of Marlins, Sailfishes, Spearfishes and Swordfishes Known to Date. FAO Fisheries Synopsis No. 125. Food and Agriculture Organization, Rome.

FAO Fisheries Department. 2009. Guidelines to Reduce Sea Turtle Mortality in Fishing Operations. Food and Agriculture Organization, Rome.

Gilman, E., E. Zollett, S. Beverly, H. Nakano, K. Davis, D. Shiode, P. Dalzell, and I. Kinan. 2006. Reducing sea turtle by-catch in pelagic longline fisheries. Fish and Fisheries 7:2–23.

Gjertsen, H. 2011. Can we improve our conservation bang for the buck? Cost effectiveness of alternative leatherback turtle conservation strategies. In: Dutton, P. H., D. Squires, and M. Ahmed (Eds.), Conservation of Pacific Sea Turtles. University of Hawaii Press, Honolulu. Pp. 60–84.

Hall, M. A., D. L. Alverson, and K. I. Metuzals. 2000. By-catch: problems and solutions. Marine Pollution Bulletin 41:204–219.

Hall, S. J., and B. M. Mainprize. 2005. Managing by-catch and discards: how much progress are we making and how can we do better? Fish and Fisheries 6:134–155.

Hanan, D. A., and A. L. Coan Jr. 1993. The California Drift Gill Net Fishery for Sharks and Swordfish, 1981–82 through 1990–91. Fish Bull. No. 175. California Department of Fish and Game, Long Beach.

Harrington, J. M., R. A. Myers, and A. A. Rosenberg. 2005. Wasted fishery resources: discarded by-catch in the USA. Fish and Fisheries 6:350–361.

Holts, D. B. 1988. Review of U.S. West Coast commercial shark fisheries. Marine Fisheries Review 50:1–8.

Holts, D. 2001. Swordfish. In: Leet, W. S., C. M. Dewees, R. Klingbeil, and E. J. Larson (Eds.), California's Living Marine Resource: A Status Report. Publication SG01-11. California Department of Fish and Game, Sacramento, CA. Pp. 322—323.

Howell, E. A., D. R. Kobayashi, D. M. Parker, G. H. Balazs, and J. J. Polovina. 2008. TurtleWatch: a tool to aid in the bycatch reduction of loggerhead turtles in the Hawaii-based pelagic longline fishery. Endangered Species Research 5:267–278.

IATTC. 2008. Fishery Status Report: Tuna and Billfishes in the Eastern Pacific Ocean in 2006. Inter-American Tropical Tuna Commission, La Jolla, CA.

ISC. 2009. Report of the Ninth Meeting of the International Scientific Committee for Tuna and Tuna-like Species in the North Pacific Ocean. Plenary Session, 15–20 July 2009, Kaohsiung, Taiwan. International Scientific Committee for Tuna and Tuna-like Species in the North Pacific Ocean. Available: http://isc.ac.affrc.go.jp/.

Jennings, S., and A. S. Revill. 2007. The role of gear technologists in supporting an ecosystem approach to fisheries. ICES Journal of Marine Science 64:1525–1534.

Kaplan, I. C. 2005. A risk assessment for Pacific leatherback turtles (Dermochelys coriacea). Canadian Journal of Fisheries and Aquatic Sciences 62:1710–1719.

Kaplan, I. C., S. P. Cox, and J. F. Kitchell. 2007. Circle hooks for Pacific longliners: not a panacea for marlin and shark bycatch, but part of the solution. Transactions of the American Fisheries Society 136:392–401.

Kato, S. 1969. Longlining for swordfish in the Eastern Pacific. Commercial Fisheries Review 31:30–32.

Kennelly, S. J., and M. K. Broadhurst. 2002. By-catch be gone: changes in the philosophy of fishing technology. Fish and Fisheries 3:340–355.

Kerstetter, D. W., and J. E. Graves. 2006. Effects of circle versus J-style hooks on target and non-target species in a pelagic longline fishery. Fisheries Research 80:239–250.

Ketten, D. R., and S. M. Bartol. 2005. Functional Measures of Sea Turtle Hearing—Final Report. 1 March 2002–30 September 2005. Technical Report, Contract No. N00014–02–1–0510. Woods Hole Oceanographic Institution, Woods Hole, MA.

McShane, P. E., M. K. Broadhurst, and A. Williams. 2007. Keeping watch on the unwatchable: technological solutions for the problems generated by ecosystem-based management. Fish and Fisheries 8:153–161.

NMFS. 1997. Final Offshore Cetacean Take Reduction Plan. U.S. National Marine Fisheries Service Southwest Regional Office, Long Beach, CA.

NMFS. 2000. Section 7 Consultation on Authorization to Take Listed Marine Mammals Incidental to Commercial Fishing Operations under Section 101(a)(5)(E) of the Marine Mammal Protection Act for the California/Oregon Drift Gillnet Fishery. U.S. National Marine Fisheries Service Office of Protected Resources, Silver Spring, MD.

NMFS. 2007. Endangered Species Act Section 7 Consultation on the Issuance of a Shallow-Set Longline Exempted Fishing Permit under the Fishery Management Plan for U.S. West Coast Highly Migratory Species Fisheries. Signed 28 November 2007. U.S. National Marine Fisheries Service, Long Beach, CA.

NMFS and PFMC. 2007. Environmental Assessment on the Issuance of an Exempted Fishing Permit to Fish with Longline Gear in the U.S. West Coast Exclusive Economic Zone. U.S. National Marine Fisheries Service and Pacific Fishery Management Council, Long Beach, CA and Portland, OR.

Peckham, S. H., D. Maldonado-Díaz, V. Koch, A. Mancini, A. Gaos, M. T. Tinker, and W. Nichols. 2008. High mortality of loggerhead turtles due to bycatch, human consumption and strandings at Baja California Sur, México, 2003–2007. Endangered Species Research 5:171–183.

Peckham, S. H., D. Maldonado Díaz, A. Walli, G. Ruiz, L. Crowder, and W. J. Nichols. 2007. Small-scale fisheries bycatch jeopardizes Pacific loggerhead turtles. PLoS One 2(10):e1041.

PFMC. 2003. Fishery Management Plan and Environmental Impact Statement for U.S. West Coast Fisheries for Highly Migratory Species. Pacific Fishery Management Council, Portland, OR.

PFMC. 2009. Status of the U.S. West Coast Fisheries for Highly Migratory Species through 2008, Stock Assessment and Evaluation. Pacific Fishery Management Council, Portland, OR.

Pitman, R. L. 1990. Pelagic distribution and biology of sea turtles in the eastern tropical Pacific. In: Richardson, T. H., J. I. Richardson, and M. Donnelly (Comps.), Proceedings of the Tenth Annual Workshop on Sea Turtle Biology and Conservation, NOAA Tech. Memo. NMFS-SEC-278. U.S. National Marine Fisheries Service, Miami, FL. Pp. 143–148.

Rausser, G., S. Hamilton, M. Kovach, and R. Stifter. 2009. Unintended consequences: the spillover effects of common property regulations. Marine Policy 33:24–39.

Ridgway, S. H., E. G. Wever, J. G. McCormick, J. Palin, and J. H. Anderson. 1969. Hearing in the giant sea turtle, *Chelonia mydas*. Proceedings of the National Academy of Sciences of the United States of America 64:884–890.

Sakagawa, G. T. 1989. Trends in fisheries for swordfish in the Pacific Ocean. In: Stroud, R. H. (Ed.), Planning the Future of Billfishes: Research and Management in the 90s and Beyond. Proceedings of the Second International Billfish Symposium, Kailua—Kona, Hawaii, 1–5 August 1988. National Coalition for Marine Conservation, Savannah, GA. Pp. 61–79.

Santora, C. 2003. Management of turtle bycatch: can endangered species be protected while minimizing socioeconomic impacts? Coastal Management 31:424–434.

Sarmiento, C. 2006. Transfer function estimation of trade leakages generated by court rulings in the Hawaii longline fishery. Applied Economics 38:183–190.

Seminoff, J. A., F. V. Paladino, and A. G. J. Rhodin. 2007. Refocusing on leatherbacks: conservation challenges and signs of success. Chelonian Conservation and Biology 6: 1–6.

Serafy, J. E., D. W. Kerstetter, and P. H. Rice. 2008. Can circle hook use benefit billfishes? Fish and Fisheries 9:1–11.

Southwood, A., K. Fritsches, R. Brill, and Y. Swimmer. 2008. Sound, chemical and light detection in sea turtles and pelagic fishes: sensory-based approaches to bycatch reduction in longline fisheries. Endangered Species Research 5:225–238.

Starbird, C. H., A. Baldridge, and J. T. Harvey. 1993. Seasonal occurrence of leatherback sea turtles (*Dermochelys coriacea*) in the Monterey Bay region, with notes on other sea turtles, 1986–1991. California Fish and Game 79(2):54–62.

Stinson, M. 1984. Biology of sea turtles in San Diego Bay, California and the northeastern Pacific Ocean. Master's Thesis, San Diego State University, San Diego, CA.

Valdemarsen, J. W., and P. Suuronen. 2003. Modifying fishing gear to achieve ecosystem objectives. In: Sinclair, M., and G. Valdimarsson (Eds.), Responsible Fisheries in the Marine Ecosystem. FAO and CABI International, Rome. Pp. 321–341.

Vojkovich, M., and K. Barsky. 1998. The California-Based Longline Fishery for Swordfish, *Xiphias gladius*, beyond the U.S. Exclusive Economic Zone. NOAA Tech. Rep. NMFS 142. U.S. National Marine Fisheries Service, La Jolla, CA. Pp. 147–152.

Ward, P., and S. Elscot. 2000. Broadbill Swordfish: Status of World Fisheries. Bureau of Rural Sciences, Commonwealth Department of Agriculture, Fisheries and Forestry, Canberra, Australia.

Watson, J. W., S. P. Epperly, A. K. Shah, and D. G. Foster. 2005. Fishing methods to reduce sea turtle mortality associated with pelagic longlines. Canadian Journal of Fisheries and Aquatic Sciences 62:965–981.

Watson, J. W., and D. W. Kerstetter. 2006. Pelagic longline fishing gear: a brief history and review of research efforts to improve selectivity. Marine Technology Society Journal 40:6–11.

Werner, T. S. Craus, A. Read, and E. Zollett. 2006. Fishing techniques to reduce the bycatch of threatened marine animals. Marine Technology Society Journal 40:50–68.

6

No "Silver Bullets" but Plenty of Options

Working with Artisanal Fishers in the Eastern Pacific
to Reduce Incidental Sea Turtle Mortality
in Longline Fisheries

MARTIN HALL, YONAT SWIMMER,
AND MARILUZ PARGA

Summary

Sea turtles are vulnerable to capture in a variety of fishing gears, including nets, trawls, purse seines, and longlines. Sea turtle bycatch is significant in the eastern Pacific Ocean, an area of heavy fishing pressure because of high biological productivity and the resulting dual presence of numerous sea turtle and target fish species. Injuries and mortality related to bycatch can have significant negative impacts on the region's sea turtle populations.

Ideally, to optimize available resources to reduce the incidental mortality of sea turtles, fisheries managers must identify which fisheries or fishing methods have the most significant impacts on sea turtle populations. This involves determining, with an acceptable level of precision, the level of incidental mortality in all relevant fisheries. In this chapter, we discuss the challenges to achieve this objective and identify opportunities for reducing fishery impacts to sea turtles in the eastern Pacific and beyond, focused primarily on longline fisheries. Specifically, we describe the concept of "lines of defense" for sea turtle fisheries interactions, which can be envisioned as layers of risk or opportunity that are unique for each fishery, with the idea that efforts must be taken to prevent turtles from interacting with fishing gear initially, to ultimately ensuring that an animal has the highest chance of surviving with minimal damage. We strongly promote the training and use of at-sea fisheries observers as well as cooperation with fishing communities to increase chances of long-term sustainability of conservation efforts.

Introduction

From the moment hatchlings emerge from their nests, sea turtles are exposed to a variety of hazards (see plates 11 and 12). One of these is the incidental capture in fisheries. Turtles are vulnerable to capture in fishing gear for a variety of reasons, including chance encounters with gill nets, trawls, purse seines, and longlines, many times because they are drawn near the fishing gear by either bait or catches as they search for food (plate 11). Depending on numerous factors associated with the fishing operation, this incidental capture can lead to mortality. We define bycatch as the individuals discarded dead or likely to die as a result of the fishing operations, which can have significant impacts for some of the sea turtle populations in the eastern Pacific Ocean. This region is an area of heavy fishing pressure because of high biological productivity and shared presence of numerous sea turtle and target fish species. Moreover, the continental shelf that lines the eastern Pacific Ocean is extremely narrow, and because much of the productivity occurs over the shelf and the shelf break, extensive fishing grounds are within close range of the artisanal and industrial fishing fleets operating out of eastern Pacific ports. Just as this region is a haven for productive fisheries, the adjacent coasts of the eastern Pacific from México to Perú host numerous large congregations of nesting sea turtles. For this reason, high turtle densities overlap with intense coastal and pelagic fishing effort, leading to unavoidable interactions.

Ideally, to optimize available resources to reduce the incidental mortality of sea turtles, fisheries managers must identify which fisheries or fishing methods have the most significant impacts on sea turtle populations. The answer is a question not simply of numbers of turtles affected, but also of which sex and age (or size) classes are most vulnerable to which types of gear. Clearly, a primary goal of sea turtle conservation is to reduce incidental capture and mortality to sustainable levels. However, we must first achieve the monumental task of determining, with an acceptable level of precision, the level of incidental mortality in all relevant fisheries. In this chapter, we discuss the challenges to achieve this objective and identify opportunities for reducing fishery impacts to sea turtles in the eastern Pacific and beyond, particularly as they relate to longline fisheries.

Fisheries of the Eastern Pacific

In the eastern Pacific, the nutrient- and biologically-rich continental shelf is very narrow (for example, compared with the Gulf of México, or the Southwestern Atlantic coasts). Even small vessels can easily reach areas with characteristics of open ocean, and a relatively benign climate in parts of the regions (fewer hurricanes or large storms than other ocean basins) allows them to venture offshore in search of pelagic species such as tunas (*Thunnus* spp., *Katsuwonus pelamis*), swordfish (*Xiphias gladius*), and sharks (e.g., *Prionace glauca*, *Sphyrna lewini*, *Alopias* spp.). Both artisanal and industrial fisheries are an important source of employment for the nations of the region, and their catch constitutes an indispensable source of protein for local consumption as well as important exports to foreign markets.

Fisheries operating from different ports or countries may share a fishing ground, and their gear technology can be similar. However, artisanal fisheries exhibit a bewildering array of modes of operation. Some boats fish most of the year with the same gear, while others change seasonally, and yet others use more than one type of gear in the same fishing trip (e.g., switch from setting longlines at the surface and bottom in successive sets, or even from longlining to gill netting). Most of the small boats (<10 m in length) operate their lines or nets manually and therefore have limitations in the amount of gear they can deploy. They also tend to deploy their gear at shallow depths, frequently less than 30 m.

On the other hand, industrial-scale longline fishing vessels, generally larger in size, deploy their lines much deeper, from approximately 100 to 400 m, to catch species such as bigeye tuna (*Thunnus obesus*). Because sea turtles spend most of their time in the upper layers (frequently at <30 m; Polovina et al. 2002; Swimmer et al. 2006; Seminoff et al. 2008), turtle bycatch rates are orders of magnitude lower in deep-set gear than in shallow-set gear. Inversely, the probability of surviving a net entanglement or longline hooking is much higher for shallow gear because the turtles have a greater chance of reaching the surface to breathe than they do in deep-set gear (Gilman et al. 2006).

Quantifying Sea Turtle Bycatch and Bycatch-Related Mortality

Despite many researchers' assertions that fisheries are a driving force in the decline of some sea turtle populations, we still are very far from having reliable estimates of the mortality caused by the different fisheries (Lewison et al. 2004) or by other factors. Similar to such estimation for any pelagic species, the range of methods available to estimate sea turtle mortality is quite limited because of costs and logistical constraints. In some populations, with knowledge of the age structure, or a series of abundance data, it is possible to estimate the total mortality and then, based on some reasonable assumption for natural mortality, to separate the component caused by the fishing operations (which can then be useful for understanding impacts at the population level). However, age determination of sea turtles has proven to be a fairly complex process (e.g., Bjorndal et al. 1998). In the future, use of electronic tags, such as passive integrated transponder (PIT) tags, which can be used on hatchlings (Rowe and Kelly 2005), may increase the accuracy of the process and greatly facilitate more accurate estimations.

Strandings of sea turtles have been used as a proxy for mortality (Epperly et al. 1996; Alava et al. 2005). However, strandings reflect only a sample of dead turtles within reach of the coasts, subject to oceanographic and atmospheric variability, and if the causes of mortality are not easy to identify (e.g., turtles washing ashore without clear hook or net scars), strandings do not allow a clear distinction between natural and fishing mortality (Epperly et al. 1996). Stranding studies could provide some valuable information on the species and sizes present in an area, but not reliable estimates of mortality, or in some cases of the causes of mortality.

These limitations bring us to the estimation of incidental mortality through sampling of fishing activities of fleets. This can be obtained through the use of fishers' logbooks, observers, or fisher surveys. Fishers' logbooks can provide valuable data on the distribution of effort and operational modes (e.g., day or night sets), but the fishers are usually dedicated to the fishing

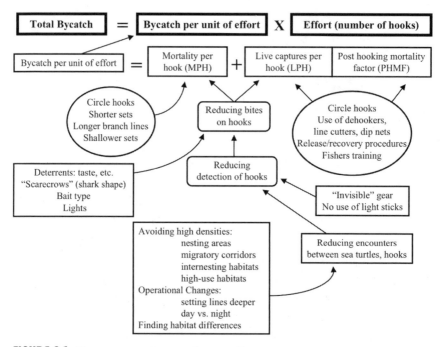

FIGURE 6.1. Components of sea turtle mortality.

operations and are not likely to keep complete records of the incidental captures. We believe that, in most cases, fisher surveys have little or no quantitative value, because the questions are never perceived in a neutral way, and answers reflect biases from either fishers or researchers. Thus, in order to effectively manage sea turtle bycatch in fisheries, it is important that efforts are undertaken to directly observe, through time and space, the extent to which gear types are more prone to interact with sea turtles (fig. 6.1).

Observer Programs

The use of observers is critical to gather the type of detailed data necessary to pinpoint seasonality, geographic range, fishing techniques, and species most associated with sea turtle bycatch. Observer programs can produce a vast amount of information that can be used to understand why bycatch happens and what affects its magnitude. If observers are well trained and collect data on a series of variables that are known or expected to affect the bycatch rates, they can be used to generate databases that have proven invaluable to reduce bycatch in many cases (e.g., Hall et al. 2001). We believe that observers are a valuable tool for the scientists searching for solutions, and they are also very important in creating and maintaining open communication links with fishers and in bringing their knowledge and feedback to scientists and managers.

Once the existence of a problem has been identified, the first step is to develop a research program to identify the technological or operational changes that could reduce bycatch mortality, while at the same time allowing fishers to continue to practice a sustainable fishery by improving the selectivity of the gear and minimizing the impact on other species, the habitat, and so on. This type of program addresses a wide range of issues, from understanding how turtles and target species sense their environment in order to devise ways to attract or reject them from fishing gear, to behavioral and ecological studies that can help us understand why bycatch happens and how to avoid it. By ensuring the participation of the fishing community in the development of the solutions (Campbell and Cornwell 2008), we ensure that the solutions will be effective and practical and will increase the chances of adoption by the fishers themselves, reducing problems of rejection or noncompliance that could later deter full use of the bycatch reduction strategy.

The Lines of Defense in Longline Fisheries

A reasonable way to approach a sea turtle bycatch mitigation program is the concept of "lines of defense" (Hall 1996), which can be envisioned as layers of risk or opportunity that are unique for each fishery. In the case of a longline, which we will use as the main example, the six lines of defense are sequential and involve preventing turtles from, first, encountering the fishing gear; second, detecting or approaching the bait; and third, biting the bait. Subsequently, efforts must be taken to, fourth, minimize the number of hookings (i.e., use hooks that slide off the mouth without setting); fifth, use hooks that are less harmful (i.e., with higher posthooking survival probability) and are easier to remove; and sixth, use the best dehooking and resuscitation techniques to improve the turtles' chances of surviving the encounter.

While the specifics of these lines of defense differ for each fishery, the sequence is generally similar. Stopping the problem in the initial lines of defense is better, because it decreases the risk of mortality. However, the problem of mortality reduction should not be seen as an all-or-nothing battle to win at any given line of defense. Rather, it should be seen as a gradual process of containment, where reductions in successive lines lead to a sustainable level of mortality based on a precautionary approach that would allow all populations to persist. Research efforts should be accompanied by programs to implement effective mitigation methods and to work with the communities to obtain their authentic support for the program and its objectives. However, it is important to note that a bycatch reduction strategy proven successful in one area or one fishery may not be universally so for all regions and fisheries. This variability calls for rigorous regional verification of what constitutes a "successful" bycatch reduction strategy.

One solution is to reduce or eliminate the fishing effort that results in the mortality. This heavy-handed approach may be impossible to enact, given the social and economic burdens it would cause in most fisheries; thus, we concentrate our attention on the ways to minimize the impacts while allowing fishing activities to continue.

First Line of Defense: Separating Sea Turtles and Fishing Gear

VERTICAL SEPARATION. Because most pelagic-stage turtles spend most of their time near the surface (with the exception of olive ridley turtles), deeper gear will have fewer interactions than shallower gear. For instance, in longline gear, deeper hooks have lower hooking rates. The Japanese industrial longline fishery targeting tunas in the eastern tropical Pacific deploys its hooks at 100–400 m of depth, and their hooking rates (in turtles per 1,000 hooks) are a fraction of those observed in shallow sets in the coastal zone (Gilman et al. 2007). In some cases, the shallower hooks from a deep-set longline may be removed to reduce encounters (Beverly et al. 2009). In other cases, bottom longlines or nets may have more interactions with benthic foraging turtles.

HORIZONTAL SEPARATION. If it were possible to define (or predict with high levels of accuracy) the route and timing of turtle migration toward the breeding and nesting areas, it may be possible to establish spatial-temporal closures to reduce gear interactions with migrating turtles. The existence of "migratory corridors" has been postulated (Morreale et al. 1996), but more recent tagging studies have shown more diffuse patterns of movement (Seminoff et al. 2008; Shillinger et al. 2008). The oceanographic correlates of the movements add a component of spatial variability (Polovina et al. 2004; Shillinger et al. 2008). When the density of sea turtles is very high, such as in the vicinity of nesting beaches, the most sensible solution is to avoid hooking and entanglement by establishing temporary fishing closures. This type of mitigation method is most likely to succeed if the closures are temporary and agreed upon by the fishing community, or if they are heavily enforced, which is rarely a viable option. Oftentimes, ambitious attempts to create marine parks and protected areas end up becoming "paper victories" without real means of enforcement or a steady stream of funding to finance them. An alternative way to identify and avoid high risk areas is the use of fleet communication among fishing vessels (Gilman et al. 2006), which we believe has great promise as a conservation tool.

HABITAT SEPARATION. When turtles and target species have different habitat preferences, it may be possible to constrain fishing operations to the target species habitat (Plotkin 2003; Canadas et al. 2005; Sasso and Epperly 2007; Kobayashi et al. 2008; Seminoff et al. 2008). This is more difficult when sea turtles and targeted fish species are associated with the same oceanographic features, such as sea surface temperatures or oceanic fronts (Polovina et al. 2000; Swimmer et al. 2010a). Attempts to delineate the habitats of both target and protected species are fraught with difficulty, yet efforts are under way to achieve this using TurtleWatch, a real-time, online system using satellite imagery to guide fishers away from predicted locations of sea turtles based on their habitat preferences (Howell et al. 2008). This represents yet another conservation tool that would be especially valuable to fishers operating in regulatory regimes with hard caps, such that successfully avoiding turtles could result in fewer fishing restrictions.

Second Line of Defense: Prevent Turtles from Detecting the Bait

Physiological and behavioral research has helped to identify the sensory mechanisms used by target fish species and sea turtles to draw them into the vicinity of the fishing gear. By understanding how animals sense their environment, we can modify fishing gear or bait so that it is less attractive to turtles (e.g., reduce its visibility via camouflaging in water, make floats transparent, mask the smell of bait) but still attractive to the target species. We can also use the information to attract turtles away from the nets or lines, to prevent them from approaching the gear. Primarily visual senses, and secondarily olfactory senses, are believed to play stronger roles than auditory senses in attracting turtles to fishing gear or bait (Constantino and Salmon 2003; Swimmer and Brill 2006; Southwood et al. 2008).

Third Line of Defense: Prevent Turtles from Biting the Bait

This approach aims to render bait unattractive or even repellent to the turtles without affecting its attraction to the target species. Bait can be presoaked in different chemicals and used the regular way, without causing additional work to fishers, which would facilitate adoption. Unfortunately, attempts to date to develop repellent scents for the turtles have not been successful (Swimmer and Brill 2006).

Another approach is to use a visual "deterrent" based on a predator–prey relationship. This model assumes that prey such as sea turtles would flee upon sight of a predator, such as a shark. Efforts are under way to test this idea, with preliminary success (Wang et al. 2010). Additionally, trials in a gill net fishery in Baja, México, have also shown that illuminating fishing gear in nighttime fisheries by adding LEDs or light sticks reduces sea turtle capture (Wang et al. 2010). A theoretical, as yet untested idea for a visual deterrent is a blinking light that relies on differences in the speed of vision between turtles and fish to repel turtles from the fishing gear.

Fourth Line of Defense: Reduce the Number of Turtles Hooked

Use of different hook sizes and shapes have been used to achieve a reduction in the number of animals that become hooked or entangled in the gear. Laboratory experiments with captive sea turtles and a variety of tested hook sizes suggest that using wider hooks could reduce the number of hooking events (Watson et al. 2005; Gilman et al. 2007; Read 2007; Piovano et al.

2009). Relatively wider circle hooks can be used instead of the straight J or Japanese-style tuna hook (fig. 6.2), which are commonly used in most longline fisheries. It is believed that circle hooks are effective for two reasons: (1) the increased width makes it more difficult to hook a turtle's mouth in the first place, and (2) because of the hook's shape, it should slide along the mouth while the fish or turtle takes the bait and lodge externally in the jaw, rather than being swallowed for a deep hooking, with a higher assumed probability of mortality (Cooke and Suski 2004). Fishing experiments have produced essentially two types of outcomes: either (1) the hooking rates of sea turtles are reduced and fewer deep hookings occur, or (2) the hooking rates remain the same but fewer deep hookings occur, so the effectiveness is essentially related to severity of injury and likelihood of surviving the encounter, and not whether turtles were caught on the line. Recent reviews of circle hooks (Gilman et al. 2006; Read 2007) reach the general conclusion that they seem to reduce bycatch of some species, including turtles. Furthermore, use of circle hooks has not shown to decrease catch rates of target

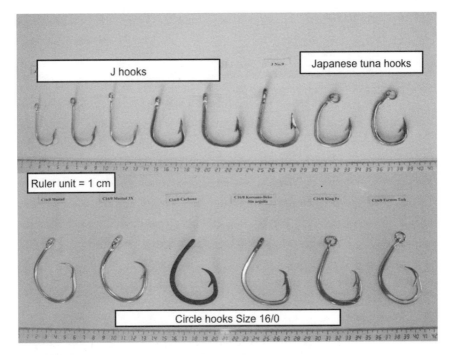

FIGURE 6.2. Different hook types typically used in longline fishing.

species, thereby improving the chances that the hooks will be more easily adopted by the fishing industry. An exception is the fishery targeting mahi-mahi (*Coryphaena hippurus*) in the Pacific off the coast of South America, which catches smaller fishes at the beginning of the season. For this range of sizes, circle hooks do not produce at the same level as the traditional J hooks. However, one could argue that the capture of a large number of small fishes is not a desirable outcome of the fishery, and the selectivity of the circle hook in this case could also be beneficial to the sustainability of the target species fishery.

Different bait types also have an impact on the hooking rates of turtles (Watson et al. 2005). Fish bait (e.g., mackerel) results in fewer hookings than does squid bait, probably because of how the turtle removes the bait from the hook. Because of the sucking action of the turtle, the squid bait plus hook is generally brought deeper into the turtle's digestive tract. In contrast, the turtle likely takes smaller bites of bait fish and thus avoids biting the metallic hook. The combination of fish bait plus circle hooks has the highest reduction of sea turtle hooking rates of all hook/bait combinations (Watson et al. 2005).

Fifth Line of Defense: Use Hooks that are Less Harmful

If the turtle is going to be hooked, a number of factors and actions can play a significant role in determining the fate of the animal. The special shape and the usually larger size of circle hooks tend to affect turtles in the mouth, as opposed to J and tuna hooks, which usually hook turtles in the esophagus and stomach. Many scientists assume that hooks left in the esophagus have a high percentage of mortality, but several authors question this belief (Tomás et al. 2001; Alegre et al. 2006; Valente et al. 2007), at least for smaller hooks. The esophagus in turtles is strong and very muscular, and in many occasions the hooks that pass through cause only local scarring that does not affect the long-term health of the animal. There is also an erroneous assumption that hookings in the mouth are generally benign. In fact, mortality resulting from hookings in the mouth most likely depends on the exact location of the hooking. For example, based on observations, a hook affecting the epiglottis will probably lead to mortality due to aspiration pneumonia, and a hook that has fractured a jawbone or is affecting the mandibular joint might stop a turtle from eating or cause a severe infection. Tests of hooks with wires attached to the shank ("appendage hooks") suggest that they could in theory be used to reduce rates of turtle capture, but they also resulted in significant declines of target species catch rates, so this method would not be acceptable to fisheries (Swimmer et al. 2010b). One should also consider that, if a sea turtle swallows this type of hook, the related mortality would increase because of the larger size of the complete hook and the transverse wire crossing the esophagus.

The importance of the location of the hook depends very much on the attitude of the fishers when they incidentally catch a turtle, and their level of desire to recover the hook. For example, Mediterranean surface longline fishers do not attempt to recover the hooks from captured turtles. Rather, even when turtles have the hook lodged externally (e.g., jaw), fishermen cut the line and let the animal go with the hook still lodged. In this case, especially if the hooks are small, it might be better for the animal to swallow the hook than to have it hooked in the mouth. However, concerns remain regarding the influence of the hooks lower in the gastrointestinal tract. Mainly for economical reasons, fishers in some regions of the eastern tropical Pacific tend to remove every hook from these animals, so the more external the hook and easier it is to remove, the better for the fishers and the turtles—removal of deeper hooks may result in mortality.

Sixth Line of Defense: Use the Best Dehooking and Resuscitation Techniques

Although the location of hooking may be the most important factor affecting the posthooking mortality of incidentally captured turtles (Ryder et al. 2006), we feel strongly that the actions taken after the hooking event to ensure a safe release to the sea may be the most critical. In most cases, the correct handling of a captured turtle coupled with a good hook-removal technique can greatly increase the probability of that animal surviving the encounter. Here are some recommendations arising from our recent experience on the subject:

1. Reduce to a minimum the pressure caused by pulling on the line with the hook lodged in a turtle; this will reduce the possibility of tears and larger lesions in the mouth or esophagus.

2. If the crew and boat can accommodate bringing the turtle on board, make sure that is done with the use of a dip net. Alternatively, hold the turtle's shell and bring it on board, if possible. Avoid grabbing the animal only by the flippers, because this can damage the tendons and ligaments of the joints, reducing the mobility of that flipper for some weeks.

3. If a turtle is very weak or appears to have drowned, never put it on its back, which restricts the animal's breathing as all the organs in the coelomic cavity push on the lungs, which are located in the dorsal area.

4. If the animal is strong and lively, and you put it on its back for easier handling, be careful when you return it to its original position. The stomach and intestines in these animals are very loose in the coelomic cavity, and it is easy to cause them to twist, which always leads to death.

5. If you cannot remove a hook, always cut the line as short as possible. Long pieces of line can get tangled up in the animal's flippers, causing constriction and eventually necrosis of that limb. If the animal swallows the line, then it will cause severe damage to both the stomach and the intestines, leading in all cases to the turtle's death after a long agony.

6. Hook removal is complicated and in many cases determines the chances of the animal surviving after release. It should only be carried out by well-trained observers, who can then train the fishers. They should have the right equipment for this job and proper protocols describing how to proceed in the different cases. Among other things, they should know the different anatomical parts of the mouth, to assess the severity of the lesions caused by the hook; how much tension to put on a line in order to remove a swallowed hook; how to use the different dehookers; and when it is best to leave a hook inside the turtle.

7. If the boat and crew are not prepared to bring the turtle on board, use a line cutter to cut as much line off the turtle while it is still in

the water. This may require a line cutter with an extension. Once again, this should be done with caution so as not to add further injury to the turtle.

Conclusion

The ideal bycatch reduction strategy is one that uses all opportunities available to reduce capture, and to improve survival if capture happens. It promotes technological or operational changes that could reduce bycatch mortality, while at the same time allowing fishers to continue to practice sustainable fishing by improving the selectivity of the gear and minimizing the impacts on other species, the habitat, and so forth. Such a program would touch on a wide range of issues, from understanding how turtles and target species sense their environment, in order to devise ways to repel or attract them to fishing gear, to behavioral and ecological studies that can help us understand why bycatch happens and how to avoid it. There should also be an exploration of sensible and effective management options that can help in the process. And of course, the practicality of implementation should also be taken into account, to maximize successful adoption and compliance. By ensuring participation of the fishing community in the development of solutions, we ensure that solutions will be effective and practical and increase chances of adoption. Achieving this goal would reduce problems of rejection or noncompliance that could later plague full use of the bycatch reduction strategy.

Perhaps the main lesson that researchers involved in bycatch reduction programs have learned is that successful programs require the full participation and engagement of the fishing communities (Hall et al. 2007; Campbell and Cornwell 2008). Success stories come from cases where scientists and managers learned to communicate and fully collaborate with fishers in the search for solutions. The time is ripe for the development of academic programs that include, in addition to biological and ecological components, exposure to social sciences that could facilitate understanding the structure and function of fishing communities, their decision-making processes, and especially some real-world experience with these communities to prepare new cohorts of scientists to address bycatch issues successfully. In the eastern Pacific Ocean, the fishing sector ranges from highly advanced industrial fleets to pirogues and numbers in the hundreds of thousands of fishers. More resources should be dedicated to understanding how these communities function (Campbell and Cornwell 2008).

Success ultimately hinges on fishers' willingness to adhere to whatever bycatch reduction strategies are promoted by shore-bound fishery managers and sea turtle conservationists. The engagement of fishers in the reduction of mortality will play a crucial role, and interaction with the fishing community is perhaps the most significant component in the process. If fishers are aware of the vulnerable condition of sea turtle populations and are willing to help mitigate fishery impacts, then putting the best technology available in their boats (e.g., hooks, and instruments to handle the turtles and remove the hooks) and teaching them the best mitigation techniques will be the best approach.

REFERENCES

Alava, J. J., P. Jiménez, M. Peñafiel, W. Aguirre, and P. Amador. 2005. Sea turtle strandings and mortality in Ecuador: 1994–1999. Marine Turtle Newsletter 108:4–7.

Alegre, F., M. L. Parga, C. Castillo, and S. Pont. 2006. Study on the long-term effect of hooks lodged in the mid-esophagus of sea turtles. In: Frick, M., L. Panagopoulo, A. F. Rees, and K. Williams (Comps.), Proceedings of the 26th Annual Symposium on Sea Turtle Biology and Conservation. International Sea Turtle Society, Athens, Greece.

Beverly, S., D. Curran, M. Musyl, and B. Molony. 2009. Effects of eliminating shallow hooks from tuna longline sets on target and non-target species in the Hawaii-based pelagic tuna fishery. Fisheries Research 96: 281–288

Bjorndal, K. A., A. B. Bolten, R. A. Bennett, E. R. Jacobson, T. J. Wronski, J. J. Valeski, and P. J. Eliazar. 1998. Age and growth in sea turtles: limitations of skeletochronology for demographic studies. Copeia 1998:23–30.

Campbell, L. M., and M. L. Cornwell. 2008. Human dimensions of bycatch reduction technology: current assumptions and directions for future research. Endangered Species Research 5:325–334.

Canadas, A., R. Sagarminaga, R. De Stephanis, E. Urquiola, and P. S. Hammond. 2005. Habitat preference modeling as a conservation tool: proposals for marine protected areas for cetaceans in southern Spanish waters. Aquatic Conservation: Marine and Freshwater Ecosystems 15:495–521.

Constantino, M. A., and M. Salmon. 2003. Role of chemical and visual cues in food recognition by leatherback posthatchlings (Dermochelys coriacea). Zoology 106:173–181.

Cooke, S. J., and C. D. Suski. 2004. Are circle hooks and effective tool for conserving marine and freshwater recreational catch-and-release fisheries? Aquatic Conservation: Marine Freshwater Ecosystems 14:299–326.

Epperly, S. P., J. Braun, A. J. Chester, F. A. Cross, J. V. Merriner, P. A. Tester, and J.

Churchill. 1996. Beach stranding as an indicator of at-sea mortality of sea turtles. Bulletin of Marine Science 59:289–297.

Gilman, E., P. Dalzell, and S. Martin. 2006. Fleet communication to abate fisheries bycatch. Marine Policy 30(4):360–336.

Gilman, E., D. Kobayashi, T. Swenarton, N. Brothers, P. Salzell, and I. Kinan-Kelly. 2007. Reducing sea turtle interactions in the Hawaii-based longline swordfish fishery. Biological Conservation 139;19–28.

Gilman, E., E. Zollett, S. Beverly, H. Nakano, K. Davis, D. Zooide, P. Dalzell, and I. Kinan. 2006. Reducing sea turtle by-catch in pelagic longline fisheries. Fish and Fisheries 7:2–23.

Hall, M. A. 1996. On bycatches. Review of Fish Biology and Fisheries 6:319–352.

Hall, M. A., M. Campa, and M. Gomez. 2001. Solving the tuna-dolphin problem in the eastern Pacific purse-seine fishery. 2003. In: Mann Borgese, E., A. Chircop, and M. McConnell (Eds.), Ocean Yearbook, Vol. 17. University of Chicago Press, Chicago. Pp. 60–92.

Hall, M. A., H. Nakano, S. Clarke, S. Thomas, J. Molloy, H. Peckham, J. Laudino-Santillán, W. J. Nichols, E. Gilman, J. Cook, S. Martin, J. P. Croxall, K. Rivera, C. A. Moreno, and S. J. Hall. 2007. Working with fishers to reduce by-catches. In: Kennelly, S. J. (Ed.), By-catch Reduction in the World's Fisheries. Methods Technol. Fish Biol. Fisheries 7. Springer, New York. Pp. 235–288.

Howell E. A, D. R. Kobayashi, D. Parker, G. H. Balazs, and J. Polovina. 2008. TurtleWatch: a tool to aid in the bycatch reduction of loggerhead turtles, *Caretta caretta*, in the Hawaii-based pelagic longline fishery. Endanger Species Research 5:267–278

Kobayashi, D. R., J. J. Polovina, D. M. Parker, N. Kamezaki, I. J. Chenge, I. Uchida, P. H. Dutton, and G. H. Balazs. 2008. Pelagic habitat characterization of loggerhead sea turtles, *Caretta caretta*, in the North Pacific Ocean (1997–2006): insights from satellite tag tracking and remotely sensed data. Journal of Experimental Marine Biology and Ecology 356: 96–114.

Lewison, R., S. Freeman, and L. Crowder. 2004. Quantifying the effects of fisheries on threatened species: the impact of pelagic longlines on loggerhead and leatherback sea turtles. Ecology Letters 7:221–231.

Morreale, S. J., E. A. Standora, J. R. Spotila, and F. V. Paladino. 1996. Migration corridor for sea turtles. *Nature* 384: 319–320.

Piovano, S., Y. Swimmer, and C. Giacoma. 2009. Are circle hooks effective in reducing incidental captures of loggerhead sea turtles in a Mediterranean longline fishery? Aquatic Conservation: Marine Freshwater Ecosystems 19: 779–785.

Plotkin, O. 2003. Adult migrations and habitat use. In: Lutz, P., J. A. Musick, and J. Wyneken (Eds.), The Biology of Sea Turtles, Vol. 2. CRC Press, Boca Raton, FL. Pp. 225–242.

Polovina, J. J., G. H. Balazs, E. A. Howell, D. M. Parker, M. P. Seki, and P. H. Dutton. 2004. Forage and migration habitat of loggerhead (*Caretta caretta*) and olive ridley (*Lepidochelys olivacea*) sea turtles in the central North Pacific Ocean. Fisheries Oceanography 13:36–51.

Polovina, J. J., E. Howell, D. M. Parker, and G. H. Balazs. 2002. Dive depth distribution of loggerhead (*Caretta caretta*) and olive ridley (*Lepidochelys olivacea*) sea turtles in the central North Pacific: might deep longline sets catch fewer turtles? Fisheries Bulletin 101(1): 189–193.

Polovina, J. J., D. R. Kobayashi, D. M. Parker, M. P. Seki, and G. H. Balazs. 2000. Turtles on the edge: movement of loggerhead turtles (*Caretta caretta*) along oceanic fronts, spanning longline fishing grounds in the central North Pacific, 1997–1998. Fisheries Oceanography 9: 71–82.

Read, A. J. 2007. Do circle hooks reduce the mortality of sea turtles in pelagic longlines? A review of recent experiments. Biological Conservation 135:155–169.

Rowe, C. R., Kelly, S. M. 2005. Marking hatchling turtles via intraperitoneal placement of PIT tags: implications for long-term studies. Herpetological Review 36(4):408–410.

Ryder, C., T. Conant, and B. Schroeder. 2006. Report of the Workshop on Marine Turtle Longline Post-Interaction Mortality. Bethesda, Maryland, USA, 15–16 January, 2004. NOAA Technical Memorandum NMFS-OPR-29. U.S. National Marine Fisheries Service, Bethesda, MD.

Sasso, C. R., and S. P. Epperly. 2007. Survival of pelagic juvenile loggerhead turtles in the open ocean. Journal of Wildlife Management 71:1830–1835.

Seminoff, J. A., P. Zárate, M. Coyne, D. G. Foley, D. Parker, B. N. Lyon, and P. H. Dutton. 2008. Post-nesting migrations of Galápagos green turtles *Chelonia mydas* in relation to oceanographic conditions: integrating satellite telemetry with remotely sensed ocean data. Endangered Species Research 4:57–72.

Shillinger G. L., D. M. Palacios, H. Bailey, S. J. Bograd, A. M. Swithenbank, et al. 2008. Persistent leatherback turtle migrations present opportunities for conservation. PLoS Biol 6(7):e171.

Southwood, A., K. Fritsches, R. Brill, and Y. Swimmer. 2008. A review of sound, chemical, and light detection in sea turtles and pelagic fishes: exploring the potential for sensory-based bycatch reduction measures in longline fisheries. Endangered Species Research Journal 4:1–14.

Swimmer, Y., R. Arauz, M. McCracken, L. McNaughton, J. Ballestero, M. Musyl, K. Bigelow, and R. Brill. 2006. Diving behavior and delayed mortality of olive ridley sea turtles Lepidochelys olivacea after their release from longline fishing gear. Marine Ecology Progress Series 323:253–261.

Swimmer, Y., and R. Brill (Eds.). 2006. Sea Turtle and Pelagic Fish Sensory Biology: Developing Techniques to Reduce Sea Turtle Bycatch in Longline Fisheries. NOAA Tech. Memo. NOAA-TM-NMFS-PIFSC-7. Pacific Islands Fisheries Science Center, U.S. National Marine Fisheries Service, Honolulu, HI.

Swimmer, Y., L. McNaughton, D. Foley, L. Moxey, and A. Nielsen. 2010a. Movements of olive Ridley sea turtles (*L. olivacea*) and associated oceanographic features as determined by improved light-based geolocation. Endangered Species Research 10:245–254.

Swimmer, Y., J. Suter, R. Arauz, K. Bigelow, A. Lopez, I. Zanela, A. Bolanos, J. Ballestero, R. Suarez, J. Wang, and C. Boggs. 2010b. Sustainable fishing gear: the case of modified circle hooks in a Costa Rican longline fishery. Marine Biology 158:757–767.

Tomás, J., A. Dominici, S. Nannarelli, L. Forni, F. J. Badillo, and J. A. Raga. 2001. From hook to hook: the odyssey of a loggerhead sea turtle in the Mediterranean. Marine Turtle Newsletter 92:13–14.

Valente, A. L. S., M. L. Parga, R. Velarde, I. Marco, S. Lavin, et al. 2007. Fishhook lesions in loggerhead sea turtles. Journal of Wildlife Diseases 43:737–741.

Wang, J. H., S. Fisler, and Y. Swimmer. 2010. Developing visual deterrents to reduce sea turtle bycatch in gill net fisheries. Marine Ecology Progress Series 408:241–250.

Watson, J. W., S. Epperly, A. Shah, and D. G. Foster. 2005. Fishing methods to reduce sea turtle mortality associated with pelagic longlines. Canadian Journal of Fishery and Aquatic Sciences 62:965–981.

7

International Instruments

Critical Tools for Conserving Marine Turtles
in the Eastern Pacific Ocean

JACK FRAZIER

Summary

Marine turtles of the eastern Pacific are "shared resources"—not the exclusive rights of any single person, community, society, or State. With worldwide concern about decimated turtle populations that are facing increasing and evermore complex threats, a variety of initiatives have been developed to promote greater cooperation among countries for the protection, conservation, management, and recovery of marine turtles and also for the protection of diverse habitats on which they depend. These accords vary from formal treaties to a wide range of "soft law" tools. Hence, a large number and variety of instruments are either directly or indirectly relevant to marine turtles and their habitats. Yet, there is a lack of basic understanding about even the fundamentals and relevance of particular provisions of even the more influential instruments. Likewise, there is a lack of understanding about how the various instruments that have been ratified, adopted, or signed by a particular country integrate with one another, at national and international levels. Many field researchers discount international instruments, organizations, and initiatives as unimportant, irrelevant, or of low priority; and they pay little attention to them. There is no question that international instruments, organizations, and initiatives are conceived, structured, and implemented in arenas foreign to many researchers and conservationists. However, in a complex world where human relations are always difficult to comprehend, much less regulate and influence, international instruments are one of the few viable options for promoting long-term survival of highly migratory species that spend significant portions of their life histories within the jurisdictions of numerous States, as well as on the high seas outside the jurisdiction of any single State. Hence, it behooves turtle biologists, conservationists, and specialists in community affairs to have at least a basic understanding of international instruments. This chapter summarizes the situation regarding international agreements as well as the associated organizations and initiatives, particularly those most directly related to the conservation of marine turtles and their habitats in the eastern Pacific Ocean. The chapter also provides basic sources for more detailed information.

Introduction

The research, conservation, and management of marine turtles and their habitats depend on far more than carrying out active field projects. As explained below, accords to promote cooperation and collaboration are absolutely essential; and there is an enormous amount of information on international accords relevant to marine turtle conservation in just the eastern Pacific, but unfortunately there are relatively few recent compilations and even fewer syntheses of these instruments. For these reasons, this chapter provides a summary of the fundamental relevance of these instruments, organizations, and initiatives, as well as some of the major points related to international accords in the eastern Pacific. Basic sources for more detailed information is also provided; in the limited space available it is not possible to give an in-depth analysis, but some general observations and recommendations are presented.

Of the various life history parameters that characterize marine turtles, their capacity to disperse and migrate over vast distances is most relevant to the central theme of this chapter. These reptiles disperse and migrate widely, with individuals of various life stages routinely moving into and out of the areas under the respective jurisdictions of numerous sovereign states, as well as the high seas, where no State has exclusive rights. They can transit and traverse ocean basins on a regular basis, and turtles from the eastern Pacific typify this characteristic as well as those from anywhere else in the world. Individuals of at least two species regularly cross from one side of the Pacific to the other: loggerhead turtles (*Caretta caretta*) make circum-Pacific movements in both the northern and southern Pacific, and leatherbacks (*Dermochelys coriacea*), depending on the population, cross the Pacific in east–west or north–south trajectories. In addition, green turtles (*Chelonia mydas*) make large-scale movements along the western coasts of the Americas, with displacements as distant as México to Chile; olive ridleys (*Lepidochelys olivacea*) make vast oceanic dispersions, also with displacements from as far as México to Chile (Frazier and Salas 1983, 1986, 1987; see chapters 1, 4, and 13). Of the five species found in the eastern Pacific Ocean, only the hawksbill turtle (*Eretmochelys imbricata*) is not known to make regular large-scale migrations, but this may be an artifact of the lack of basic information on this species (see chapter 10).

This highly developed migratory habit has enormous implications on conservation and management activities (Giordano 2002); it means that during their life cycles marine turtles are resources that are shared by different interested parties, and not the exclusive rights of any one person, community, society, or country. People of many communities and nations have traditions, rights, and responsibilities regarding activities that affect marine turtles, both negatively and positively. These practices may be contemplated both in traditional (unwritten) laws and in a variety of formal (written) legal regimes. To academic specialists in institutional relations and property rights, it means that these reptiles are "shared resources"; indeed, marine turtles are classic examples of this concept (Frazier 2004). While this is intellectually fascinating and challenging, it means that conservation activities must be designed to deal with this complication—the best conservation program in the world can be rendered useless if the animals disperse and migrate out of an area where protection is provided and move into other areas where they suffer high mortality and/or low reproductive success. The classic example of this dilemma is popularly known as the "Tragedy of the Commons" (Harding 1968; Frazier 2004).

In this light, international laws *should* promote sovereign states to collectively recognize these reptiles as "common property" of all countries, but in the world of law it is not correct to call marine turtles "common property," for this legal term refers to joint ownership at all times. Marine turtles, however, move from jurisdiction to jurisdiction, and a particular State actually has sole ownership during the time that the turtle is in its jurisdiction. This is a function of sovereignty and a related concept known as "permanent sovereignty" over natural resources. Except when on the high seas (when no Party has sole ownership), marine turtles are essentially subject to sequential sole ownership. Even the term "shared resource" has no legal status. The term "commons resource" would relate to species on the high seas; that is, something not subject to ownership until reduced to possession (Wold 2002). As a result, one of the great challenges of international law is to create an accepted basis for clearly regulating marine turtles as common property, and controlling the rights of any one person, community, society or country to exercise private control over a resource that is sequentially shared by many interested parties.

In addition to the migratory habit, which results in these legal and management complications, other characteristics of marine turtle life histories further complicate conservation and management initiatives. During different stages of its life cycle, a turtle lives in, and depends on, a great diversity of environments, ranging from the terrestrial high beach, where nests are dug and eggs deposited, to the high seas, where many years can be spent growing from a hatchling to a subadult. The open oceans are also transited during migrations between reproductive and feeding areas, which are commonly in coastal and inshore waters, although leatherbacks and olive ridleys may spend most of their immature and adult lives on the high seas of the East Pacific (see chapters 4 and 13). Hence, the protection of just a single turtle requires adequate management of beaches, inshore waters, coastal waters, and the high seas: terrestrial, estuarine, benthic, neritic, and pelagic habitats. Combined with these complex spatial-environmental considerations are challenging temporal issues. Marine turtles are slow maturing and long-lived (Musick and Limpus 1997; Crouse 2000), so protection of just a single turtle requires that adequate policies be implemented not for a few years, which is the normal horizon for politicians—as well as for countless research projects—but for many decades, up to a century. But wildlife conservation requires the protection and management of populations, not just individuals. In other words, the time frame required for effective marine turtle conservation programs—decades or centuries—conflicts with the time frame for political systems, which are rarely more than a few years; this is not to mention the time frame in which contemporary economic systems operate, sometimes only a few weeks or days, or less. Because economic systems normally drive political systems, which in turn drive management systems, tremendous inherent conflicts between different systems must be overcome: effective marine turtle conservation depends on adequate protection and management of a wide variety of human activities in a diversity of environments over extended periods of time, in contrast to many other—dominating—human activities. To deal adequately with these many special and temporal considerations is a daunting challenge.

These highly complex, interrelated, and interacting features emphasize even further that marine turtles are shared resources or "common property," although in the nonlegal sense, and that they should logically be recognized as common property at an international level. Hence, diverse forms of cooperation and collaboration, at multiple geographic, temporal, disciplinary, social, and political levels, are fundamental for understanding and conserving these animals. As a result, there have been various calls for developing regional and international cooperation through diverse mechanisms at different levels (Frazier and Salas 1983, 1986, 1987; Frazier 1984a, 2002; Hykle 1999, 2002; Trono and Salm 1999; Wold 2002). Indeed, marine turtles could be used as index species for assessing the success of international cooperation (Frazier 1981).

Initiatives to meet these challenges vary from collaborative research efforts in gathering information of diverse kinds—biological, cultural, ecological, economic, historic, and so on—to politically complex international meetings to develop legally binding agreements among sovereign states. In both cases, it is essential to bear in mind that these are tools for managing the resources in question—not end goals. Without a doubt, developing and implementing cooperative agreements are far more complex and exigent than information gathering, a lesson that this author has learned, again and again, from personal experience. Decades ago my focus was on gathering and exchanging information on natural history and conservation in the eastern Pacific region, including surveys of geographic distribution and syntheses of biological, historical, and anthropological information, with determined attempts to promote greater in-country interest and commitment, particularly through publications in Latin American journals and conferences (e.g., Frazier 1979, 1982, 1983, 1984b, 1985a, 1985b, 1990, 1992a, 1992b, 1993, 1994, 1995; Frazier and Salas 1982, 1983, 1984, 1986, 1987; Frazier and Brito Montero 1990; Frazier and Hártasanchez 1992; Rosano Hernández et al. 1996, 1998; Frazier and Bonavia 2000). However, in recent years, with the realization that no amount of "scientific information" necessarily results in effective conservation activities without effective policies

(Meffe and Viederman 1995; Frazier 2005), I have made a concerted effort to promote cooperative initiatives under the aegis of different forms of international accords (e.g., Frazier 1984a, 1997, 1999, 2000a, 2000b, 2000c, 2000d, 2002a, 2002b, 2004, 2006; Frazier and Bache 2002; Bache and Frazier 2006). For anyone trained in, and dedicated to, field biology and conservation—as will be the case for most readers of this volume—entry into the worlds of law and policy means learning and managing very different sets of rules and procedures, often with perfectly clear words having very different meanings in the different arenas; and it is with this "crossover" perspective in mind that this chapter is organized.

A Diversity of Instruments and Organizations Relevant to Marine Turtle Conservation

Treaties to promote international collaborative efforts for marine turtles date back at least to 1916 (Giordano 2002); and unwritten agreements between different members of "traditional" or "indigenous" societies are likely to date back millennia (e.g., societies with strong fishing traditions [Dyer and McGoodwin 1994], such as the Comcáac in the Gulf of California [Nabhan 2003], the Vezo of southwestern Madagascar [Lilette 2006], and the Meriam of Torres Strait, Australia [Smith and Bliege Bird 2000; Bliege Bird et al. 2001]). Given the diversity of environments on which these reptiles depend, and the multiplicity of human actions that result in direct and indirect impacts, an enormous variety and number of written instruments, intergovernmental organizations, and initiatives are germane to the conservation and management of marine turtles and their habitats (see plates 13 and 14). This conclusion is underscored by initiatives that at first seem to have no direct relevance to marine turtles. For example, the Global International Waters Assessments for the east Pacific region, coordinated with the United Nations Environment Programme (UNEP 2006a, 2006b), occasionally mention marine turtles in a superficial way, but the various themes discussed in the assessments, as well as the regional mechanisms for coordinating management efforts, are highly relevant to root issues in the protection, management, rehabilitation, and conservation of diverse habitats critical to the survival and status of marine turtles.

On the geopolitical scale, instruments and organizations range from local to national, binational, regional, and international. Instruments also vary according to the principal topic that they address, varying from a direct focus on species and habitat protection, to industrial/fisheries/shipping regulatory systems, to more general ocean governance and even international trade. Hence, there are a great number and an enormous diversity of instruments and organizations, and accompanying information and analyses, relevant to marine turtle conservation.

Ideally, given this wide range of instruments that States must employ effectively, each country should produce detailed analyses of its domestic laws and policies as well as its political acceptance (e.g., ratification) of the various international legal instruments, organizations, and initiatives relevant to the conservation of marine turtles and their habitats. Such an analysis of policies and legal regimes provides a framework and synthesis for understanding what regulations and norms exist, how they interrelate with each other and with environmental issues, which administrative jurisdictions apply, and what major gaps need attention. Regrettably, although marine turtles occur in nearly every coastal state on the planet, very few countries have national studies of the legal regimes available to protect these marine resources. A notable exception are Central American countries, where thanks to the vision and support of the Caribbean Conservation Corporation (now called the Sea Turtle Conservancy), detailed compilations and syntheses of the legal regimes for each of the seven States were commissioned at end of the 1990s: Belize (Ellis 1997), Costa Rica (Espinoza and Piskulich 1997), El Salvador (Álvarez 1997), Guatemala (de Noack 1997), Honduras (Galindo 1997), Nicaragua (Ruiz and Jarquín 1997), and Panamá (Tulio 1997); and a regional synthesis was also produced (Espinoza 1997). Other countries that have studies of the legal regime relevant to marine turtles include Argentina (Tonelli 2006), Colombia (Oeding 2007), Venezuela (Omaña Parés 2005), and India (Upadhyay and Upadhyay 2002). Also important, but more limited in reach, are detailed analyses of primary instruments. For example, Namnum (2002) evaluated how effectively the Inter-American Convention for the Protection and Conservation of Sea Turtles is integrated within the Mexican legal regime. However, in all cases these fundamental studies have been conceived, designed, and executed by students, academics, and nongovernmental organizations (NGOs); government investment and support—and sometimes even approval—have been conspicuously absent. There are also several evaluations of how marine turtle conservation is supported by major treaties such as the United Nations Convention on the Law of the Sea (UNCLOS), the Convention on International Trade in Endangered Species of Wild Fauna and Flora (CITES), and the Convention on the Conservation of Migratory Species of Wild Animals (CMS), as well as other international agreements and certain fisheries accords (Bache 2002; Giordano 2002; Hykle 2002; Wold 2002; Bache and Frazier 2006; Cajiao Jiménez et al. 2006; Murphy 2006).

Although not specific to marine turtles, national compilations of marine or aquatic legislation provide valuable information on legal foundations. The Permanent Commission for the South Pacific (Comisión Permanente del Pacífico Sur, CPPS) promulgated national compendia of marine legislation beginning more than three decades ago, including the four countries of the commission: Chile (CPPS 1972), Colombia (CPPS 1997a, 1997b), Ecuador (CPPS 1974, 1986), and Perú (CPPS 1973). Although somewhat out of date, these compendia are still valuable information sources. Recently the Parties to the Lima Convention (including the four countries of CPPS as well as Panamá; see below) commissioned compilations and syntheses of national and international legislation to evaluate compliance and concordance of legislation of each country regarding their obligations to the Action Plan for the Protection of the Marine Environment and Coastal Areas of the Southeast Pacific, the backbone of the Lima Convention: Chile (Bermúdez Soto 2008), Colombia (Ramos 2008), Ecuador (Hernández et al. 2008), Perú (Iturregui 2008), and Panamá (Díaz 2008). Because the status of marine turtles depends intimately on the condition of various marine and coastal environments, extending from beaches to the open sea, these national compilations provide invaluable sources of information relevant to the nuts and bolts of identifying, coordinating, constructing, and improving instruments and organizations for national, regional, and international cooperation.

Other countries in the Americas that have recent syntheses of marine legislation and policies include Costa Rica (Cajiao Jiménez 2003), Cuba (Speir 1999), Guatemala (de Noack et al. 2005), and Venezuela (Omaña Parés 2007); and Cajiao Jiménez et al. (2006) provide summaries and syntheses for the four countries of the Eastern Tropical Pacific Marine Corridor (see below): Colombia, Costa Rica, Ecuador, and Panamá. One of the most exhaustive compilations was done by the Marine Mammal Commission of the United States (Wallace 1994, 1997; Weiskel et al. 2000); this five-volume compendium, of almost 6,000 pages, contains information on more than 650 multi- and bilateral instruments that are relevant to marine mammals, animals that live in many of the same environments as marine turtles, facing many of the same risks. Only rarely are these sorts of publications available in Native American languages; a notable exception is a Spanish-*Q'eqchi'* presentation of the general fisheries and aquaculture law of Guatemala (de Noack et al. 2005).

Categorization of Instruments, Organizations, and Initiatives

A primary categorization is between legally binding agreements, such as treaties (hard law), on the one hand, and nonbinding accords, such as memoranda of understanding (soft law), on the other. As mentioned above, instruments can be grouped according to geographic coverage as well as geopolitical hierarchy, from local to global. Although an individual accord may often cover diverse themes, all instruments can be grouped according to their principal focus, for example, species protection, habitat protection, fisheries, ocean governance, pollution, and trade. The instruments that routinely get the most attention, and command the most political consideration, are legally binding treaties; these are also the most complex, for they regularly involve detailed, tedious, and often contentious multinational negotiations, which must then be followed by a series of bureaucratic and political steps to pass through the internal political review and approval processes within each respective State en route to ratification (or accession), a lengthy domestic process that might (or might not) result in the respective country becoming a Contracting Party. On the other hand, nonbinding instruments, such as memoranda of understanding, international principles and guidelines, technical action plans, or regional initiatives promoted by NGOs, are usually simpler for countries to negotiate and sign (Hykle 2002). In these cases, the intention is that Signatory States, while not legally bound, will be "morally compelled" by the soft law to cooperate with other Signatory States and implement the measures, resolutions, and recommendations of the agreement. Action plans and technical agreements, often promoted by technical agencies of governments (and not involving ministries of foreign affairs), or even developed by NGOs, are far less onerous to negotiate and sign than are the more formal accords, and while these sorts of instruments may have less political clout, they can still be highly effective in promoting certain levels of international cooperation.

Among the examples of soft law instruments are several international principles, such as the Precautionary Principle and the Principle of Due Diligence. While these do not provide specific measures for marine turtles, they help to establish a common legal foundation that can be used in future negotiations and proceedings (Hunter 1998); these in turn can have direct bearing on measures highly relevant to the conservation of marine turtles and their habitats. An especially important example of a soft law tool for marine turtles is the "Guidelines to Reduce Sea Turtle Mortality in Fishing Operations." These guidelines were developed by the Food and Agriculture Organization of the United Nations (FAO) through an expert consultation followed by a technical consultation (FAO 2004a, 2004b). They were then discussed and approved at the twenty-sixth meeting of the FAO Committee on Fisheries (FAO 2005), and while these guidelines are "voluntary and not intended to affect trade" (FAO 2005), they come with the unanimous approval of an extremely powerful global body. Hence, soft law has enjoyed considerable support from numerous organizations and legal specialists (Lugten 2006), often with the hope that the instruments will eventually evolve to include obligatory measures.

On the other hand, simplifying the developmental process of an agreement by using soft law can have its drawbacks. Typically, governments evaluate the importance of complying with agreements in direct relationship to the degree of official governmental involvement in the negotiation and development of the instrument. Hence, heavy dependence on NGOs and their actors can have dire costs if not handled with sensitivity and imagination. Despite the fact that most governments, including those of the industrialized North, do not have the human and financial resources to adequately implement environmental agreements—especially for complex issues like marine turtles—nongovernmental actors (including academics) are frequently viewed with suspicion by official delegations, even if these people are recognized technical experts.

The political importance of instruments and organizations also tends to increase with geographic scope: the more countries and territory involved, the greater attention paid to the accord, with the belief that it has greater importance and impact. However, some veteran specialists have observed numerous cases where regional instruments have been more effective than global ones, especially in the conservation arena (T. Young, written communication, 14 January 2009). Likewise, the complexity in development, administration, and implementation of an international accord increases with the number of participating States and expanse of territory. At the same time, instruments that deal with higher-priority issues in the political arena, such as security, sovereignty (or States' rights), territory, trade, and the flow of fiscal resources, are taken much more seriously by governments than are issues like species conservation. Hence, in recent years there has been a growing interest in linking environmental and conservation concerns to trade issues and in that way attaching considerable political importance to what may otherwise be an issue of ephemeral or limited significance in the political arena. Thus, although the polemics of trade may seem distant and unrelated to marine turtle conservation, in fact they have been highly relevant, as shown by the much cited case of the threat of a trade embargo linked to marine turtle conservation (e.g., Shaffer 1999; Frazier and Bache 2002; Howse 2002; Bache and Frazier 2006). Indeed, there are regional environmental accords that stem from trade agreements (Giordano 2002) and that have injected considerable importance into conservation programs for marine turtles, for example, leatherback turtles in the eastern Pacific (Commission for Environmental Cooperation 2005). Trade–environment relations are a major topic in their own right, beyond the scope of this chapter, but it is essential to understand that international accords relevant to marine turtle conservation in the eastern Pacific are not limited to just those that focus specifically on turtle conservation.

In this light, it is important to appreciate that there are different reasons and motivations for States to become signatories or Parties to international accords: access to environmental resources, access to fiscal resources, access to trade markets, access to other forms of help and cooperation, demands from powerful citizens and lobby groups, and pressure from more powerful countries or neighboring States (Danaher 1999). These different underlying motivations bear directly on the level of commitment that a State will have in implementing the measures of the accord, especially if it is a soft law agreement.

Instruments of Particular Importance to Marine Turtles in the Eastern Pacific

For this volume, the eastern Pacific is defined as the area from the western shores of North, Central, and South America, west to 100° W longitude (fig. 7.1); hence, in addition to a considerable area of high seas, it comprises fifteen States that have terrestrial and maritime territories in the eastern Pacific, including three from North America (Canada, the United States, and México), six from Central America (Guatemala, El Salvador, Honduras, Nicaragua, Costa Rica, and Panamá), four from South America (Colombia, Ecuador, Perú, and Chile), and one European State (France [Clipperton Island]); this excludes British Overseas Territories in the Pacific, as well as independent island States of the Pacific. Each of the countries named above is Party or Signatory to a variety of international instruments that relate, in different ways, to the conservation of marine turtles and the habitats on which they depend (fig 7.1). Discussions on how some of these instruments are relevant to marine turtles are available in publications cited previously (e.g., Bache 2002; Frazier 2002a; Hykle 2002; Wold 2002; Bache and Frazier 2006; Cajiao Jiménez et al. 2006) and are not treated in detail herein. However, four instruments especially important to the conservation of marine turtles and their habitats in the eastern Pacific are summarized to illustrate the diverse types of accords that can be directly relevant to marine turtle conservation: a stand-alone, hemispheric treaty; a regional convention that is part of the United Nations Environment Programme (UNEP), with its associated protocols as well as a regional program for marine turtles; a regional fisheries management organization (RFMO); and a nonbinding regional accord promoted by an international NGO—all but the last are legally binding.

Although the Convention on International Trade of Endangered Species of Wild Fauna and Flora (CITES) is perhaps the best-known international instrument relevant to wildlife conservation, this treaty—which focuses on international trade—is not discussed herein. There are, however, extensive discourses on CITES (e.g., Giordano 2002; Wold 2002;

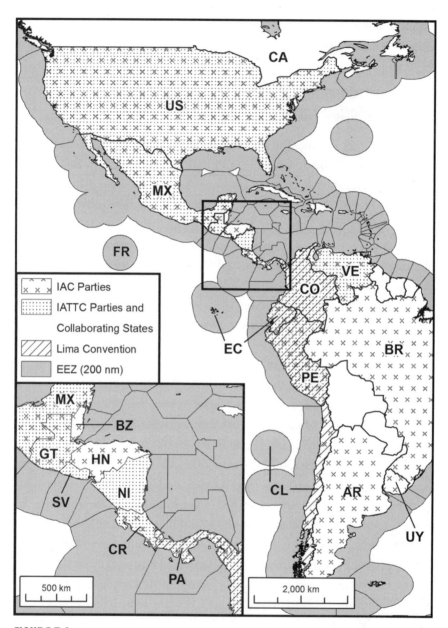

FIGURE 7.1. Schematic of the Americas, showing countries that border on the eastern Pacific and their respective relationships with three important international instruments: (1) Parties to the Inter-American Convention for the Protection and Conservation of Sea Turtles (IAC), (2) Members of the Inter-American Tropical Tuna Commission (IATTC), and (3) Parties to the Lima Convention. The Exclusive Economic Zones (EEZs), which extend out 200

Bache and Frazier 2006; Murphy 2006), as well as discussions and critiques of its limitations, particularly as related to marine turtle conservation (e.g., Danaher 1999; Carpenter 2006).

The Inter-American Convention for the Protection and Conservation of Sea Turtles (IAC)

The text of this treaty was adopted on 5 September 1996, and it entered into force on 2 May 2001. At the date of writing there are fifteen Parties. Nine of the IAC Parties border on the eastern Pacific. The objective of the IAC is "to promote the protection, conservation and recovery of sea turtle populations and of the habitats on which they depend, based on the best available scientific evidence, taking into account the environmental, socioeconomic and cultural characteristics of the Parties" (Article 2). This is the only *legally binding* international instrument focused specifically in marine turtle conservation. The language and content of the IAC have attracted the attention and approval of many specialists in turtle conservation as well as experts in international instruments, among other reasons because research, education, community participation, bycatch mitigation, habitat protection, and other measures are specifically mandated. Both a consultative committee and a scientific committee are established to support the IAC in compiling and evaluating the best scientific information available for decision making by the Parties.

During the decade after the IAC entered into force, five Conferences of the Parties have been convened, as well as one extraordinary meeting of the Parties, four meetings of the Consultative Committee, and eight meetings of the Scientific Committee. These meetings have resulted in nearly two dozen resolutions, four of which are on conservation issues: promotion of the conservation of leatherback turtles (Resolution COP2/2004/

---

nautical miles, are also shown for each country. Clipperton Island, an overseas territory of France that lies in the eastern Pacific, is also shown in this schematic. Country codes are as follows: AR, Argentina; BR, Brazil; BZ, Belize; CA, Canada; CL, Chile; CO, Colombia; CR, Costa Rica; EC, Ecuador; FR, France; GT, Guatemala; HN, Honduras; MX, México; NI, Nicaragua; PA, Panamá; PE, Perú; SV, El Salvador; US, United States; UY, Uruguay.

R-1), promotion of the conservation of hawksbill turtles (Resolution COP3/2006/R-1), mitigation of negative impacts of fisheries interactions (Resolution COP3/2006/R-2), and adaptation of marine turtle habitats to climate change (Resolution CIT-COP4-2009-R5). An annual report is mandated by the treaty, but it has suffered from an overly ambitious format that results in dozens, or scores, of pages for each Party—when countries comply (which has happened less than half the time, and even so, some information provided has been contradictory and inaccurate). Nevertheless, in several States that are Parties to this treaty, reference to the IAC has been useful in national legal and policy documents for advancing marine turtle conservation at higher domestic policy levels. The text of this treaty has served as a model for other instruments that were developed afterward to promote marine turtle conservation, such as the Memorandum of Understanding concerning Conservation Measures for Marine Turtles of the Atlantic Coast of Africa and the Memorandum of Understanding on the Conservation and Management of Marine Turtles and Their Habitats of the Indian Ocean and South-East Asia (see Frazier 2002a), and some of the turtle-specific materials of the Lima Convention (see below).

Initially, conservationists were enthusiastic that this unique instrument would become a major force for promoting international cooperation for marine turtle conservation in the Western Hemisphere (e.g., Frazier 1997, 1999, 2000a), but there is now widespread disenchantment; NGOs and observers that initially participated actively in meetings of the Conferences of the Parties and committees have all but abandoned this treaty in recent years. Despite the expenditure of more than a million dollars and hundreds of thousands of person-hours in meetings and other related activities, at the time of this writing the Parties to the IAC have still not resolved several fundamental issues. These include the location and staffing of a permanent secretariat, financial and logistic support to cover basic operational needs, and the implementation of rudimentary measures that are stipulated by the convention, for example, the regular submission of an annual report by each Party, not to mention more complex and contentious measures such as the effective mandatory use of turtle excluder devices (TEDs) in shrimp trawls and the development of management programs through regional consultation when turtles and/or their eggs are subject to legal, directed exploitation.

Hence, despite the laudable objective and language of the treaty and the high marks it has attracted from specialists in multilateral environmental agreements, in practice the IAC has languished. Rather than advancing turtle conservation, it has become known for great confusion and polemic. The excessively politicized atmosphere in part is a vestige from the genesis of the treaty; it stems from the threat of an embargo mandated by a U.S. law passed in 1989 that essentially calls for the compulsory use of TEDs by shrimp exporting nations (Frazier 1997; Frazier and Bache 2002; Bache and Frazier 2006). The treaty has not been able to shake this stigma of a trade restriction unilaterally imposed by a superpower whose foreign policies in Latin America have created a long history of grief and distrust (e.g., Galeano 2004). It is worth noting, however, that international treaties are notorious for taking decades to develop and mature to the point where they begin to be effective.

The Convention for the Protection of the Marine Environment and Coastal Area of the South-East Pacific (Lima Convention)

The Lima Convention, together with its associated Action Plan for the Protection of the Marine Environment and Coastal Areas of the South-East Pacific, was developed as one of the Regional Seas Programme of the UNEP; the region includes the waters under the jurisdictions of Chile, Colombia, Ecuador, Panamá, and Perú, extending from the northern border of Panamá to the southern border of Chile, and including oceanic islands such as Galápagos and Easter Island. The convention text and action plan were adopted in Lima on 12 November 1981 and entered into force on 19 May 1986. During the first decade after this convention entered into force, the Parties adopted an accord and three protocols focused on contamination issues, as well as one protocol on protected areas (CPPS 2007).[1] Numerous other initiatives of this convention are relevant to marine conservation, addressing marine mammals and the regional network of marine and coastal areas. For example, the protection of nesting beaches of marine turtles was one of the three most conspicuous priorities described for the creation of marine protected areas in continental Ecuador (Terán et al. 2006).

In 2001, national workshops on marine turtle biology and conservation were carried out in all five countries of the convention, and between 2006 and 2007 several regional workshops prepared the way for the Regional Program for the Conservation of Marine Turtles in the South-East Pacific to finally be adopted at the fourteenth meeting of the Contracting Parties (30 November 2007). The regional program has five main components: (1) research and monitoring, (2) sustainable management, (3) environment education and community participation, (4) information and dissemination, and (5) institutional strengthening and international cooperation. A regional scientific committee for the turtle program was created on 30 April 2008, composed of two people from each of the five countries. Three priorities were established: (1) development and strengthening of national committees/networks, (2) regional workshop on standardization of methodologies, and (3) national diagnostics.

A major goal of the committee's work plan is to coordinate closely with the IAC and also the Protocol Concerning Specially Protected Areas and Wildlife in the Wider Caribbean Region (SPAW Protocol) of the Cartagena Convention for the Protection of the Marine Environment in the Greater Caribbean. Harmonization between the IAC and the Regional Program for the southeast Pacific is an obvious need, especially because at the time of writing four of the five Parties to the Lima Convention are also Parties to the IAC (Chile, Ecuador, Panamá, and Perú). Unfortunately, nearly four years after the creation of the Regional Scientific Committee, there has been little advance. This is a common symptom of committees that are all-volunteer, meet only once a year—if not less frequently—are tasked with communicating and coordinating through correspondence, and have no regular budget.

On the other hand, the Lima Convention, together with its Action Plan, as part of the Regional Seas Programme of the UNEP, has international credibility and linkages. Moreover, the Parties to the Lima Convention enlisted the Secretariat of the Permanent Commission of the South Pacific (CPPS)[2] to administer the convention's Action Plan on the Protection of the Marine Environment and Coastal Areas. After more than half a century of involvement in marine sovereignty, resources, fisheries, and research issues, with at least fifteen different internal accords, CPPS offers unique administrative, procedural, research, and political experience to the Lima Convention and includes nearly the same countries. In this capacity, the General Secretary of CPPS is also the Executive Secretary of the Lima Convention, and the two secretariats share the same offices, which provide the Lima Convention with valuable administrative and political support. However, as a result of this arrangement, there is frequently confusion between the two instruments, although in fact they have separate legal status, budgets, and objectives, and the contracting Parties are not identical.

The Inter-American Tropical Tuna Commission (IATTC)

The IATTC was created pursuant to the Convention for the Establishment of an Inter-American Tropical Tuna Commission (adopted in Washington, DC, 31 May 1949, in force 3 March 1950).[3] Its primary objective is "conservation and management of tuna and other marine resources in the eastern Pacific Ocean," particularly to develop collaborative mechanisms for managing tuna fisheries in this region. As of the date of writing, the IATTC comprised twenty-one Members and one Cooperating non Party; it includes twelve States with jurisdictions in the East Pacific (Canada, United States, México, Guatemala, El Salvador, Nicaragua, Costa Rica, Panamá, Colombia, Ecuador, Perú, and France). For nearly half a century the IATTC was focused on the purse seine tuna fishery, and apart from managing the exploitation of tuna stocks, the primary conservation issue was the bycatch and mortality of dolphins in purse seines. By the twenty-first century, however, all Parties finally acknowledged that there are other serious management issues to be addressed: some fisheries in the eastern tropical Pacific other than purse seining cause significant bycatch—particularly longlining—and various marine wildlife species besides dolphins are significantly affected by these activities, including especially marine turtles. As a regional fisheries management organization (RFMO), the IATTC exists primarily to keep the fishing industry in business. Thus, in order to retain discussions on the mitigation of bycatch of endangered marine turtles within a fisheries forum, and not have this contentious issue escape into a conservation forum that would be more difficult to control with the interests of the fisheries industry, IATTC Parties became actively involved with marine turtle bycatch issues.

As a result, between 2003 and 2007, meetings of different working groups developed proposals for mitigating bycatch problems; and meetings of the Parties to the IATTC adopted a series of resolutions, or modifications to resolutions, critical to mitigating marine turtle bycatch.[4] Noteworthy are resolutions adopted at the seventy-second and seventy-fifth IATTC meetings. Two resolutions passed at the seventy-second meeting of the IATTC (14–18 June 2004, Lima, Perú) relevant to turtles include (1) Consolidated Resolution on Bycatch C-04-05, which includes a dozen specific measures directed at governments, fishermen, IATTC staff, and the FAO to strengthen mitigation measures for turtles caught in purse-seine and longline fisheries, and (2) Resolution C-04-07, establishing a three-year program to mitigate the impact of tuna fishing on marine turtles. The program consisted of five main components: collection and analysis of information, mitigation measures, industry education, capacity building in coastal developing countries, and reporting. The seventy-fifth meeting (25–29 June 2007, Cancun, México) adopted Resolution C-07-03 to mitigate the impact of tuna fishing vessels expressly on marine turtles, which consists of more than a dozen specific measures, including implementing the FAO guidelines to reduce the bycatch, injury, and mortality of marine turtles in fishing operations (see above); reporting on progress in implementation of the FAO guidelines; collaboration with other actors; implementation of observer programs; requiring fishermen to avoid catching, handle properly if caught, release, and resuscitate turtles; conduct research on gear modification for reducing capture; and review progress. Based on the information generated under IATTC-required reporting, the eighth and ninth meetings of the commission's Permanent Working Group on Compliance (21 June 2007, Cancun, México and 19 June 2008, Panamá, respectively) presented details on marine turtle captures reported by the onboard observer programs of the purse-seine fishery. It is important to appreciate that, in theory, at least, these compliance reports can result in sanctions on the implicated IATTC Party; and it is noteworthy that two key countries in the eastern Pacific and Member States of IATTC (as well as Parties to the IAC and Lima Convention), Ecuador and Panamá, had by far the highest levels of "violations of sea turtle release requirements recorded by IATTC observers" in 2006 and 2007. However, as yet there seem to have been no sanctions, other than possible embarrassment that the two respective delegations might have felt. As is clear, the IATTC Parties continue to refine and modify regulations related to reducing the bycatch, injury, and mortality of marine turtles in fishing operations.

In addition, IATTC staff have participated actively, since its inception in 2003, in a regional program that extends from México to Perú, to evaluate the effects of circle hooks in mitigating marine turtle bycatch. Hence, between 2003 and 2007, this fisheries management organization became ever more involved in marine turtle bycatch issues, and the IATTC has had the strongest marine turtle conservation measures of any regional fisheries management organization on the planet. It would be advisable and more cost-effective if other instruments such as the IAC and Lima Convention were to base their respective recommendations, resolutions, and actions related to turtle bycatch on IATTC advances and develop functional coordination between the instruments and their implementation. In conclusion, it is remarkable that this regional fisheries management organization has taken the lead over other multilateral environmental agreements regarding turtle bycatch mitigation, and it has been far more active and effective than the two specific turtle agreements in the eastern Pacific—the IAC and the Regional Program for the Conservation of Marine Turtles in the South-East Pacific of the Lima Convention.

The Eastern Tropical Pacific Marine Corridor

In 1997 Costa Rica's Ministry of Environment and Energy (MINAE) (now the Ministry of Environment, Energy, and Telecommunications [MINAET]) and the Ecuadorian Institute of Forestry and Natural Areas (INEFAN) (which was absorbed two years later into the Ministry of Environment, which itself had been created in 1996) signed the San José Declaration. This declaration was to promote cooperation and coordination in conservation initiatives focused on Costa Rica's Cocos Island and the Galápagos Islands of Ecuador: the Exclusive Economic Zones around these islands make Costa Rica and Ecuador neighboring states. In 2002 the accord was expanded to include the island of Coiba, Panamá, as well as Malpelo and Gorgona islands, Colombia, forming the Eastern Tropical Pacific Marine Corridor (Cajiao Jiménez et al. 2006). This is a voluntary, nonbinding agreement among the governments of Colombia, Costa Rica, Ecuador, and Panamá to coordinate activities in their respective jurisdictional seas, particularly surrounding the respective islands, all of which have special conservation significance. This agreement involves a very different approach from the last three instruments described. It was not a plenipotentiary process, involving ministries of foreign affairs and/or other high-level governmental officers, but rather stems from agreements between more technical actors in the respective governments. As a nonbinding agreement, it is more flexible than a formal treaty or convention, hence the relative ease and speed with which the number of countries doubled and the area greatly increased.

The total area of the marine corridor covers more than 10,000 square kilometers. There have been various specialist workshops to coordinate research and conservation of flagship species, primarily sharks, marine mammals, and marine turtles. Because a significant part of the original program was to restrict fishing activities in an area that is intensively exploited by several industrial fisheries, there has been considerable resistance by certain members of the fishing community. However, some specialists feel that a significant impediment to the development of the corridor proposal has been a lack of adequate promotion and consolidation of the accord by the interim secretariat (M. Ribadeneira Sarmiento, written communication, 18 March 2009). The development of the agreement and follow-up workshops have been supported by several governmental organizations and NGOs; among them, Conservation International, one of the largest, best-endowed, and most powerful conservation organizations, has been especially active (e.g., Wallace 2010).

In some circles, especially where fisheries issues are prominent, the fact that a large NGO has played a leading role in this initiative has been used to argue against the accord. Nonetheless, there is no question that NGOs have a critical role to play in governance of environmental issues in general (Oberthür et al. 2002), and the initiative for the Eastern Tropical Pacific Marine Corridor would never have achieved what it has without NGO support. Moreover, with the active lobbying of NGOs, the initiative has been adopted by the UNESCO World Heritage Program as the Eastern Tropical Pacific Seascape. While the involvement of UNESCO should be an asset, there are concerns that administrative and bureaucratic complications could increase dramatically (M. Ribadeneira Sarmiento, written communication, 18 March 2009). It is unclear how much official support the initiative will receive from the respective governments. At the time of this writing, one of the two initial signatories to the San José Declaration, Ecuador, has withdrawn official support (reportedly because of pressure from the national fishing industry). In addition, Panamá briefly reopened some of the previously protected area surrounding Coiba National Park to tuna purse seine fishing, until the decision was reconsidered, and intense slaughter (finning) of sharks has been reported from the waters around Malpelo Island, Colombia. This shows that although soft law instruments are much easier to negotiate, they are also much easier for countries to ignore, or even walk away from. To date, it is not clear if the marine corridor has served a unique role for promoting turtle conservation or, rather, is simply an accessory to other existing instruments.

Discussion

While international law and politics are a far cry from the usual stuff of conservation biology, and many of the concepts and details discussed above may seem too strange or irrelevant to be useful to marine turtle conservation, there are pressing reasons for conservationists to be better informed and more involved in these initiatives. Several attempts have been made over the past years to promote a greater understanding of, and interest in, this topic among marine turtle biologists and conservationists (e.g., Frazier 2002a, 2004, 2005, 2006), with several area-specific summaries, such as for Latin America and the Caribbean (Frazier 2000; Young 2001) and the Indian subcontinent (Upadhyay and Upadhyay 2002; Bache and Frazier 2006). However, international wildlife law is not generally well understood, even among legal specialists (Giordano 2002); and the level of understanding among marine turtle specialists is remarkably poor (Tiwari 2002). Add to this the minimal level of professional engagement of turtle conservationists in international policy issues. Indeed, even government employees responsible for promoting national policies for international instruments are often loath to invest time, effort, and resources, much preferring simpler tasks, such as technical activities and bilateral arrangements, that require little integration with political actors and less complex policy development. This often means that the more complex instruments are given less attention and resources than is needed to make them function adequately. Also, while there is a tendency to treat certain initiatives as outcomes (e.g., the organization of meetings; the production of meeting reports, guidelines, and recommendation; and the development of different types of agreements), it is absolutely essential to bear in mind that these "deliverables" of which bureaucrats are so enamored are in fact steps, or tools, on the way to reaching practical, effective conservation and management objectives.

Bilateral relationships, because of far fewer actors, are much easier to negotiate; and when there is an unequal relationship between the two parties, the more powerful can further its agenda more easily. Technical meetings routinely involve people who are oriented toward the same goal, despite differences in nationalities and institutions. However, a functioning agreement between two countries, or a successful technical meeting, is a far cry from effective collaboration between all the countries and institutional actors that impinge upon the life history of a marine turtle. Moreover, to be effective, bilateral, regional, and international agreements should not exist as isolated, stand-alone initiatives but must integrate into wider national policies, and these in turn must be coordinated at regional and international levels. The need for an integrated policy is especially clear for marine turtles, shared resources that not only crisscross ocean basins but during various stages of their extended life history depend on a variety of different habitats, including terrestrial beaches, coastal waters, and the open ocean. Because marine turtles are subject to countless direct and indirect impacts originating from diverse sources ranging from land-based to coastal to oceanic, the conservation of these reptiles and their habitats requires the investigation, evaluation, monitoring, control, and mitigation of an incredible variety of anthropogenic activities; this can be effective only if conducted in a coordinated way. Indeed, the vast amount of information relevant to marine turtle conservation, from diverse sources, summarized in the present chapter for just the eastern Pacific, underscores the pressing need for cooperation and coordination in this field. Despite the large diversity and number of accords, as well as their interrelated, and overlapping objectives, there is relatively little coordination or cooperation among instruments and the institutions that manage them. Indeed, it is not uncommon for delegations to different instruments to have different positions, even though they are from the same Party.

However, coordination does not come about just because it is decreed; it requires consistent, dedicated, competent efforts, which in turn require adequate human, institutional, and financial resources. This is true not only of coordination among countries but also of coordination within countries. In the case of marine turtles with their complex life histories, various government ministries are directly involved, including those that are responsible for environmental protection, fisheries, coastal development, education, research, marine affairs, law enforcement, security, international relations, and more. In addition, governments from local to state/provincial to national levels are involved in different ways with issues that directly affect marine turtles and their habitats. Hence, the mechanisms and procedures for coordination will vary depending on the governance structure, practices, and culture of each institution and country. Development and maintenance of the levels of coordination required is time-consuming and expensive (not to mention tedious and frustrating); and when resources and/or political will are inadequate, coordination will be inadequate, and as a result—despite the best research and scientific information—conservation policies and actions will be inadequate and ineffective. When raised to the level of interactions among sovereign states, the challenge of coordination—and the allocation of adequate human, institutional, and financial resources—is even more acute.

Criticisms of turtle accords abound; among biologists, social scientists, and conservationists they are frequently regarded as "political," "just talk and no action," or out of touch with theoretical constructs that are now in fashion. A lack of clarity in convention texts is easy to find. For example, Article IV(3)(a) of the IAC allows exceptions from strict protection "to satisfy economic subsistence needs of traditional communities," but there is no attempt to define either "subsistence needs" or "traditional communities" (a general problem that is also found in much larger treaties such as the UNCLOS [Johnson 2004]). Wording of accords is often out of step with recent theory, and numerous other imperfections unquestionably warrant critique. However, prolonged condemnation, repeating again and again the same criticisms (e.g., Campbell et al. 2002; Richardson et al. 2006; Campbell 2007) without offering pragmatic alternatives for how to deal with highly migratory, common pool resources that are subjected to diverse threats, is not constructive—except perhaps for academic purposes. International instruments, like other complex human ideas such as democracy, are far from perfect, but they seem to be preferable to other alternatives that have been used throughout history, such as military invasion, colonization, imposition of anarchy, or other forms of domination of one State over another.

Anyone who has worked in committees—including committees in academia—will have experienced the frustration of processes and products that emanate from group work, typified by inconsistencies and lapses in logic, not to mention certain individuals and institutions that unabashedly drive personal agendas. It is very naive to expect the process to be simpler in politically charged negotiations of international instruments where, despite the noble objective of multilateral cooperation, each delegation will have different, even contradictory, interests and objectives; this is more evident in cases involving binding, international law. Hence, by the very nature of their genesis, no instrument can be perfect, and the most realistic outlook is to view each as a work in progress, no matter how long it has been in existence or how much expert involvement it has experienced (Johnson 2004).

In this light, there are a number of excellent syntheses that provide numerous details of international environmental law and policy and discuss how international instruments are directly relevant to marine conservation; and they also provide basic information and suggestions on how to make the best use of policy and legal tools (e.g., Kimball 2001; Johnson 2004; Kuru-kulasuriya and Robinson 2006; Schiffman 2008). These works also include detailed national studies, such as for México (Brañes 2000; Carmona Lara 2000), which serve as valuable models for many other nations in the eastern Pacific region. There are also numerous academic studies that debate diverse aspects of environmental law, definitions, and theoretical underpinnings (e.g., Maes 2007). There is even the *Multilateral Environmental Agreement Negotiator's Handbook* (Carruthers et al. 2007), designed to help develop more effective instruments. It is important to emphasize that in the phases of both development and implementation, active participation and technical inputs from scientists and conservationists can make a world of difference in the objectives, focus, structure, measures, follow-through, and other critical elements of the instrument in question.

Beyond the many theories and discussions about international law, there are several fundamental, pragmatic questions that must be asked of any instrument's merit: Have the much-debated and carefully worded resolutions or convention articles expounding marine turtle conservation actually been implemented? Has the instrument under consideration made significant contributions to "on the ground" conservation of marine turtles (e.g., through recovered populations or truly protected areas of critical habitat—not just more and more meetings and reports)? What are the most effective policies for promoting marine turtle conservation at an international level; that is, how are shared resources managed for diverse societies and institutions? Because of the complex, lengthy life histories of marine turtles, meaningful evaluations of success must be based on periods of at least a decade or more. Hence, even if there were straightforward procedures for evaluating complex policy initiatives—which there are not—it is too early to be able to answer the fundamental questions. Nonetheless, some preliminary conclusions are warranted. Of the four instruments described above in detail, only the IATTC seems to have the potential to have promoted measures that could reduce fisheries related mortality on marine turtles over a large region. While the IAC, the Regional Marine Turtle Program of the Lima Convention, and the Eastern Tropical Pacific Marine Corridor will have promoted greater awareness of marine turtles and encouraged some measures at different national levels, there is no clear evidence that any of these instruments have resulted in any substantive advances at a regional level. Several critical characteristics of the IATTC stand out: half a century of experience, directly related to an activity (fisheries) that represents significant financial investment and gain, the possibility of sanctions, and a staff of trained personnel. Nonetheless, clearly some IATTC Parties are not implementing some of the mitigation measures that have been adopted by the commission. While personalities are rarely drawn into discussions of this nature, it is essential to understand that the success of multilateral agreements, like any other institutional activity, depends heavily on key players. When secretariat staff are inexperienced, as has been the case of the IAC, basic learning and office organization will—in the least—absorb significant amounts of time and resources at the cost of resolving the many complexities of international coordination. The relatively greater flexibility of the agreement for the marine corridor shows that it matters less what kind of instrument is involved than whether there is political will, ability, and funding on all sides of the instrument. At the same time, soft law initiatives, such as the marine corridor, can be more easily ignored by collaborating countries.

In summary, it is essential that marine turtle biologists and conservationists understand—and appreciate—that there is an enormous body of knowledge and experience that focuses on international environmental law and the diverse instruments that derive from initiatives to promote different forms of cooperation and collaboration. These fields have their own theoretical and disciplinary constructs. In this respect, it is important to note that in Latin America, law and policy are often referred to as "sciences," showing a certain detachment from the common arrogance of the physical and natural sciences (see Brown 1994; Nader 1996). To be truly effective at promoting the conservation of marine turtles and the diverse habitats on which they depend, biologists and conservationists must learn how to make the best use of these tools. At the same time, it is essential that specialists in marine turtle biology and conservation engage actively with policy makers and specialists outside the natural sciences to help provide technical and scientific information and also to make it relevant within forums where such information is not known or not easily understood. If policies are to be effective within a democratic context, they must be based on the best available information, and the onus is on the purveyors of such information—scientists and conservationists—to provide and explain it to a wide audience.

ACKNOWLEDGMENTS

Various colleagues provided useful comments on earlier drafts of this chapter: David Balton, Anna Cederstav, Fernando Félix Grijalva, Douglas Hykle, Aimee Leslie, Gladys Martínez, Samantha Namnum, Gustavo Omaña, María Cristina Puente, Marina Ruíz Slater, Mónica Ribadeneira Sarmiento, and Howard Schiffman; especially detailed and thought-provoking comments were provided by Alexis Gutierrez, Tomme Young, and Chris Wold.

NOTES

1. Accords and protocols adopted by the Lima Convention include the Accord on Regional Cooperation to Combat Against Contamination of the Southeast Pacific by

Hydrocarbons and Other Hazardous Substances in Cases of Emergency (Lima, 1981, in force 1986); the Complementary Protocol of the Accord on Regional Cooperation to Combat Against Contamination of the Southeast Pacific by Hydrocarbons and Other Hazardous Substances in Cases of Emergency (Quito, 1983, in force 1987); the Protocol for the Protection of the South-East Pacific Against Pollution from Land-Based Sources (Quito, 1983, in force 1986); the Protocol for the Conservation and Management of Protected Marine and Coastal Areas of the South-East Pacific (Paipa, 1989, in force 1995); and the Protocol for the Protection of the South East Pacific Against Radioactive Contamination (Paipa, 1989, in force, 1995) (CPPS 2007a).

2. The Permanent Commission for the South Pacific (CPPS) was established by agreement between Chile, Ecuador, and Perú in 1952 (the Santiago Declaration of 18 August 1952), and Colombia joined in 1979 (CPPS 2007a).

3. Since this chapter was first drafted in mid-2008, the Convention for the Strengthening of the Inter-American Tropical Tuna Commission Established by the 1949 Convention, Between the United States of America and the Republic of Costa Rica (Antigua Convention) entered into force on 27 August 2010. It was developed between October 1998 and 27 June 2003, when the text was adopted by the delegations of fourteen States at the seventieth meeting of the Inter-American Tropical Tuna Commission. The Antigua Convention was adopted to strengthen and replace the 1949 Convention establishing the IATTC. At the time of this writing, there are fourteen Parties to the new convention, eight of which have jurisdiction in the eastern tropical Pacific. One of the more important changes is that the Antigua Convention provides a formal mechanism for fishing entities to participate directly in the convention, essentially as if they were national delegations. In addition, the new convention reflects more recent ideas in fisheries management, such as the United Nations Fish Stocks Agreement. Although the Antigua Convention replaces the 1949 Convention, the Secretariat and operational aspects are still centered at the IATTC headquarters in La Jolla, California; and it continues to build on the advances made under the 1949 Convention. (For more information, see www.iattc.org/IATTCdocumentationENG.htm.)

4. IATTC meetings relevant to mitigating marine turtle bycatch include (1) the fourth meeting of the IATTC Permanent Working Group on Compliance (June 2003, Antigua, Guatemala), which proposed amendments to the Consolidated Resolution on bycatch (Resolution C-03-08); (2) IATTC seventieth meeting (26–28 June 2003, Antigua, Guatemala), which adopted Recommendation C-03-10 to strengthen measures for protecting marine turtles that interact with the purse seine fishery; (3) the fourth meeting of the Bycatch Working Group (14–16 January 2004 in Kobe, Japan), which included presentations of various working documents, including (i) "Review of the status of sea turtle stocks in the eastern Pacific"; (ii) "Interactions of sea turtles with tuna fisheries and other impacts on turtle populations: (a) Interactions in the purse-seine fishery and (b) Interactions in the longline fisheries"; (iii) "Proposed amendments to the consolidated resolution on bycatch"; and (iv) "Resolution C-03-08, Consolidated Resolution on bycatch"; (4) IATTC seventy-second meeting (14–18 June 2004, Lima, Perú), which passed two resolutions: Consolidated Resolution on Bycatch C-04-05, and Resolution C-04-07 on a three-year program to mitigate the impact of tuna fishing on marine turtles; (5) IATTC seventy-third meeting (20–24 June 2005, Lanzarote, Spain), which adopted modifications to the Consolidated Resolution on Bycatch C-04-05, which further strengthened certain mitigation measures; (6) IATTC

seventy-fourth meeting 26–30 June 2006, Pusan, Korea), which adopted modifications to Consolidated Resolution on Bycatch C-04-05, which further strengthened certain mitigation measures; and (7) IATTC seventy-fifth meeting (25–29 June 2007, Cancun, México), which adopted Resolution C-07-03 to mitigate the impact of tuna fishing vessels on marine turtles.

REFERENCES

Álvarez, G. 1997. Cuestionario: Régimen Institucional y Legal de las Tortugas Marinas. Fundación Salvadoreña de Derecho Ambiental (FUNDASALDA), San Salvador, El Salvador.

Bache, S. 2002. Turtles, tuna and treaties: strengthening the links between international fisheries management and marine species conservation. In: Frazier, J. (Ed.), International Instruments and Marine Turtle Conservation (Special Issue). Journal of International Wildlife Law and Policy 5:49–64.

Bache, S., and J. Frazier. 2006. International instruments and marine turtle conservation. In: Shanker, K., and B. C. Choudhry (Eds.), Sea Turtles on the Indian Subcontinent. Universities Press, Hyderabad, India. Pp. 324–353.

Bermúdez Soto, J. 2008. Análisis de la concordancia de los instrumentos nacionales y regionales con la legislación internacional sobre el manejo sostenible y utilización del medio marino y costero [Chile]. Available: www.cpps-int.org/spanish/planaccion/reunion/consultoria/0808.INFORME.CHILE.pdf.

Bliege Bird, R., E. A. Smith, and D. W. Bird. 2001. The hunting handicap: costly signaling in human foraging strategies. Behavioral Ecology and Sociobiology 50:9–19.

Brañes, R. 2000. Manual de derecho ambiental mexicano, 2nd ed. Fundación Mexicana para la Educación Ambiental, Fondo de Cultura Economómica, Mexico City.

Brown, A. C. 1994. Is biology a science? Transactions of the Royal Society of South Africa 49(2):141–146.

Cajiao Jiménez, M. V. 2003. Régimen legal de los recursos marinos y costeros de Costa Rica. Instituto de Pesquerias IPECA (Editorial Ivan Pérez Castillo), San José, Costa Rica.

Cajiao Jiménez, M. V., M. Flórez, A. González, P. Hernández, C. Martans, N. Porras, and J. A. Zornoza. 2006. Manual de legislación ambiental para los países del Corredor Marino de Conservación del Pacífico Este Tropical. Fundación Marviva, San José, Costa Rica.

Campbell, L. M. 2007. Local conservation practice and global discourse: a political ecology of sea turtle conservation. Annals of the Association of American Geographers 97:313–334.

Campbell, L. M., M. H. Godfrey, and O. Drif. 2002. Community-based conservation via global legislation? Limitations of the Inter-American Convention for the Protection and Conservation of Sea Turtles. In: Frazier, J. (Ed.), International Instruments and Marine Turtle Conservation (Special Issue). Journal of International Wildlife Law and Policy 5:121–143.

Carmona Lara, M. D. C. 2000. Derechos en relación con el medio ambiente. Universidad Autónoma Nacional de México, México, DF.

Carpenter, A. I. 2006. Conservation convention adoption provides limited conservation benefits: the Mediterranean green turtle as a case study. Journal for Nature Conservation 14:91–96.

Carruthers, C., Y. Le Bouthillier, A. Daniel, J. Bernstein, and D. McGraw. 2007. Multilateral Environmental Agreement Negotiator's Handbook, 2nd ed. Department of Law, University of Joensuu, Joensuu, Finland.

Commission for Environmental Cooperation. 2005. North American Conservation Action Plan: Pacific Leatherback Sea Turtle, *Dermochelys coriacea*. Commission for Environmental Cooperation, Montreal, Quebec, Canada. Available: www.cec.org/Storage/59/5167_NACAP-Leatherback-SeaTurtle_en.pdf.

CPPS. 1972. Legislación marítima y pesquera vigente de Chile Secretaría General. Comisión Permanente del Pacífico Sur, Quito, Ecuador.

CPPS. 1973. Legislación marítima y pesquera vigente del Perú Secretaría General. Comisión Permanente del Pacífico Sur, Quito, Ecuador.

CPPS. 1974. Legislación marítima y pesquera vigente y otros documentos referentes al derecho del mar Ecuador Secretaría General. Comisión Permanente del Pacífico Sur, Santiago, Chile.

CPPS. 1986. Legislación marítima y pesquera del Ecuador Secretaría General. Comisión Permanente del Pacífico Sur, Bogota, Colombia.

CPPS. 1997a. Legislación marítima de Colombia Secretaría General. Comisión Permanente del Pacífico Sur, Lima, Perú.

CPPS. 1997b. Legislación marítima de Colombia Secretaría General. Comisión Permanente del Pacífico Sur, Lima, Perú.

CPPS. 2007a. Convenios, acuerdos, protocolos, declaraciones, estatuto y reglamento de la CPPS, 3rd ed. Secretaría General. Comisión Permanente del Pacífico Sur, Guayaquil, Ecuador.

CPPS. 2007b. XIV Reunión de la Autoridad General del Plan de Acción para la Protección del Medio Marino y Áreas Costeras del Pacífico Sudeste; Guayaquil-Ecuador, 30 de Noviembre de 2007. Acta Final. Comisión Permanente del Pacífico Sur, Guayaquil. Available: www.cpps-int.org/spanish/planaccion/reunion/informes/Informe%20AG%202007.pdf.

Crouse, D. T. 2000. The consequences of delayed maturity in a human-dominated world. American Fisheries Society Symposium 23:195–202.

Danaher, M. 1999. Nature conservation, environmental diplomacy and Japan. Asian Studies Review 23(2): 247–270.

de Noack, J. 1997. Legislación Nacional e Internacional que Afecta a Las Tortugas Marinas en Centroamérica, Caso Guatemala. IDEADS (Environmental Law and Sustainable Development Institute), Guatemala City.

de Noack, J., E. Cifuentes, and R. Vásquez. 2005. Ley General de Pesca y Acuacultura. Decreto No. 80-2002 del Congreso de la República Versión Español Q'eqchi'. Centro de Acción Legal Ambiental y Social de Guatemala, Editorial Batres, Guatemala City.

Díaz, M. P. 2008. Informe de consultoría. Análisis de la concordancia de los instrumentos nacionales y regionales con la legislación internacional sobre el manejo sostenible y utilización del medio marino y costero de acuerdo con la decisión no. 12 de la XIII Reunión de la Autoridad General del Plan de Acción para la Protección de Medio Marino y Áreas costeras del Pacífico Sudeste, realizada en Guayaquil, Ecuador, el 31 de

Agosto de 2006 [Panamá]. Available: www.cpps-int.org/spanish/planaccion/reunion/consultoria/0808.INFORME.PANAMA.pdf.

Dyer, C. L., and J. R. McGoodwin (Eds.). 1994. Folk Management in the World's Fisheries. Lessons for Modern Fisheries Management. University Press of Colorado, Nowot.

Ellis, Z. 1997. National and International Legislation Which Affects Marine Turtles in Central America—The Belize Case. BELPO (Belize Institute of Environmental Law and Policy), Belmopan City, Belize.

Espinoza E., L. 1997. Estudio Regional sobre Tortugas Marinas (Regulaciones Normativas). Centro de Derecho Ambiental y de los Recursos Naturales, San José, Costa Rica.

Espinoza E., L., and C. Z. Piskulich. 1997. Políticas y Régimen Legal e Institucional de las Tortugas Marinas en Costa Rica. Centro de Derecho Ambiental y de Recursos Naturales, San José, Costa Rica.

FAO. 2004a. Papers Presented at the Expert Consultation on Interactions between Sea Turtles and Fisheries within an Ecosystem Context; Rome, 9–12 March 2004. FAO Fisheries Report No. 738, Supplement. Food and Agriculture Organization of the United Nations, Rome.

FAO. 2004b. Report of the Technical Consultation on Sea Turtles Conservation and Fisheries; Bangkok, Thailand, 29 November–2 December 2004. FAO Fisheries Report No. 765. Food and Agriculture Organization of the United Nations, Rome.

FAO. 2005. Report of the Twenty-Sixth Session of the Committee on Fisheries; Rome, 7–11 March 2004. FAO Fisheries Report No. 780. Food and Agriculture Organization of the United Nations, Rome.

Frazier, J. 1979. Marine Turtles in Peru and the East Pacific. Smithsonian Institution, Washington, DC.

Frazier, J. 1981. Marine turtles as index species for international conservation. In: Ambasht, R. S., and H. N. Pandey (Eds.), Ecology and Resource Management in the Tropics. Abstracts. International Society of Tropical Ecology, Ecology and Resource Management in the Tropics; Varanasi, India. International Society of Tropical Ecology, Department of Botany, Banaras Hindu University, Varanasi, India. Pp. 66–67.

Frazier, J. 1982. Crying "wolf" at La Escobilla. Marine Turtle Newsletter 21:7–8.

Frazier, J. 1983. The marine turtle situation in the east Pacific. CDC Newsletter 2:7–10.

Frazier, J. 1984a. Contemporary problems in sea turtle biology and conservation: the urgent need for regional cooperation. Bulletin of the Central Marine Fisheries Research Institute 18:77–91.

Frazier, J. 1984b. Análisis estadístico de la tortuga golfina *Lepidochelys olivacea* (Eschscholtz) de Oaxaca, México. Ciencia Pesquera 4[1983]:49–75.

Frazier, J. 1985a. Misidentification of marine turtles: *Caretta caretta* and *Lepidochelys olivacea* in the East Pacific. Journal of Herpetology 19:1–11.

Frazier, J. 1985b. Biology of the olive ridley sea turtle. National Geographic Research Reports 21:175–179.

Frazier, J. 1990. Marine turtles in Chile: an update. In: Richardson, T. H., J. I. Richardson, and M. Donnelly (Comps.), Proceedings of the Tenth Annual Workshop on Sea Turtle Biology and Conservation. NOAA Tech. Memo. NMFS-SEFC-278. Department of Commerce, U.S. National Marine Fisheries Service, Miami, FL. Pp. 39–41.

Frazier, J. 1992a. Update on Playa Grande, Costa Rica and Troubles on French Guiana beaches. Marine Turtle Newsletter 59:12–14.

Frazier, J. 1992b. La tortuga marina: recurso del pueblo o pelota política? In: Resumenes del IX Congreso Nacional de Oceanografía. Secretaría de Pesca, Mexico City. P. 171.

Frazier, J. 1993. Educación ambiental: una respuesta al reto de los retos. In: II Reunión Regional de Educadores Ambientales del Sur-Sureste de México: Memorias. División Académica de Ciencias Biológicas, Universidad Juárez Autónoma de Tabasco, Villahermosa. Pp. 154–158.

Frazier, J. 1994. La tortuga marina: ¿Dios, seducción, excusa, o recurso? Boletín de la Sociedad Herpetológica Mexicana 6:9–14.

Frazier, J. 1995. ¿Es la educación ambiental realmente educación? In: Memorias de la III Reunión Regional de Educadores Ambientales del Sur-Sureste de México. Pronatura, Chiapas, AC. Pp. 13–37.

Frazier, J. 1997. The Inter-American Convention for the Protection and Conservation of Sea Turtles. Marine Turtle Newsletter 78:7–13.

Frazier, J. 1999. Update on the Inter-American Convention for the Protection and Conservation of Sea Turtles. Marine Turtle Newsletter 84:1–3.

Frazier, J. 2000a. Advances with the Inter-American Convention for the Protection and Conservation of Sea Turtles. Marine Turtle Newsletter 90:1–3.

Frazier, J. 2000b. Building support for regional sea turtle conservation in Indian Ocean region: learning from the Inter-American Convention for the Protection and Conservation of Sea Turtles. In: Pilcher, N., and G. Ismail (Eds.), Sea Turtles of the Indo-Pacific: Research, Conservation and Management. ASEAN Academic Press, London. Pp. 277–306.

Frazier, J. 2000c. Actualización sobre la Convención Interamericana para la Protección y Conservación de las Tortugas Marinas. Memorias del IV Taller Regional para la Conservación de las Tortugas Marinas en Centroamérica. Asociación ANAI San José, Costa Rica. Pp. 79–85.

Frazier, J. 2000d. Annex V: brief assessment of some international instruments that are of major relevance to the conservation of marine turtles and marine turtle habitats in the western hemisphere. In: Frazier, J. (Ed.), Developing an Action Plan for Marine Turtles in Latin America and the Caribbean. World Wildlife Fund for Nature, Washington, DC. P. 8.

Frazier, J. (Ed.). 2002a. International instruments and marine turtle conservation (Special Issue). Journal of International Wildlife Law and Policy 5(1–2):1–208.

Frazier, J. 2002b. Marine turtles and international instruments: the agony and the ecstasy. Journal of International Wildlife Law and Policy 5:1–10.

Frazier, J. 2004. Marine turtles: whose property? Whose rights? In: Frazier, J. (Comp.), Marine Turtles: A Case Study of "Common Property" from the "Global Commons." Proceedings of the Tenth Biennial Conference of the International Association for the Study of Common Property (IASCP): The Commons in an Age of Global Transition: Challenges, Risks and Opportunities, 9–13 August 2004, Oaxaca, México. Available: http://dlc.dlib.indiana.edu/dlc/bitstream/handle/10535/242/Frazier_Marine_040531_Paper547b.pdf?sequence=1.

Frazier, J. 2005. Science, conservation, and sea turtles: what's the connection? In: Coyne, M. S., and R. D. Clark (Comps.), Proceedings of the 21st Annual Symposium on Sea Turtle Biology and Conservation. NOAA Tech. Memo. NMFS-SEFSC-528. Department of Commerce, U.S. National Marine Fisheries Service, Miami, FL. Pp. 27–29.

Frazier, J. (Ed.). 2006. Instrumentos Internacionales y la Conservación de las Tortugas Marinas. Abyayala, Quito, Ecuador.

Frazier, J., and S. Bache. 2002. Sea turtle conservation and the "big stick": the effects of unilateral US embargoes on international fishing activities. In: Mosier, A., A. Foley, and B. Brost (Comps.), Proceedings of the 20th Annual Symposium on Sea Turtle Biology and Conservation. NOAA Tech. Memo. NMFS-SEFSC-477. Department of Commerce, U.S. National Marine Fisheries Service, Miami, FL. Pp. 118–121.

Frazier, J., and D. Bonavia. (2000). Prehispanic marine turtles in Peru: where were they? In: Abru-Grobois, F. A., R. Briseño-Dueñas, R. Márquez-Millán, and L. Sarti-Martínez (Comps.), Proceedings of the Eighteenth International Sea Turtle Symposium. NOAA Tech. Memo. NMFS-SEFSC-436. Department of Commerce, U.S. National Marine Fisheries Service, Miami, FL. Pp. 243–245.

Frazier, J. G., and J. L. Brito Montero. 1990. Incidental capture of marine turtles by the swordfish fishery at San Antonio, Chile. Marine Turtle Newsletter 49:8–13.

Frazier, J., and I. Hártasanchez. 1992. Educando a los Educadores: el reto de los retos. In: Memoria del 1er Taller Nacional de Educación Ambiental para la Protección y Conservación de las Tortugas Marinas. Subsecretaria de Ecología, SEDUE, México, DF. Pp. 23–26.

Frazier, J., and S. Salas. 1982. Ecuador closes commercial turtle fishery. Marine Turtle Newsletter 20:5–6.

Frazier, J., and S. Salas. 1983. Tortugas marinas en la Costa Peruana. Boletin de Lima 5(30):13.

Frazier, J., and S. Salas. 1984. Tortugas marinas en Chile. Boletin del Museo Nacional de Historia Natural de Santiago, Chile, 39[1982]:63–73.

Frazier, J., and S. Salas. 1986. Tortugas marinas del Pacífico oriental: ¿El recurso que nunca acabrá? In: Symposio de Conservación y Manejo de Fauna Silvestre Neotropical (IX Congreso Latinoamericano de Zoología PERU) (1983) Arequipa, Perú. Pp. 87–98.

Frazier, J., and S. Salas. 1987. La Situación de las Tortugas Marinas en el Pacífico Este. In: Aguirre, S. G. (Ed.), VII Simposio Latinoamericano Sobre Oceanografía Biológica. 15–19 November 1981, Acapulco, Mexico. Secretaría de Pesca, Mexico City. Pp. 615–624.

Galeano, E. H. 2004. Las venas abiertas de América Latina, 21st ed. Siglo XXI, Buenos Aires.

Galindo, F. M. G. 1997. Legislación Nacional e Internacional que Afecta a las Tortugas Marinas en Centro América. Centro de Derecho Ambiental de Honduras, Tegucigalpa, Honduras.

Giordano, M. 2002. The internationalization of wildlife and efforts towards its management: a conceptual framework and the historic record. Georgetown International Environmental Law Review 14:1–24.

Harding, G. 1968. The tragedy of the commons. Science 162:1243–1248.

Hernández, R. P., M. C. Puente, and J. G. Auz. 2008. Concordancia de los instrumentos nacionales y regionales con la legislación internacional sobre manejo sostenible y utilización del medio marino y costero. In: Evaluación y Observancia de los Convenios del Plan de Acción para la Protección del Medio Marino y las Áreas Costeras del Pacífico Sudeste y sus Instrumentos. Comisión Permanente del Pacífico Sur, Guayaquil, Ecuador. Available: www.cpps-int.org/spanish/planaccion/reunion/consultoria/0808 .INFORME.ECUADOR.pdf.

Howse, R. 2002. The appellate body rulings in the shrimp/turtle case: a new legal baseline for the trade and environment debate. Colombia Journal of Environmental Law 27:491–521.

Hunter, D., J. Salzman, and D. Zaelke. 1998. International Environmental Law and Policy, USA. University Case Book Series. Foundation Press, New York.

Hykle, D. 1999. International conservation treaties, In: Eckert, K. L., K. A. Bjorndal, F. A. Abreu-Grobois, and M. Donnelly (Eds.), Research and Management Techniques for the Conservation of Sea Turtles. IUCN/SSC Marine Turtle Specialist Group Publication No. 4. IUCN/SSC, Washington, DC. Pp. 228–231.

Hykle, D. 2002. The Convention on Migratory Species and other international instruments relevant to marine turtle conservation: pros and cons. Journal of International Wildlife Law and Policy 5:105–119.

Iturregui, P. 2008. Aplicación nacional de los Acuerdos del Plan de Acción para la Protección del Medio Marino y las Áreas Costeras del Pacífico Sudeste por parte del Perú. Available: www.cpps-int.org/spanish/planaccion/reunion/consultoria/0808.INFORME .PERU.pdf.

Johnson, L. S. 2004 Coastal State Regulation of International Shipping. Oceana, Dobbs Ferry, NY.

Kimball, L. A. 2001. International ocean governance: using international law and organizations to manage marine resources sustainably. International Union for Conservation of Nature, Gland, Switzerland.

Kurukulasuriya, L., and N. A. Robinson (Eds.). 2006. Training Manual on International Environmental Law. United Nations Environment Programme, Nairobi, Kenya.

Lilette, V. 2006. Mixed results: conservation of the marine turtle and red-tailed tropicbird by Vezo semi-nomadic fishers. Conservation and Society 4(2):262–286.

Lugten, G. L. 2006. Soft law with hidden teeth: the case for a FAO International Plan of Action on sea turtles. Journal of International Wildlife Law and Policy 9:155–173.

Maes, F. 2007. Los principios de derecho ambiental, su naturaleza y sus relaciones con el derecho internacional marítimo. Un cambio para los legisladores nacionales. Anuario Mexicano de Derecho International. 7:189–255. Originally published in: Sheridan, M., and L. Lavrysen (Eds.), Environmental Law, Principles in Practice. Bruyllant, Brussels, 2007. Available: http://revistas.unam.mx/index.php/amdi/article/download/ 16021/15213.

Meffe, G. K., and A. Viederman. 1995. Combining science and policy in conservation biology. Wildlife Society Bulletin 23:327–332.

Murphy, J. B. 2006. Alternate approaches to the CITES "non-detriment" finding for Appendix II species. Environmental Law 36:531–563.

Musick, J. A., and C. J. Limpus. 1997. Habitat utilization and migration in juvenile sea turtles. In: Lutz, P. L., and J. A. Musick (Eds.), The Biology of Sea Turtles. CRC Press, New York. Pp. 137–163.

Nabhan, G. P. 2003. Singing the Turtles to Sea: The Comcáac (Seri) Art and Science of Reptiles. University of California Press, Berkeley.

Nader, L. (Ed.). 1996. Naked Science: Anthropological Inquiry into Boundaries, Power, and Knowledge. Routledge, New York.

Namnum, S. 2002. The Inter-American Convention for the Protection and Conservation of Sea Turtles and its implementation in Mexican law. Journal of International Wildlife Law and Policy 5:87–103.

Oberthür, S., M. Buck, S. Müller, S. Pfahl, R. G. Tarasofsky, J. Werksman, and A. Palmer. 2002. Participation of Non-governmental Organisations in International Environmental Governance: Legal Basis and Practical Experience. Ecologic: Institute for International and European Environmental Policy; and FIELD: Foundation for International Environmental Law and Development, Berlin and London.

Oeding, M. 2007. Análisis de los instrumentos jurídicos internacionales relacionados con el tema de tortugas marinas WWF Colombia. Santiago de Cali, Colombia.

Omaña Parés, G. A. 2005. Régimen tutelar de las tortugas marinas en Venezuela. DOCTUM 8:181–198.

Omaña Parés, G. A. 2007. Legislación acuática de Venezuela. Legis, Caracas, Venezuela.

Ramos, A. 2008. Gobernabilidad en la aplicación, observancia e implementación nacional de los acuerdos del Plan de Acción para la Protección de Medio Marino y las Áreas costeras del Pacífico sudeste—Capítulo Colombia. Available: www.cpps-int.org/spanish/planaccion/reunion/consultoria/08.08.INFORME.COLOMBIA.pdf.

Richardson, P. B., A. C. Broderick, L. M. Campbell, B. J. Godley, and S. Ranger. 2006. Marine turtle fisheries in the UK overseas territories of the Caribbean: domestic legislation and the requirements of multilateral agreements. Journal of International Wildlife Law and Policy 9:223–246.

Rosano Hernández, M. C., T. Arqueta V., and J. G. Frazier. 1996. Factores que pueden afectar la sobrevivencia de embriones de tortuga golfina Leidochelys olivacea en la playa Escobilla, Tonameca, Oaxaca: observaciones en el campo. In: Memorias 1er Encuentro Regional sobre Investigación y Desarrollo Costero: Guerrero, Oaxaca, Chiapas. Universidad del Mar, Puerto Angel, Oaxaca, México. P. 4.

Rosano Hernández, M. C., T. Arqueta V., and J. Frazier. 1998. Return of the beetles: observations at La Escobilla. In: Proceedings of the Seventeenth International Symposium on Sea Turtle Biology and Conservation. NOAA Tech. Memo. NMFS-SEFSC-415. Department of Commerce, U.S. National Marine Fisheries Service, Miami, FL. Pp. 257–260.

Ruíz, G. A., and L. E. Jarquín C. 1997. Legislación Nacional que Afecta la Tortuga Marina en Nicaragua. Centro de Derecho Ambiental y Promoción para el Desarrollo, Managua, Nicaragua.

Schiffman, H. 2008. Marine Conservation Agreements: The Law and Policy of Reservations and Vetoes. Martinus Nijhoff-Brill, Leiden.

Shaffer, G. 1999. United States–import prohibition of certain shrimp and shrimp products. American Journal of International Law 93(2):507–514.

Smith, E. A., and R. L. Bliege Bird. 2000. Turtle hunting and tombstone opening: public generosity and costly signalling. Evolution and Human Behavior 21: 245–261.

Speir, J. (Ed.). 1999. Cuban Environmental Law: The Framework Environmental Law and an Index of Cuban Environmental Law. Tulane Law School and Center for Marine Conservation, New Orleans, LA, and Washington, DC.

Terán, M. C., K. Clark, C. Suárez, F. Campos, J. Denkinger, D. Ruiz, and P. Jiménez. 2006. Análisis de vacios e identificación de áreas prioritarias para la conservación de la biodiversidad marino-costero en el Ecuador continental: resumen ejecutivo. Ministerio del Ambiente, Quito, Ecuador.

Tiwari, M. 2002. An evaluation of the perceived effectiveness of international instruments for sea turtle conservation. Journal of International Wildlife Law and Policy 5:145–156.

Tonelli, M. 2006. Acciones a ser implementadas para el establecimiento de una buena cooperación internacional entre la República Argentina, República Federativa do Brasil y la República Oriental del Uruguay para la protección y conservación de tortugas marinas. Universidad Internacional de Andalucía, Sede Antonio Machado de Baeza, Baeza, España.

Trono, R. B., and R. V. Salm. 1999. Regional collaboration. In: Eckert, K. L., K. A. Bjorndal, F. A. Abreu-Grobois, and M. Donnelly (Eds.), Research and Management Techniques for the Conservation of Sea Turtles. IUCN/SSC Marine Turtle Specialist Group Publication No. 4. IUCN/SSC, Washington, DC. Pp. 224–227.

Tulio H., M. 1997. Aspectos Legales y Manejo de las Tortugas en Panamá. Asociación Legal y Administrativa de Panamá, Panama City.

Upadhyay, S., and V. Upadhyay. 2002. International and national instruments and marine turtle conservation in India. Journal of International Wildlife Law and Policy 5:65–86.

UNEP (United Nations Environment Programme). 2006a. Permanent Commission for the South Pacific (CPPS). Humboldt Current, GIWA [Global International Waters Assessment] Regional Assessment 64. University of Kalmar, Kalmar, Sweden.

UNEP (United Nations Environment Programme). 2006b. Permanent Commission for the South Pacific (CPPS). Eastern Equatorial Pacific, GIWA [Global International Waters Assessment] Regional Assessment 65. University of Kalmar, Kalmar, Sweden.

Wallace, B. (Ed.). 2010. Informe sobre el Estado de las Tortugas marinas del Pacífico este tropical ETPS: Perspectivas Regionales, Informes Nacionales—Costa Rica, Panamá, Colombia, Ecuador. Vol. 1. Conservation International-Ecuador, Quito, Ecuador.

Wallace, R. L. (Comp.). 1994. The Marine Mammal Commission Compendium of Selected Treaties, International Agreements, and Other Relevant Documents on Marine Resources, Wildlife, and the Environment, vols. 1–3. U.S. Government Printing Office, Washington, DC.

Wallace, R. L. (Comp.). 1997. The Marine Mammal Commission Compendium of Selected Treaties, International Agreements, and Other Relevant Documents on Marine Resources, Wildlife and the Environment: First Update. U.S. Government Printing Office, Washington, DC.

Weiskel, H. W., R. L. Wallace, and M. M. Boness (Comps.). 2000. The Marine Mammal Commission Compendium of Selected Treaties, International Agreements, and Other Relevant Documents on Marine Resources, Wildlife, and the Environment: Second Update. U.S. Government Printing Office, Washington, DC.

Wold, C. 2002. The status of sea turtles under international environmental law and international environmental agreements. Journal of International Wildlife Law and Policy 5:11–48.

Young, T. 2001. Multi-lateral agreements for conservation of hawksbill turtles. International Instruments Relevant to the Conservation of Hawksbill Turtles [and their Habitats]. Convention on International Trade in Endangered Species of Wild Fauna and Flora. First CITES Wider Caribbean Hawksbill Turtle Dialogue Meeting, Mexico, April 2001, Doc. 8. Convention on International Trade in Endangered Species of Wild Fauna and Flora, Geneva, Switzerland.

PART THREE

Field–Based Conservation and Signs of Success

8

Leatherbacks in the Balance

Reconciling Human Pressures and
Conservation Efforts in Pacific Costa Rica

BRYAN P. WALLACE AND
ROTNEY PIEDRA CHACÓN

Summary

For decades Costa Rica has been regarded as the model for biodiversity conservation in Central America, primarily because of its network of protected areas spanning diverse terrestrial ecosystems. Likewise, Costa Rica recently has emerged as a regional leader in efforts to repeat these conservation successes in the marine realm. In particular, Costa Rica is home to several well-developed sea turtle conservation efforts, specifically along its Pacific coast. One of the best-known examples of the struggles, complexities, and successes of sea turtle conservation in the eastern Pacific Ocean has occurred at Parque Nacional Marino Las Baulas (PNMB), in Guanacaste Province, northwest Costa Rica. The leatherback nesting colony at PNMB has been reduced more than 90 percent over the past two decades, mirroring the regionwide precipitous decline of eastern Pacific leatherbacks, which is considered to be one of the most urgent sea turtle conservation issues globally. However, unlike most leatherback nesting beaches in the eastern Pacific, historic threats (e.g., harvest of eggs) to nesting leatherbacks at PNMB have been eradicated through comprehensive protection of nesting females and their eggs and hatchlings by integrated efforts of park rangers, scientists, local communities, and volunteers. For these reasons, the leatherback nesting colony at PNMB has been identified as the most likely site for recovery of leatherbacks in the eastern Pacific. In this chapter, we review how the relationships between humans and leatherbacks and their nesting habitat have changed over time, how the current scenario of conflicting as well as convergent human interests is affecting the continued threat of extinction confronting eastern Pacific leatherbacks, and how current conservation efforts and scientific research focusing on leatherbacks in Costa Rica are being leveraged to achieve broader national and international conservation aims. We end by presenting three possible scenarios that might portray the future of human interests and conservation of leatherbacks and natural areas in PNMB.

Brief Background on Costa Rica Conservation Issues

Costa Rica is unique with respect to the rest of Central America for several reasons. For example, Costa Rica has enjoyed peace (it has no standing army) as well as relative political and economic stability (representative democracy and a market-based, capitalist economy) during the same period that several of its neighbors have been dramatically affected by civil wars, governmental turmoil and dysfunction, and severe poverty. Indeed, the 2010–2011 United Nations Human Development Index, which assesses the achievements in health, knowledge, and standard of living for countries around the world, ranked Costa Rica among countries with "High Human Development" (UNDP 2007).

These fortuitous characteristics have also allowed biodiversity conservation to flourish in Costa Rica as a movement, as the engine of a highly lucrative tourism industry, as a well-developed national parks system, and as a source of national pride and identity for Costa Ricans. As such, Costa Rica has earned special recognition as a model for conservation in Latin America, owing to its landmark network of terrestrial protected areas that encompasses a wide array of ecosystems and associated biodiversity. In fact, according to the Ministerio del Ambiente, Energía y Telecomunicaciones (MINAET) and the Sistema Nacional de Áreas de Conservación (SINAC), approximately 26 percent of Costa Rica's land area is afforded some form of protection (SINAC 2011).

Likewise, in the marine realm, Costa Rica recently reinforced its regional leadership in conservation when former president Abel Pacheco declared the nation's intention to protect 25 percent of its marine exclusive economic zone under some type of management scheme; currently 16 percent of Costa Rica's marine areas have some form of protected status (SINAC 2011). While suffering from challenges faced by protected area systems worldwide, which consist predominantly of resource deficiencies (e.g., numbers of trained staff, money, equipment), the system of protected areas in Costa Rica has achieved significant success in biodiversity conservation (Boza 1993).

Not only has Costa Rica's investment in biodiversity conservation increased over the past few decades, but revenue from tourism now constitutes more than 20 percent of all of Costa Rica's exports, or more than double the revenue generated by coffee and banana exports combined (Instituto Costarricense de Turismo 2005). Not surprisingly, this influx of tourism and investment in tourism-related development projects is not always in harmony with conservation of natural resources and protected areas. Thus, biodiversity conservation in Costa Rica currently finds itself at the mercy of often competing human interests. However, because the tourism industry ultimately depends on healthy, protected ecosystems and biodiversity, integration of sound conservation strategies into land-use and resource management plans and development projects is imperative for both lucrative tourism and vibrant biodiversity to persist in Costa Rica.

Because conservation is typically local in scale, each struggle between competing interests transpires according to its unique circumstances and players. It is in this context that several efforts to conserve populations of sea turtles in Pacific Costa Rica are taking place.

Sea Turtles in Costa Rica: Overview of Species and Special Nesting Sites

Faced with the daunting challenge of creating a system of protected areas from scratch in a developing country, Don Mario Boza, Don Alvaro Ugalde, and others decided to highlight areas where strong scenic, historic, and natural values coincided to generate interest and support for the idea of protected area conservation (Boza 1993). Interestingly, but perhaps not surprisingly, of the first four national parks created in Costa Rica in 1970–1971, three—Tortuguero, Cahuita, and Santa Rosa—included important sea turtle nesting beaches. Currently, all six of the eleven Conservation Areas within the framework of SINAC that have marine coastlines include protected areas in which sea turtles occur. Five of the world's seven sea turtle species occur in Costa Rica: the loggerhead (*Caretta caretta*), the hawksbill (*Eretmochelys imbricata*), the green turtle, also called the East Pacific green turtle or black turtle (*Chelonia mydas*), the olive ridley turtle (*Lepidochelys olivacea*), and the leatherback turtle (*Dermochelys coriacea*) (see plates 1–10). The loggerheads (see chapter 12), hawksbills (see chapter 10), black turtles (see chapter 11), and olive ridleys (see chapter 13) in the eastern Pacific are covered elsewhere in this volume.

Costa Rica (Tortuguero National Park, specifically) is considered by many to be the cradle of sea turtle conservation and research, owing to the legacy of famed sea turtle researcher Archie Carr. Dr. Carr's pioneering work started in the 1950s and included studies on nesting ecology and reproduction as well as at-sea movements and migrations. His books, including *The Windward Road* (1956) and *So Excellent a Fishe* (1967), offer detailed and fascinating perspectives into Dr. Carr's life and work. In the decades since Dr. Carr raised the international profile of sea turtle conservation, many new sea turtle conservation projects involving countless researchers, volunteers, and local communities have taken root in Costa Rica. Sea turtles are now iconic species for Costa Rica's biodiversity conservation efforts and ecotourism industries.

A microcosm of the complex and challenging interactions among people, sea turtles, and natural areas they share in the eastern Pacific region is Parque Nacional Marino Las Baulas (PNMB, or Leatherback National Marine Park) on the Pacific coast of Costa Rica (plate 1). PNMB has been the setting for a remarkable story, starting with the relationship between a small local community and thousands of leatherbacks each year, thirty years ago, to a burgeoning residential and tourist population and fewer than a hundred leatherbacks per year at present (fig. 8.1). Creating a balance among the survival of leatherbacks, the livelihood of the local communities, and the prosperity of developers and investors has been elusive, and it will require effective cooperation among all parties and interests involved.

Playa Grande in the Early Days: Lots of Turtles, Few People

The number of human players and the nature of their interactions with leatherbacks and leatherback nesting habitat within present-day PNMB have changed tremendously during the past five decades. During the 1950s

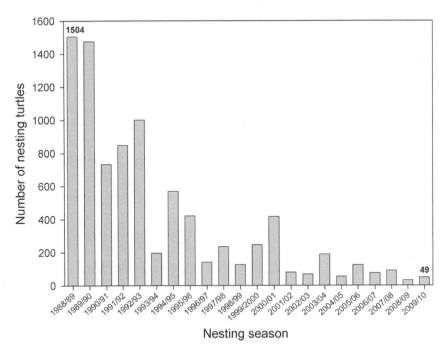

FIGURE 8.1. Number of nesting leatherback turtles in Parque Nacional Marino Las Baulas, Costa Rica, from 1988 through 2010. Annual nesting has declined from approximately 1,504 leatherbacks in 1988/1989 to 49 individuals in 2009/2010. Figure courtesy of M. Santidrián Tomillo, F. V. Paladino, and J. R. Spotila.

through the early 1970s, the area including Playas Grande, Ventanas, Tamarindo, and Langosta and surrounding mangrove estuaries and maritime forests was inhabited by a just handful of families, because it was accessible from outside areas only on horseback or by foot. Local families made trips to the beach at night, gathering in small *fiestas*, as parents brought their children to Playa Grande to watch the leatherbacks come ashore to lay their eggs. As turtles lumbered back to the sea after nesting, children would literally ride turtles down to the surf and jump off before being carried into the dark ocean horizon.

People who lived in the area recall the increase in leatherback numbers on Playa Grande (the primary nesting beach in the system) from relatively few in the early 1950s to more than 100–150 individual turtles *nightly* during the December–January peak of the nesting season in the late 1960s and early 1970s (Wallace and Saba 2009). Witnesses likened these incredible numbers of nesting leatherbacks on the beach to swarms of *hormigas*, or ants—an amusingly illustrative comparison considering the enormous size of adult leatherbacks (~1.5 m long and ~400 kg in mass). At that time, local families engaged in subsistence harvest of leatherback eggs, typically using no more than one or two nests per week. Thus, leatherback eggs supplemented local families' typical food intake, which included agriculture (e.g., rice, fruits, and livestock) and other local sources (e.g., deer from the dry forest and clams, crabs, and fish from the ocean and estuaries). This level of egg harvest was sustainable and by itself would have had a negligible impact on the leatherback population.

What Led to the Collapse: No Eggs Left Behind

By the late 1970s, however, Playa Grande was a very different place. Several people who lived in the area at the time corroborated that the main turning point was the construction of the road that led directly to the beach. Because people could reach Playa Grande by motorized vehicles, subsistence egg harvest gave way to systematic egg harvest that resulted in more than 90 percent of all eggs being removed from the beach each night (Santidrián Tomillo et al. 2007). This eventually would be a primary driver for the overall population decline at PNMB, coupled with unsustainably high mortality due to interactions with fisheries (Santidrián Tomillo et al. 2008; reviewed by Wallace and Saba 2009).

The eggs were largely sold to far-off markets in the Central Valley of Costa Rica (home to the capital, San José, and surrounding towns) and Limón, rather than remaining within the local communities. In previous years, parents had brought their children to the beach to watch the *hormigas*, and the number of turtles was always greater than the number of people on the beach at any given time. In contrast, during this period of comprehensive egg harvest, the beach was occupied from one end to the other with people collecting entire clutches of the billiard-ball-sized eggs from nearly every nesting leatherback. Tracts of beach roughly 100 m in length were reserved by individuals who claimed the "rights" to all the clutches laid by turtles in their section. The eggs were then sold to middlemen, who transported them by truck to the distant markets. Instead of only coming from communities adjacent to Playa Grande, the people who reaped the bounty of leatherback eggs at Playa Grande came from all over Guanacaste Province and the rest of Costa Rica, and from various walks of life. For example, one anecdote recalls policemen from Liberia (the provincial capital, ~50 km from Playa Grande) coming directly to the beach after their shifts—still in uniform—to partake in the harvest.

Because leatherbacks continued to nest in extremely large numbers during this period of intense egg harvest, local people generally did not foresee the imminent population collapse that would result from essentially removing a generation of new recruits from the leatherback population. In fact, an ironic (if not chronologically accurate) local legend maintained by a few old-timers has it that it was the formation of the national park and the scientific activities of the monitoring studies (e.g., tagging, weighing, measuring, etc.)—not the population impacts of comprehensive egg harvest—that drove the turtles away because they were averse to the change in treatment from the caring, respectful handling during the harvest years to the callous manipulation by scientists under the national park regime.

Egg harvest as a main factor in declines of sea turtle populations is not unique to Playa Grande. Comprehensive egg harvest contributed to the collapse of the Terengganu, Malaysia, leatherback population in the western Pacific Ocean (Chan and Liew 1996), as well as to the decline of leatherbacks nesting on the Pacific coast of México (Sarti Martínez et al. 2007; see chapter 9 in this volume). Sea turtle eggs are consumed worldwide for subsistence and are used for baking, but they are also considered delicacies and are widely believed to be aphrodisiacs (Spotila 2004; see plate 12). The black market for sea turtle eggs remains strong in Costa Rica as well, where local bars throughout Guanacaste and elsewhere continue to offer *tragos* or shots of raw sea turtle egg yolks that accompany sugar cane liquor drinks (*guaro*) or beers. Income generated by the legal egg harvest of the first 36 hours of olive ridley *arribadas* at Playa Ostional Wildlife Refuge has resulted in numerous community development projects (Campbell 1998, 2007); however, this egg harvest program's existence and execution remain controversial (Spotila and Paladino 2004). Nonetheless, as the example of Playa Grande's leatherbacks demonstrates, unchecked, comprehensive, unsustainable egg harvest eventually results in declines in numbers of nesting female sea turtles (Santidrián Tomillo et al. 2008).

The Shift from Exploitation to Protection: The National Park

As is the case in many local conservation stories, the paradigm shift from unsustainable human behavior threatening biodiversity to responsible human behavior to conserve and protect it began as a combination of outside influences and internal transitions. Doña Esperanza Rodríguez was one the "landlords" of Playa Grande who not only collected eggs but also "rented" to the "tenants" of the 100-m sections of beach, patrolling each night's activities on horseback. In the late 1980s, Doña Esperanza began to accompany a biologist from San José, Doña María Teresa Koberg, on María Teresa's daily censuses of turtle tracks and nests poached, because Esperanza wanted to ensure María Teresa's safety. Through her efforts, and those of others like Dr. Peter Pritchard (see the foreword to this volume), María Teresa was the first to begin the shift from egg harvest to beach protection.

Over time, María Teresa befriended Esperanza to the point that when María Teresa had to return San José, she enlisted Esperanza to continue counting turtle tracks and the proportion of nests whose eggs had been taken and sold. Esperanza began performing morning beach censuses on horseback using two hand counters—one to tally the number of turtles that nested the previous night, and the other to tally the number of nests that were poached. Upon returning from her censuses, Esperanza gave the counters to record keepers who transcribed each night's counts for each season and each year; Esperanza could neither read nor write. Interestingly, the data recorded by Esperanza in the early years (1988–1992) actually composed the baseline for analyses published in a seminal study in one of the world's premiere scientific journals, *Nature*, that projected the eventual extinction of leatherbacks at Playa Grande based on the exponential decline in the number of nesting turtles from 1988 through 1999 (Spotila et al. 2000).

It was her involvement in the monitoring efforts and her friendship with María Teresa that eventually led Esperanza to eschew her occupation as egg harvester and "landlord" at Playa Grande in exchange for a position of protection of nesting females and their nests. While some affectionately refer to Esperanza as "the first national park ranger" at Playa Grande (fig. 8.2), she was not alone in the momentum to mitigate the threats to the leatherback population and shift the focus from exploitation to conservation.

FIGURE 8.2. Doña Esperanza Rodríguez on horseback, donning her Park Guard uniform shirt and cap (photo courtesy of B. P. Wallace). Note the counter in her left hand, which she used to do daily nest tallies on Playa Grande and Playa Ventanas. Her morning censuses provided the baseline data for tracking the population trend for nesting leatherbacks at Parque Nacional Marino Las Baulas.

At the same time that María Teresa and Esperanza began to work together, the first significant conservation actions by the Costa Rican government occurred in 1987, with the creation of the Tamarindo National Wildlife Refuge, which included Playa Grande, Playa Ventanas, and the Tamarindo mangrove estuary. This mangrove system was declared a wetland of international significance by the RAMSAR Convention in 1993. The creation of the Parque Nacional Marino Las Baulas by executive decree in 1991 and later by law in 1995 was the result of many years of efforts by several people, including María Teresa Koberg, Peter Pritchard, Louis Wilson, Mario Boza, Clara Padilla, Jim Spotila, and Frank Paladino. For a more thorough account of this process and the important players, see Spotila and Paladino (2004).

Unfortunately, despite the protection afforded to the leatherbacks, their eggs, and their nesting habitat by the creation of PNMB, the latent effects of the historic egg harvest, compounded with high levels of mortality due to fisheries bycatch, manifested in the inevitable collapse of the population from more than 1,000 individual females annually in the late 1980s to fewer than 100 individuals in recent years (fig. 8.1; Santidrián Tomillo et al. 2007). However, the establishment of the national park, administered by officials and park rangers of the Área de Conservación Tempisque (ACT) and SINAC/MINAET, and a long-term research and monitoring program led by Spotila and Paladino at Playa Grande and by Elizabeth Vélez and Rotney Piedra at Playa Langosta (Piedra et al. 2007) have ensured the comprehensive protection of leatherbacks and their nesting habitat to prevent complete extinction of this population (Santidrián Tomillo et al. 2008).

Meanwhile, the leatherback population nesting on the Pacific coast of México has suffered a similar decline in numbers due to the effects of similar threats of egg harvest and fisheries bycatch (plates 11 and 12), but also including harvest of nesting females (Sarti Martínez et al. 2007; see chapter 9 in this volume). In contrast to PNMB, leatherback nesting beaches in México are extremely long and difficult to patrol and police. Without the infrastructure for enforcement of conservation regulations provided by a national park in a discrete area, threats to leatherbacks have been much more difficult to address on Mexican nesting beaches than in PNMB, despite enormous, admirable efforts by Mexican biologists and conservationists for the past two decades (Sarti Martínez et al. 2007; see chapter 9 in this volume). For these reasons and because of the integrated efforts of the people described below, PNMB is recognized as the most plausible site for eventual recovery of leatherbacks in the eastern Pacific.

The New Threat: Coastal Development

Despite PNMB's advantages, it is an anomaly in many ways: it includes a fairly large marine component extending twelve nautical miles from shore, a coastal mountain, and a complex of mangrove estuaries, but the protected nesting beach zone extends only 125 m from the mean high tide line, and development is fast encroaching on existing open beachfront land. The land within this narrow strip of beach and bordering vegetation is currently the subject of an intense struggle between conservationists and developers, pitting the future of the endangered leatherbacks that nest there against the economic interests of the tourism and development industries in Costa Rica.

In recent years, unsustainable coastal development has replaced egg harvest as the principal threat to the present and future sanctity of critical leatherback nesting habitat at PNMB. In the town of Tamarindo, located across Tamarindo Bay from Playa Grande, impacts associated with beachfront development, including pollution from artificial light as well as solid and chemical wastes, unsustainable water consumption, increased number of scavenging domestic and feral animals (e.g., dogs), beach erosion, deforestation, and loss of mangrove estuary habitat, have occurred within the past two decades. Playa Tamarindo once hosted significant leatherback nesting in the 1970s and 1980s but now hosts only numerous hotels directly on the beach, as well as thousands of residents and tourists. Not a single leatherback has nested at Playa Tamarindo since the beach was robbed of its sand for construction of paved roads and new buildings and became bathed in artificial light about fifteen years ago.

In stark contrast to the brightly lit Tamarindo beachfront, much of Playa Grande's beach habitat remains relatively pristine, maintaining a "green curtain" that creates a dark, natural backdrop for nocturnal leatherback nesting. However, in response to the thousands of tourists that visit PNMB each year, development is increasing rapidly at the northern and southern ends of Playa Grande, as well as in the park's buffer zone and im-

FIGURE 8.3. Juxtaposition of high levels of coastal development in Tamarindo, where leatherbacks no longer nest (left), and essentially pristine nesting habitat on Playa Langosta, where leatherbacks continue to nest (right). Lights and noise from Tamarindo flood Langosta at night, because the two are separated by the narrow mouth of the Langosta mangrove estuary.

mediately surrounding areas; dozens of hotels, houses, and restaurants are already in use, and there are new plans to construct more than 300 additional housing units in the upcoming years. This specter of development, having already extirpated leatherback nesting from Playa Tamarindo and other beaches in the region, has cast a new shadow on the prospect of successful recovery and persistence of leatherbacks at PNMB. Figure 8.3 illustrates this stark contrast: Tamarindo is just across the narrow mouth of the Langosta mangrove estuary from Playa Langosta, and lights and noise from Tamarindo flood the otherwise wild beach habitat of Langosta at night.

In 2002, an environmental impact statement was sought by an enormous development project proposing to build 186 condominium residences in the middle of Playa Grande, also the principal nesting hotspot for leatherbacks in the park, as part of its requirements to obtain building permits from the Costa Rican government. This proposal, with its potentially disastrous results for leatherbacks and their nesting habitat, was rejected because of strong opposition from local communities, park rangers, scientists, and conservationists. This episode served as a wake-up call for several groups involved in conservation at PNMB, including the PNMB administration, community groups, scientists, and nongovernmental organizations (NGOs).

Beginning in 1991 with the establishment of PNMB, Costa Rica declared its commitment to the protection of the leatherback and its critical nesting habitat. Today, many years after the park's creation, its consolidation has become the flashpoint for an intense struggle among developers and investors, conservationists and scientists, the local community and the government. To facilitate consolidation of the open land within the national park, Drs. Spotila and Paladino formed the Leatherback Trust in 2000, a nonprofit organization registered both in the United States and in Costa Rica, including administration and participation by Costa Rican citizens. One of the first actions of this organization was to fund the development of the management plan for PNMB by the Tropical Science Center and the ACT. The Leatherback Trust's Costa Rican arm, directed by Mario Boza, has also supported park activities and various local communities through capacity-building initiatives and environmental education projects. In addition, the Leatherback Trust has executed an energetic public relations campaign to keep the public informed about how various activities and actions will help or hurt PNMB. The primary activity of the trust has been to raise funds to facilitate the Costa Rican government's expropriation of contested open land within PNMB. The Leatherback Trust and other allies of PNMB, including PRETOMA (Programa Restauración de Tortugas Marinas) and the Costa Rican National Network for the Conservation of Marine Turtles, have persisted in the urgent need to consolidate the PNMB against intense and relentless resistance from investors and developers with interests in the area.

Despite monumental legal and philanthropic efforts by the Leatherback Trust, as well as ample available funds, most of the expropriations have not been undertaken at the time of this writing. This lack of action has occurred for many reasons, including insufficient government funds, but principally because expropriation of land from private landowners, despite the land being within a national park area, is a complicated, delicate, tenuous, and extremely contentious process for the government to execute, especially for a democratic country friendly to foreign investment like Costa Rica. However, in 2008 the Costa Rican Supreme Court ruled that because national law calls for absolute protection of biodiversity, the government is obligated to proceed with the acquisition of this land to carry out the consolidation of PNMB. Despite the complications, the expropriation process is continuing, with the imminent acquisition of eleven plots of land (pending a final judicial resolution) in critical areas for leatherback nesting within PNMB.

In addition to the terrestrial consolidation efforts, complementary efforts by other governmental organizations and NGOs, such as UNESCO, Conservation International, and MarViva, have strengthened the network of marine protected areas under the jurisdiction of the governments of Costa Rica, Panamá, Colombia, and Ecuador, within the Eastern Tropical Pacific Seascape (Shillinger 2005). Fortunately, the consolidation of PNMB is also being bolstered by the collective work of a diverse group of conservation-minded people also working to ensure a future for PNMB and its leatherbacks.

Equipo Baulas: An Integrated Team for Leatherback Conservation at PNMB

Successful conservation in PNMB has occurred because of a dynamic, collaborative relationship between groups of people with different backgrounds and experiences. Among the players are park rangers, who are well trained and experienced; local community guides for turtle tours, who work day jobs and take tourists onto the beach at night to see nesting leatherbacks; biologists, who patrol the nesting beach nightly for nesting leatherbacks and conduct scientific research on all aspects of leatherback ecology to inform conservation efforts; and a broad spectrum of others, including local residents and business owners (specifically Carlos Enrique "Kike" Chacón and Yanira Vargas), volunteers from Costa Rica and other countries, administrators, lawyers, and tourists. Although getting all of the pieces to work together effectively has taken several years and continues to require substantial efforts, these seemingly disparate groups currently comprise an effective *equipo*, or team, focusing on leatherback conservation at PNMB.

Park Rangers

Park rangers manage all activities within PNMB each night during the nesting season, including vigilance of tourists for turtle viewing and tours, control of entry to the beach during nocturnal hours, and oversight of research activities. Several of the park rangers are themselves biologists, not only offering support and advice to the permanent research team but also conducting research and monitoring in the park. Since 2006, park rangers have also conducted patrols of the marine sector of PNMB, which is a no-take reserve, in cooperation with MarViva, a nonprofit organization that supports marine protected area management throughout the eastern Pacific. During this time, a cooperative agreement between MarViva and MINAET established marine patrols that involve MarViva personnel and resources, PNMB park rangers, and Costa Rican Coast Guard officials. Conservation International and UNESCO have facilitated the acquisition of a new boat and radar for ACT to allow patrols to continue in the marine sector of PNMB. Through these combined efforts, fishing and boating impacts have been dramatically reduced within PNMB. Additionally, this initiative has led to collaborative research in ecosystem function in PNMB between park rangers and Costa Rican National University researchers.

In addition to the oversight and enforcement of national park rules during turtle nesting on the beach and in the marine sector, park rangers also deal with issues of land use (e.g., construction) in and around the park. PNMB rangers document and cite violators of illegal vegetation clearance and filling of mangrove estuaries for purposes of building. Rangers also assist in development and implementation of environmental education programs, solid waste management, community outreach, and general protection efforts. Thus, the cumulative responsibilities of a PNMB ranger are year-round, rather than just during leatherback nesting season, as is the case for all of the other groups involved in leatherback conservation in PNMB. Unfortunately, due to the aforementioned shortage of people and funding available to PNMB, park rangers are typically stretched thin and are unable to prevent all infractions at all times. In fact, budget limitations at the beginning of 2009 resulted in a decrease in PNMB staff from six or more park rangers to only four, which presents yet another significant challenge for effective management of the complete protected area.

Local Community Guides

In addition to the strong influence and important roles played by people from outside Playa Grande and outside of Costa Rica, many members of the local community also have recognized the importance of conserving the leatherbacks nesting on the beaches in their "backyards." Before the formation of the national park, Don Idanuel Contreras, Don Santos Arrieta Arrieta, and Don Luzgardo Rosales Gutiérrez recognized that a healthy population of nesting leatherbacks represented more consistent income for the community than an extirpated one, and they formed a local guide association based in the town of Matapalo, about 7 km from Playa Grande. Today, the Matapalo Local Guide Association is composed of long-term members who had previously participated and benefited economically from harvesting eggs, as well as their children and grandchildren, members of a younger generation with a perception of turtles as a source of income only through tourism, not egg harvest. Similarly, another local guide association from Tamarindo also leads turtle and mangrove tours in PNMB. The Tamarindo Guide Association, originally created by the merger of three local associations committed to conservation in the PNMB area, has been led by Warren Chacón, Franklin Barrantes, Gerardo Santana, and Enrique Chavarría, and like the Matapalo guide association, its membership includes former egg harvesters as well as younger community members. This paradigm shift holds tremendous promise for a sustained conservation effort from the local community for the leatherbacks and PNMB.

Together, the Matapalo and Tamarindo guide associations administer to thousands of tourists per nesting season, both from Costa Rica and from several other countries and continents, using the income garnered from tourist fees for community development projects. Examples of such projects in Matapalo include a fence around the elementary school to keep children out of street traffic, a public address system for the town church, and renovation of the town meeting hall. In addition, nearly half of the Matapalo Local Guide Association's members are women, which is an intentional attempt to take advantage of the disproportionately large influence women have on children in Costa Rican society and thereby effectively instill a conservation ethic in the next generation. The guide associations have also broadened their scope to raising environmental awareness and promoting sustainable resource use with respect to other relevant local issues, such as slash-and-burn agriculture and water consumption.

In addition to the Matapalo Local Guide Association, the Matapalo Women's Association was created in 2004 in support of the PNMB. The women's association performs public awareness campaigns in their town, informing their neighbors about the plight of the leatherbacks and the benefits of the national park to conservation and to their community. The association also provides home-cooked food and coffee as well as souvenirs and other items for sale to the tourists who wait for their chance to go onto Playa Grande to view a nesting turtle. This group is another example of opportunities for community empowerment provided by the existence of PNMB.

Biologists

Boza (1993) and Spotila and Paladino (2004) noted the important role that foreign scientists and conservation groups have played in international conservation issues, especially in Costa Rica. PNMB's history is no exception. The permanent research and monitoring presence began in 1988, when Spotila and Paladino, on a recommendation from Boza, began conducting physiology studies at Playa Grande and Playa Langosta. Within three years, Spotila and Paladino had initiated a population monitoring project, supported by volunteers from the Earthwatch Institute's Center for Field Research. Today, Spotila and Paladino continue to oversee the Playa Grande Leatherback Conservation and Research Project, as dozens of Earthwatch volunteers assist the team of field biologists, mostly undergraduate and graduate students, in nightly patrols and daily activities on Playa Grande and Playa Ventanas during the leatherback nesting season.

The third nesting beach in PNMB, Playa Langosta, is monitored by biologists, park rangers, and volunteers associated with PNMB and is in close contact with the project at Playa Grande. The monitoring at Langosta began in 1991–1992, when Anny Chaves and colleagues from the University of Costa Rica recorded the first leatherback nesting data for the site. The monitoring temporarily ceased until 1994, when Stanley Rodriguez, Guiselle Monge, Elizabeth Vélez, and Rotney Piedra (the latter three biolo-

gists from the National University of Costa Rica) established nest protection efforts. Two years later, Monge, Vélez and Piedra initiated the research
and monitoring project with support from Paladino and Spotila and the
Wildlife Conservation Society-Costa Rica. Since 1997, the project has been
supported by Dr. Peter Dutton from the U.S. National Marine Fisheries
Service and Earl Possardt from the U.S. Fish and Wildlife Service, and
since 2008 financial assistance from the U.S. Marine Turtle Conservation
Act's Conservation Grant Program has also played an important role. Vélez
and Piedra have continued to manage the research and conservation project
at Playa Langosta since 1998, and the help of countless volunteers has contributed greatly to the success of this project (Piedra et al. 2007).

The work of the biologists consists of monitoring and protection of
nesting females, their eggs, and their hatchlings during nightly foot patrols
and daytime activities such as nest excavations and beach hatchery work. In
addition to the monitoring and protection efforts, the biologists conduct
research studies on aspects of leatherback life history, nesting ecology, at-
sea behavior and movements, physiology, and population demography.
Over the course of almost 20 years, the research project has produced more
than thirty graduate and honors students and at least fifty scientific publications, and it has become one of the best-known and respected sea turtle research projects in the world. In addition, the information generated by the
biologists is disseminated not only in peer-reviewed scientific literature but
also to park rangers and local guides in information workshops and written
reports. Like the park rangers and local guides, the biologists play an inex-
tricable role in the overall conservation work at PNMB, especially because
of the long-term nature of the research efforts and the valuable information
they have generated.

The Role of Research and Monitoring in Conservation Efforts at PNMB

Research and monitoring are strategically integrated with efforts to
establish and enforce priorities for management within PNMB. At its core,
the research conducted at PNMB allows continuous assessment of the
leatherback population's status. A saturation tagging program began in
1993–1994 (Steyermark et al. 1996) and has resulted in identification (via

passive integrated transponders, or PIT tags) of more than 1,700 individual turtles (Santidrián Tomillo et al. 2007), which has enhanced the understanding of the dynamics of this nesting population (Reina et al. 2002; Santidrián Tomillo et al. 2007, 2008). For example, Spotila et al. (1996) combined data from PNMB with nesting population information from other beaches around the world and concluded that the global leatherback population had declined by almost 70 percent in fewer than twenty years. Moreover, Spotila et al. (2000) used a simple population demographic model based on monitoring data from PNMB to project extinction of Pacific leatherbacks within a few decades if threats continued unabated. These dire projections have justified complete protection of the nesting beach, nesting females, and their eggs and hatchlings. Among these efforts is an egg relocation program and beach hatchery that have been supported by MINAET and the park for almost a decade, resulting in 21,000 hatchlings produced from nests that had been laid below the natural high tide line and presumably would not have survived without intervention (Santidrián Tomillo et al. 2007). However, because protection of eggs and the resulting increased production of hatchlings have limited ability to strengthen the resilience of PNMB's leatherbacks to extinction (Santidrián Tomillo et al. 2008), continued vigilance at PNMB and effective reduction of at-sea threats to leatherbacks are necessary long-term conservation strategies. Thus, information from long-term monitoring at PNMB has allowed biologists, park rangers, and local guides to embrace a strategy of consistency, patience, and optimism with respect to the nesting beach conservation efforts because of the time necessary for the leatherback population to recover.

Although nesting beach conservation has received the majority of the research attention, research on environmental conditions (see chapter 2), at-sea movements, behavior, and migration have also contributed greatly to understanding of leatherback biology and to conservation strategies. Over four seasons from 2004 through 2007, a team of biologists, park rangers, and volunteers led by Dr. George Shillinger, then of Stanford University, deployed satellite transmitters on forty-six leatherbacks (Shillinger et al. 2008, 2010). This research aimed to elucidate a previously described leatherback "migration corridor" (Morreale et al. 1996), as well as spatial habitat use by leatherbacks during their internesting periods in and around PNMB.

This satellite telemetry program has been one of the most successful research initiatives ever realized at PNMB for several reasons. Importantly, the results have directly informed leatherback conservation strategies at local, national, and multinational scales. Because leatherbacks tend to spend much, but not all, of their internesting periods within the boundaries of PNMB (Shillinger et al. 2010), park managers have solidified support for consistent and continued boat-based patrols and have successfully reduced threats to leatherbacks (and other marine animals) from small-scale fishing practices and boat traffic within the marine sector of the park. Furthermore, MINAET and ACT have leveraged the findings that leatherbacks utilize marine habitat along the entire coast of the Nicoya Peninsula during the nesting season to promote the consolidation of a network of marine conservation areas that include existing official marine protected areas as well as coastal communities that will voluntarily participate in marine conservation focused on leatherbacks. The purpose of this initiative is to coordinate management of fishing activities within key marine areas in the region.

In addition to the conservation applications of the internesting habitat use, the new data confirmed and expanded on previous reports of a relatively defined migration route for leatherbacks (fig. 8.4). This detailed description of leatherback migration through the eastern Pacific Ocean underscores the importance of multinational conservation frameworks, such as the Eastern Tropical Pacific Seascape Initiative (Shillinger et al. 2008). Leatherbacks are considered a flagship species of the region, which makes PNMB a particularly important site in the region, considering the eastern Pacific's broad geographic scope and high marine biodiversity.

Additionally, the satellite telemetry work in 2007 represented the most integrated research initiative ever conducted at PNMB with respect to the composition of the research team. Biologists and park rangers first held training workshops to share methods for attaching tracking instruments to sea turtles, especially satellite transmitters to leatherbacks. Biologists then also provided information from all tagging studies conducted at PNMB and engaged park rangers in preparation of research permits and reports. In addition, biologists delivered information to local guides about the tagging work and past results and made presentations to tourists at PNMB. During transmitter deployments, tagging teams always consisted of biologists and park rangers, and sometimes worked in front of tourists and local guides.

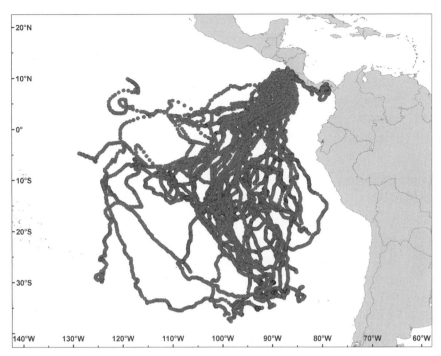

FIGURE 8.4. Postnesting migrations of forty-six leatherbacks from PNMB during 2004–2007. Figure courtesy of G. L. Shillinger.

This level of interaction and cooperation would have been unthinkable in the first years of the research project, but in recent years the close communication between research and management has provided the glue to keep integrated conservation efforts together in PNMB.

A New Group in the Mix: Local Residents of Playa Grande and Playa Ventanas

A recent and encouraging development in the struggle to consolidate the national park has been the organization of a motivated and concerned group of local community members toward effective conservation of this protected area and the biodiversity—especially leatherbacks—that it hosts. This group of residents has formed the Comité Bandera Azul Ecológica (Ecological Blue Flag Committee) to support the PNMB administration in management of the protected area, in large part because they recognize the value of responsible development and conservation of natural resources in and around the protected area. Specifically, the committee concentrates on four issues: (1) an action plan for controlling domestic animals (e.g., dogs) on the park's beaches, especially during the leatherback nesting season; (2) a plan for reforestation and restoration of altered areas, particularly beachfront areas lacking natural vegetation, or "green curtains," using native plant species; (3) development of a program for managing solid wastes; and (4) a plan for comprehensive signage throughout the park to improve orientation and information delivery to visitors. Through these activities, this group of local community members is trying to join forces with the park administration to maintain community development in harmony with the national park and conservation of its natural resources, and, in particular, its turtles. In addition, the Playa Grande Association for Community Development was formed in early 2010 to promote community improvement projects and is becoming increasingly integrated with activities of the Comité Bandera Azul Ecológica and the park administration. In a concrete example of this collaboration, the park administration together with the Comité Bandera Azul Ecológica, local residents, biologists, and other local groups participated in a joint social event in June 2010 during which participants planted native vegetation in front of existing houses on Playa Grande, marking the beginning of an ongoing reforestation initiative for PNMB.

Equipo Baulas: The Leatherback Team

Although understanding that people with roles as distinct as park rangers, local guides, local community members, and biologists share a common goal of protecting leatherbacks and their habitat is easy in concept, integrating these roles in a cohesive, effective way is much more difficult in practice. Since the park's inception, the relationship between these groups has fluctuated between tension and overt conflict, indifference toward one another, and even harmonious cooperation. However, after time and through much effort, each group has come to understand and embrace its unique role, leading to the current situation in which the collective energy and enthusiasm for conservation is palpable.

How did this *equipo*, or team, come to be? As in many comparable cases, open dialogue, information dissemination, and joint participation in various activities facilitated the cohesion of the different groups around the common conservation cause. To strengthen the working relationship, the park rangers have asserted their role and increased their visibility, responsiveness, and effectiveness as administrators. The biologists, being foreigners predominantly from the United States, have increasingly reached out to the park rangers, local community, and local guides to earn the trust and support of their Costa Rican colleagues. For example, workshops are held before the start of the nesting season where the biologists provide a status report on the leatherback population, results from the previous season's research, and plans for the present season's research objectives to the local guides and park rangers. Similar meetings take place intermittently throughout the season as well, for all groups to discuss progress and solve any problems that might have arisen. This important work, which started in earnest in the early to mid-1990s by Tony Steyermark, Richard Reina, and Philippe Mayor, has been continued by Pilar "Bibi" Santidrián Tomillo, Vince Saba, Bryan Wallace, Gabriela Blanco, and Tera Dornfeld, among many others.

In addition to work-related efforts, various social activities have also taken place. Several *fiestas* to celebrate the end of the nesting season, holidays, and other special events have involved all groups in informal, more personal settings. Additionally, the annual Leatherback Festival, organized by Elizabeth Vélez and park ranger Carlos Díaz, involves the local communities, especially schoolchildren, private property owners, local businesses, NGOs, artists, universities, and visitors to PNMB to celebrate the turtles and the efforts to protect them. The social networking efforts even manifested in a PNMB soccer team composed of park rangers, local guides, residents, and biologists that has participated in a full season as well as several weekend tournaments since 2004. Through this combination of work-related and social activities, the *Equipo Baulas* has consolidated itself as several people from diverse backgrounds and experiences have recognized and rallied around the common purpose of protecting leatherbacks and leatherback habitat for future generations—turtle and human.

The Struggle for the Land Rages On: The Future of PNMB

PNMB, with its collaboration among park rangers, local guides, local communities, and scientists, has successfully achieved elimination of egg harvest, absolute protection of nesting leatherbacks, and a safe haven for leatherbacks in the ocean within the national park's marine sector. It is a place where scientific research and conservation as well as burgeoning tourism thrive. However, given the imminent—but not inevitable—threat of unsustainable coastal development, the future of leatherbacks and their nesting habitat at PNMB is uncertain. We envision three possible scenarios, and urge careful consideration of their respective advantages and disadvantages.

The Rise of Unchecked Development and the Downfall of PNMB

Under the first scenario, the government undertakes no further expropriations to consolidate the open land within the PNMB, which leads to continued construction of houses, condominiums, hotels, and restaurants. Developers and real estate agents make enormous profits, and the residential (and largely foreign) population of Playa Grande swells to several hundreds, even thousands. Tourists flock to the area in the thousands to take advantage of its many amenities and beautiful scenery. However, as development increases, so do the number of lights, the amount of waste produced, and the volume of freshwater consumed. The water table suffers saltwater intrusion, drinking water becomes contaminated by human wastes, and local communities' artisanal wells dry up. (All of the above have already happened in the area, particularly in Tamarindo.) Loss of beach vegetation related to beachfront construction leads to increased beach erosion and light pollution, and an increased number of domestic and feral animals (e.g., dogs) freely roam the beach eating leatherback eggs and hatchlings. Leatherback nesting continues to decrease due to this litany of impacts; tourism for observation of leatherback nesting declines in tandem. Soon, Playa Grande resembles its neighbor to the south, Playa Tamarindo: lots of lights, lots of hotels, lots of people, lots of tourism, but no turtles and no associated benefits to local communities. Regrettably, Costa Rica suffers significant damage to its image as a model for biodiversity conservation because of its failure to prevent the extirpation of one of its flagship species.

Unfortunately, there have been indicators of this first scenario becoming reality. Toward the end of President Oscar Arias's second presidential term (2006–2010), a new proposal was put forth—largely by people with development interests—with the intention of redefining the current boundaries and land use regulations of the protected area and the disputed land. This maneuver followed previous failed attempts to introduce proposals in the Costa Rican Legislative Assembly to reduce or change current PNMB boundaries and development regulations. Specifically, this new proposal would have freed the privately owned land within the current limits of the park from the expropriation process, and instead—in a change from previous proposals—would convert the land's official status to Mixed Category Wildlife Refuge, which would include a 150-m-wide strip of land running parallel to the 50-m national park boundary along the beaches. Thus, this revised protected area would have included the public (i.e., national park) and public–private (i.e., mixed refuge) sections, but it would also theoretically have grown in total area with the inclusion of the private land. According to the proposal's proponents, owners of these properties would keep their rights to develop these properties but could only do so under guidelines to maintain low-density, low-impact development.

At first glance, this proposal might have appeared to be an attempt to resolve the conflict over open land within park boundaries that has plagued PNMB for several years. However, careful scrutiny reveals that this scenario would drastically jeopardize the sanctity of PNMB and the persistence of leatherbacks, and could inadvertently and wrongly set a dangerous precedent for diminishing protected status of other national parks in Costa Rica, especially in other situations where private in-holdings within park boundaries have yet to be compensated. In addition, PNMB park rangers, local guides, biologists, NGOs, and some current residents and neighbors of the Playa Grande area all voiced strong opposition to this proposal. These stakeholders object for several reasons: the proposal lacks support of scientific studies of environmental impacts; it is not a solution that considers the ecosystem in its entirety; it fails to adequately address the potential negative impacts of high-density, unsustainable development in the area; and rare coastal dry forest areas would lose their protected status.

In a surprising—and alarming—chain of events, the proposal reached Costa Rica's Legislative Assembly floor for a vote just days before the new administration of President Laura Chinchilla took over, despite the numerous legal precedents clearly in favor of consolidation of PNMB and against previous attempts to downgrade its status. Fortunately, the vote was deferred due to lack of sufficient information about environmental impacts and other concerns. The proposal now falls to the new legislative period, during which we hope that further analysis of existing arguments against it will force it to be shelved once and for all. However, the simple fact that the proposal made it through the legislative process to the point of an imminent floor vote underscores the urgency and high stakes of the current threats to PNMB represented by the first scenario.

Consolidation of PNMB: A Victory for Leatherbacks and Conservation in Costa Rica

Alternatively, under a second scenario, the Costa Rican government, led by MINAET's efforts, demonstrates its dedication to biodiversity conservation as well as to Costa Rican communities by executing the expropriations of the remaining open land in PNMB, thereby consolidating the park in perpetuity. With the threat of development within PNMB removed, the powerful integration of other human components, including national park administrators, international and national researchers and conservationists, local community members (e.g., ecotourism guides, local associations, other local residents), and volunteers from within and outside of Costa Rica, continues to create hospitable conditions for recovery of leatherbacks in the eastern Pacific. Development in areas adjacent to PNMB is strictly regulated to low-density, low-impact projects that assure protection of natural areas, resources, and biodiversity and thus preserve ecological integrity within the region. (In 2009, the Constitutional Chamber of the Costa Rican Supreme Court issued a resolution ordering the government to study the potential environmental impacts of development within the buffer zone of PNMB. The results of this effort could be extremely beneficial for promoting harmony between future development and the environment in and around PNMB.) Over the next few decades, leatherback nesting numbers increase steadily. The strength and permanence of the national park and its leatherbacks continue to provide opportunities for empowerment of the local community through turtle tours and sales of souvenirs and crafts and authentic foods and drinks to the consistent tourist influx, generating reliable revenue for local communities.

An Effective Compromise: Adaptive Management of a Consolidated PNMB

Considering the dire consequences of the first scenario, as well as the persistent obstacles (e.g., legal actions, lack of resources) to realizing the second scenario, it is worth considering a third scenario that would still achieve the most fundamental protections for PNMB while not executing the full expropriation and consolidation process. A tiered strategy could be adopted that addresses the differences in current land use within PNMB; in addition to the undeveloped land areas, at least thirty-three houses and two hotels exist within park boundaries. Under this third scenario, because the Costa Rican government lacks the monetary resources to undertake the expropriations envisioned under the second scenario, open or undeveloped areas become the primary target of PNMB's consolidation. Furthermore, this scenario would exclude from expropriation land with buildings, but these areas would be managed within a strict regulatory framework to minimize environmental impacts on the nesting beaches and adjacent natural areas. Clearly, this adaptive management scenario would require a rigorous analysis incorporating many factors, such as existing and potential future environmental impacts, mitigation of such impacts, the management process, the conditions of such an agreement, and costs of its implementation. Moreover, it would require implementation of a new legal precedent for the PNMB, due to the fact that Costa Rican law currently does not permit any private property within national parks, which are under complete protection and administration by the government.

Along these lines, and in response to the developers' proposal described above, local residents have articulated that, while they support some form of a compromise approach of a protected area that would safeguard the sanctity of the national park, they do not support unchecked development within PNMB. However, the official stance of the local Association for Community Development maintains that complete consolidation of all properties within PNMB boundaries—although an ideal outcome—is unlikely due to a lack of resources to undertake expropriations, thus making a more tenable solution necessary. Therefore, multiple opinions of how best to balance development and conservation interests are present within the Playa Grande community, clearly making resolution of this situation extremely challenging. Nonetheless, a compromise scenario that safeguards the park and the natural areas and species it protects and reflects realities of the current impediments to consolidation is emerging as the most likely candidate to break the impasse regarding PNMB.

Conclusions

We emphatically contend that the first scenario—which was nearly realized through the developers' proposal—would be disastrous for the PNMB and other national parks, for the turtles, for local communities who currently benefit directly from the relative serenity, scenic beauty, and associated tourism of PNMB, and for Costa Rica's reputation as a global leader in conservation. In contrast, we vigorously support efforts to realize the second scenario, in which PNMB is consolidated to protect leatherback nesting habitat and associated coastal ecosystems in perpetuity. However, we acknowledge efforts to explore a compromise to the first two scenarios, wherein open disputed land would be expropriated and consolidated as part of the national park while existing structures within the current boundaries of the PNMB are allowed to remain under a special management category, but must comply with certain strict conditions that minimize their impact on the protected area and biodiversity (e.g., beach vegetation, minimization of light pollution, etc.). While the Costa Rican government's lack of financial resources has been mentioned as an obstacle to executing PNMB's consolidation, a coordinated, strategic international initiative involving NGOs, foundations, and other partners, all with necessary governmental oversight, could be carried out to raise and implement sufficient funds to finance the consolidation of the open land within park boundaries.

Future discussions must include full participation of stakeholders with various interests, including government (e.g., MINAET, members of the Costa Rican Legislative Assembly), scientists, conservation, the local community, and the private sector, to ensure a solution that will benefit conservation efforts and human interests, but without the latter coming at the expense of the former. In any case, we emphasize that complete protection of leatherback turtles and their nesting habitat should remain the primary objective when considering *any* development in the area. While the future remains unclear, the situation for leatherbacks at PNMB, as for sea turtles in the eastern Pacific, depends on how the human groups involved respond to the challenges and opportunities. In light of the long history of humans and leatherbacks in PNMB, we hope that future generations of people and turtles will continue to call the area home.

ACKNOWLEDGMENTS

We acknowledge *Equipo Baulas* for its hard work and dedication to a world in which people and leatherbacks thrive together. We thank the pioneers and champions of PNMB, without whose efforts there would be no national park and no leatherbacks: Fabricio Álvarez, Santos Arrieta Arrieta, Doña Ligia, Franklin Barrantes, Gabriela Blanco, Don Mario Boza, Guillermo Briceño, Ricardo Calderón, Carlos Enrique Chacón and Yanira Vargas, José Miguel Carvajal, Luis Castro, Enrique Chavarría, Idanuel Contreras, Bernal Cortéz, Roberto and Luis Corea, Mariela Cruz, Carlos Díaz, Tera Dornfeld, Peter Dutton, Álvaro Fonseca, María Teresa Koberg, Alexander Madrigal, Phillippe Mayor, Cecilia Mesen, Guiselle Monge, Jennifer and Penelope Neeve, Clara Padilla, Frank Paladino, Earl Possardt, Peter Pritchard, Patrick Opay, Mauricio Ramírez, Richard Reina, Esperanza Rodríguez, Norma Rodríguez, Ademar Rosales, Genaro Rosales, Laura Jaen Rosales, Luzgardo Rosales, Teodora Rosales, Vince Saba, Gerardo Santana, Bibi Santidrián Tomillo, Jim Spotila, Tony Steyermark, Elizabeth Vélez, Louis Wilson, Judy Zabriskie, and the many other park rangers, local guides, members of the local community, and volunteers who have worked in PNMB.

REFERENCES

Boza, M. 1993. Conservation in action: past, present, and future of the national parks system of Costa Rica. Conservation Biology 7:239–247.
Campbell, L. M. 1998. Use them or lose them? Conservation and the consumptive use of marine turtle eggs at Ostional, Costa Rica. Environmental Conservation 25:305–319.

Campbell, L. M. 2007. Understanding the human use of olive ridleys: implications for conservation. In: Plotkin, P. T. (Ed.), Biology and Conservation of Ridley Sea Turtles. Johns Hopkins University Press, Baltimore, MD, USA. Pp. 23–44.

Chan, E. H., and H. C. Liew. 1996. Decline of the leatherback population in Terengganu, Malaysia, 1956–1995. Chelonian Conservation and Biology 2:196–2003.

Instituto Costarricense de Turismo. 2005. Tourism Statistical Yearly Report. Instituto Costarricense de Turismo, San José, Costa Rica.

Morreale, S. J., E. A. Standora, J. R. Spotila, and F. V. Paladino. 1996. Migration corridor for sea turtles. Nature 384: 319–320.

Piedra, R., E. Veléz, P. H. Dutton, E. Possardt, and C. Padilla. 2007. Nesting of the leatherback turtle (*Dermochelys coriacea*) from 1999–2000 through 2003–2004 at Playa Langosta, Parque Nacional Marino Las Baulas de Guanacaste, Costa Rica. Chelonian Conservation and Biology 6:111–117.

Reina, R. D., P. A. Mayor, J. R. Spotila, R. Piedra, and F. V. Paladino. 2002. Nesting ecology of the leatherback turtle, *Dermochelys coriacea*, at Parque Nacional Marino Las Baulas, Costa Rica: 1988–1989 to 1999–2000. Copeia 2002:653–664.

Santidrián Tomillo, P., E. Vélez, R. D. Reina, R. Piedra, F. V. Paladino, and J. R. Spotila. 2007. Reassessment of the leatherback turtles (*Dermochelys coriacea*) nesting population at Parque Nacional Marino Las Baulas, Costa Rica: effects of conservation efforts. Chelonian Conservation Biology 6:54–62.

Santidrián Tomillo, P., V. S. Saba, R. Piedra, F. V. Paladino, and J. R. Spotila. 2008. Effects of illegal harvest of eggs on the population decline of leatherback turtles in Parque Nacional Marino Las Baulas, Costa Rica. Conservation Biology 22:1216–1224.

Sarti Martínez, L., A. R. Barragán, D. García Muñoz, N. García, P. Huerta, and F. Vargas. 2007. Conservation and biology of the leatherback turtle in the Mexican Pacific. Chelonian Conservation and Biology 6:70–78.

Shillinger, G. L. 2005. The Eastern Tropical Pacific Seascape: an innovative model for transboundary marine conservation. In: Mittermeier, R., et al. (Eds.), Transboundary Conservation, A New Vision for Protected Areas. Cementos Mexicanos (CEMEX), Conservation International, Agrupacion Sierra Madre, Washington, DC. Pp. 320–331.

Shillinger, G. L., D. M. Palacios, H. Bailey, S. J. Bograd, A. M. Swithenbank, P. Gaspar, B. P. Wallace, J. R. Spotila, F. V. Paladino, R. Piedra, S. A. Eckert, and B. A. Block. 2008. Persistent leatherback turtle migrations present opportunities for conservation. PLoS Biology 6(7):e171.

Shillinger, G. L., A. M. Swithenbank, S. J. Bograd, H. Bailey, M. R. Castelton, B. P. Wallace, J. R. Spotila, F. V. Paladino, R. Piedra, and B. A. Block. 2010. Identification of high-use internesting habitats for eastern Pacific leatherback turtles: role of the environment and implications for conservation. Endangered Species Research 10:215–232.

SINAC (Sistema Nacional de Areas de Conservacion). 2011. Sistema Nacional de Áreas Protegidas. Available: www.sinac.go.cr/planificacionasp.php (accessed 18 November 2011).

Spotila, J. R. 2004. Sea Turtles: A Complete Guide to Their Biology, Behavior, and Conservation. Johns Hopkins University Press, Baltimore, MD.

Spotila, J. R., A. E. Dunham, A. J. Leslie, A. C. Steyermark, P. T. Plotkin, and F. V. Paladino. 1996. Worldwide population decline of *Dermochelys coriacea*: are leatherback turtles going extinct? Chelonian Conservation and Biology 2:209–222.

Spotila, J. R., and F. V. Paladino. 2004. Conservation lessons from a new national park and from 45 years of conservation of sea turtles in Costa Rica. In: Frankie, G. W., A. Mata, and S. B. Vinson (Eds.), Biodiversity Conservation in Costa Rica. Learning the Lessons in a Seasonal Dry Forest. University of California Press, Berkeley. Pp. 194–209.

Spotila, J. R., R. D. Reina, A. C. Steyermark, P. T. Plotkin, and F. V. Paladino. 2000. Pacific leatherback turtles face extinction. Nature 405: 529–530.

Steyermark, A. C., Williams, K., Spotila, J. R., Paladino, F. V., Rostal, D. C., Morreale, S. J., Koberg, M. T., and R. Arauz. 1996. Nesting leatherback turtles at Las Baulas national park, Costa Rica. Chelonian Conservation and Biology 2:173–183.

UNDP (United Nations Development Programme). 2007. Human Development Report 2007/2008. Available: http://hdr.undp.org/en/media/HDR_20072008_EN_Complete .pdf.

Wallace, B. P., and V. S. Saba. 2009. Environmental and anthropogenic impacts on intraspecific variation in leatherback turtles: opportunities for targeted research and conservation. Endangered Species Research 7:11–21.

9

Nesting Beach Conservation in the Mexican Pacific

The Bridge between Sea Turtles and People

ANA REBECA BARRAGÁN

Summary

For millennia, sea turtles have had an intimate relationship with people living close to the ocean. This is especially true for the populations of the Mexican Pacific, both human and turtle. When the balance of that relationship was broken by commercial capture and extensive commerce of eggs (plate 12), both sides were in trouble: sea turtle populations declined, and human populations were deprived of an important natural resource. Early conservation efforts aimed at recovering the depleted nesting populations were organized by several institutions that focused on protection of eggs and females. In the long term, it made no sense to exclude the local coastal residents—turtles were part of their culture and environment. Moreover, given the wide geographic range of most sea turtle species, the protection projects at a few important beaches were clearly not enough to recover sea turtle populations. This chapter tells the stories of significant conservation projects in nesting beaches of the Mexican Pacific, the trends of the sea turtle populations they work with, and the principle behind their success. In the end, it is clear that networking is the key to saving both sea turtles and coastal communities.

Introduction

Imagine yourself being on a field trip to the Mexican coast during the 1970s. Had you paid a night visit to any sea turtle nesting site, you'd witness a common sight: the beach looked like a turtle egg flea market. Dozens of poachers collected eggs on horseback and sold them to buyers that waited by their trucks; women and children walked the beach selling coffee and bread to the men. No doubt, the local economy heavily depended on the easy harvest; no wonder the sea turtle populations were declining since no nest was left untouched. The female turtles, oblivious to the activity around them, kept emerging from the ocean, driven by the same ancient need to reproduce that has kept their species alive for millions of years.

With time, and having gained trust from the local people, you would learn how close the relationship of these communities with the sea turtles actually was. Most towns were originally settled as close as they could to the beach, in order to have easy access to the turtles. In some areas, each community had the right to ask for a period of "exclusivity" in case of need; nobody else could harvest turtle eggs for that period. The revenues were used for construction of public buildings and other community improvements. If you are a nature lover, you would have felt helpless and sorry for the fate of the magnificent marine reptiles.

But this story is actually more complex than it might seem. Stories like this have come to show us that a beach is, above all, an interface between two worlds. The sea turtles, mainly oceanic creatures, come to the beach to fulfill a small but extremely important fraction of their life cycle: reproduction in the form of nesting, egg incubation, and emergence of hatchlings. Humans, mainly terrestrial creatures, come to the beach in search for much valued resources: food, materials for housing, and even aesthetic values when the turtle shells and bones were used for jewelry and decoration (plate 12). It is natural that for millennia the great majority of the interactions between these worlds, for good and bad, occurred at the nesting beaches.

The beaches of México are very rich in sea turtle diversity: all but one of the world's species can be found there, and México's Pacific beaches, in particular, host important nesting populations of four species: the leatherback (*Dermochelys coriacea*), the olive ridley (*Lepidochelys olivacea*), the east Pacific green (*Chelonia mydas*), and the hawksbill (*Eretmochelys imbricata*) (plates 1–9). A fifth species, the loggerhead (*Caretta caretta*), forages in the Mexican Pacific waters off Baja California after an impressive trans-Pacific migration from its nesting beaches in Japan (Peckham et al. 2008; plate 10).

The Mexican Pacific coast is also rich in cultural traditions related to sea turtles. Many nesting beaches in the area have at least one human community associated with them, and in many cases this is not mere coincidence, but instead a result of the dependence on marine resources (i.e., turtle products) by humans. From the beginning of human settlement in the country, people were attracted to the abundant (and easy to harvest) food supply in the form of turtle eggs and nesting females. Carapaces and bones became household appliances, and objects used in trade, medicine, and art; sea turtles quickly became part of the human (i.e., material) world and cosmic view of life. For example, some coastal indigenous cultures, such as the Comcáac (Seri) in Sonora and the Huaves in Oaxaca, view sea turtles as integral in their traditions regarding the creation of the world. We find evidence of the importance of sea turtles not only in the coastal areas but also as far inland as the Central Plateau of México, heart of the Aztec empire. Ancient writings give testimony of trade between the Aztecs and the people from the coast, who provided shells from the "yellow turtle" and the "tiger turtle"; coastal towns paid an annual tribute to the Aztec emperor, which was composed of all sorts of marine creatures, including sea turtles (León-Portilla 1972).

There are few records of the level of harvest during these ancient times, but the size of the nesting populations at the start of recorded history suggests that the turtles could cope with the pressure just as they did with other predators. However, as human populations grew, so did the demand of sea turtle products that were already part of traditional cultural heritage. The situation reached a turning point in the 1960s, when the Mexican government banned the commerce of crocodile skin for use in the leather industry, and the item was substituted by sea turtle skin. And once a country-wide turtle fishery was established, it was not long before a market for other turtle products such as meat and turtle oil developed, both nationally and internationally. Although the olive ridley was the species that largely sustained the turtle fisheries, all the hard-shelled species suffered from the increase of the new-found industrial-scale demand. The industrial exploitation added to the widespread traditional use of sea turtle products in small coastal communities throughout the Mexican coast, and especially in the states of Oaxaca and Guerrero in the Pacific, which have many ethnic groups with a strong cultural attachment to sea turtles. Turtle populations crashed, and these coastal groups saw a very valuable resource start to disappear. At the same time, perturbations to the ecosystem functions of beaches affected nesting sea turtles. In many cases, land adjacent to nesting beaches was sold to tourist developers or denuded of native vegetation to grow grasslands for cattle ranching. Ultimately, the traditional way of life of communities changed; people struggled to keep up with the changes and, by doing so, frequently further damaged the relationship they had with the ancient reptiles.

While these dramatic declines among sea turtle populations and the increasing human encroachment on their habitats were occurring, there was a strong commitment by the Mexican government to protect the turtles as a natural resource, beginning in 1966, when the first turtle protection camps were established. Several conservation initiatives from universities and non-governmental organizations (NGOs) began when it became evident to the Mexican scientific community that the turtle populations were in a state of collapse.

Although all the various conservation efforts are relevant and deserve attention, some stand out for their particular regions and species of focus, because of their success in rekindling the eroded relationship between coastal communities and sea turtles.

The Leatherback Turtle

The leatherback is a unique turtle in almost every way and comes first in many turtle categories: the largest, the fastest swimmer, the deepest diver, and the coldest-water dweller (plate 2). People have always been fascinated by the leatherback, from ancient cultures to modern scientists. This is why it is almost beyond belief that the major nesting sites for the eastern Pacific population (plate 3) were among the last-discovered nesting hotspots for sea turtles in México, when all the other species had conservation projects up and running. Peter Pritchard himself (see the foreword to this volume) wrote that no areas of high nesting concentrations were known for the eastern Pacific leatherback in the early 1970s (Pritchard 1971).

But those high nesting concentration areas in México were well known to egg poachers. Unfortunately for the species, leatherback eggs are the largest in size and demand better prices on the market. It was the seizing of a large truck transporting 50,000 leatherback eggs out of Punta Maldonado, Guerrero, in 1976 that led to the discovery of Tierra Colorada as a major rookery (Márquez et al. 1981). Coincidentally, in that same year René Márquez wrote the first report of leatherbacks nesting in the Mexican Pacific, mentioning the beach of San Juan Chacahua in Oaxaca as the most important nesting site, with 2,000 females nesting each season (Márquez 1976a).

A few years later, Pritchard conducted an aerial survey along the Pacific coast of México, which allowed him to estimate the size of the leatherback nesting population in the region and the world (Pritchard 1982). He concluded that, collectively, the leatherback nesting beaches of México hosted the world's largest leatherback nesting population, with 75,000 females, or roughly 65 percent of the world estimate at the time.

While Pritchard was engaged in this population census, a group of Mexican biology students from Universidad Nacional Autónoma de México (UNAM, México's National University) were part of a field biology course that had winter field trips at a beautiful bay known as Caleta de Campos in Michoacán. As they worked on their assignments, they heard rumors from the local community of a beach 8 km to the north, at which they could see a "very amazing" creature. When they decided to follow this lead, they saw hundreds of leatherbacks nesting as far as the eye could see, crawling to or from the sea like little bulldozers, or nesting high on the beach. The students learned two things: the beach was named Mexiquillo, and the turtles were not alone—the surrounding communities made heavy use of the plentiful resource. The beach was divided in parcels, each owned by a family that camped by a fire and extracted and sold every single egg they could find—no nest was spared.

The students felt that something had to be done. They teamed up and created a leatherback conservation and research project that later evolved into the Sea Turtle Laboratory at UNAM. Over the next decade, many students powered the Mexiquillo project, patrolling the beach with limited resources, protecting nests, doing research, and talking to locals about the importance of their leatherbacks. And over that period, it became clear that this nesting population was declining. Today, under the guidance of Laura Sarti, who was among those first students, the ongoing work at Mexiquillo stands as the oldest uninterrupted conservation program for leatherbacks in México.

I was lucky enough to have the opportunity to join this team in 1988, when I was a biology student at UNAM. We lived at the beach for months, and together our team became a familiar sight for the local communities. Some of the children that used to hang around the turtle camp in the early days now have children of their own and admit that the everyday contact with the biologists and their work changed their view of turtles and of life. In turn, our contact with the local people and the turtles changed our own lives.

In 1993, the project was gearing up for the nesting season, students were ready, the Mexiquillo field camp was set up, but something unexpected and worrisome happened: the leatherbacks failed to come. The nesting numbers dropped from more than a thousand nests during the 1992–1993 season to fewer than a hundred the following year. What happened? Were our leatherbacks dying? The prevailing feeling among both biologists and local people was fear for the loss of this fantastic and valuable animal. We contacted personnel from conservation projects at other important leatherback nesting beach projects, such as those in Tierra Colorada, Chacahua, and Barra de la Cruz, and learned that the same situation was going on in these areas as well. Several explanations were suggested: the leatherbacks could be moving to nest somewhere else, were just suffering a natural fluctuation on their reproductive biology, or were truly dying and the population was about to disappear. One thing was clear: none of the field teams was going to solve the mystery working independently. So the leaders of all the leatherback nesting beach conservation projects decided to join forces, and in 1995 a new coordinated conservation effort emerged as Proyecto Laúd (Project Leatherback). In this first-ever species-specific turtle network in México, everybody agreed to adopt standardized methods to monitor the nesting population size and to protect the reproductive effort of the leatherback in México through a coordinated research and management plan (Sarti et al. 2007).

With the leadership from the UNAM team, the first question addressed was the possible movement of females to unprotected beaches. It was apparent that the fastest and most efficient way to survey the full 6,000 km of the Pacific coast of México was to do what Pritchard had done a decade before: aerial surveys. After careful planning, the first flight took off on 16 January 1996, surveying from Manzanillo, Colima, south to the Guatemalan border, a total distance of about 2,000 km that comprised the known nesting range of the species in the country. It was an exciting adventure in many ways, documenting the same observations Pritchard published and finding that leatherbacks nested all along the Mexican Pacific, concentrating in specific areas. It was also a painful reminder of the conservation challenges that lay ahead, as we also observed many dead females on some beaches, found as if they were still trying to nest, and obviously killed while nesting by hurried poachers who refused to wait until the turtles laid their eggs.

There were some notable positive findings as well. For example, we documented one beach with an unexpected high nesting density, located at the northwestern end of Oaxaca State just south from Tierra Colorada and later named Cahuitán for the town closest to it. Could it be that the females were shifting their nesting activity to new beaches? After the survey was completed, Laura Sarti and Cuauhtémoc Peñaflores from Centro Mexicano de la Tortuga returned to the area by land, following the GPS signal, trying to find the new beach. And they did, driving through tiny dirt roads and crossing several rural towns. But after talking to the local people, they realized that it was not a new leatherback rookery but one that had been known by town elders for decades. However, even they commented that there are not as many as there used to be. Cahuitán immediately became among the most important nesting areas in México, and after spending the 1996 season regularly visiting the beach, doing evaluation track counts and night patrols, confirming that this rookery was one of the largest in the Mexican Pacific, we decided to establish a nesting beach conservation camp there. During night patrols we could witness the same heavy poaching that we saw at Mexiquillo several years before, and a local economy may be even more dependent on the harvest of leatherbacks and their eggs. However, the people from Llano Grande, La Culebra, and Cahuitán, the communities adjacent to the nesting beach, were well aware of the decline of the population, and despite their wacky theories to explain it (e.g., former president Salinas sold all the leatherbacks to the Japanese), they recognized that something had to be done. A significant and emotive moment was when we created the new hatchery in Cahuitán—that night, sixty-year-old Tobias, an egg poacher all his life, brought us a freshly laid turtle clutch, saying, "Here you are, for good luck."

The new millennium found a strengthened Proyecto Laúd; we had

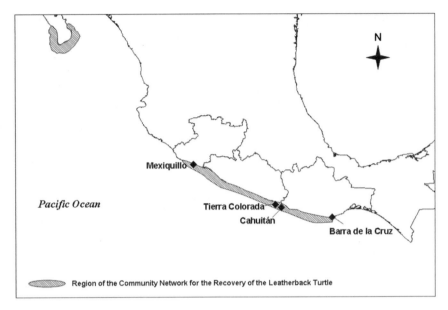

FIGURE 9.1. Location of the major nesting beaches for the leatherback turtle in the Mexican Pacific.

identified four index beaches (Mexiquillo, Tierra Colorada, Cahuitán, and Barra de la Cruz; fig. 9.1), and the local communities at all of them, concerned for their leatherbacks, wanted to participate in the conservation efforts. The style of participation has been as diverse as the number of communities involved: in Tierra Colorada, the community agreed that only three members will participate in the night patrols while the rest avoid consuming turtle products; in Cahuitán, all the adult men patrol the beach along with the biologists and equally distribute any economic gain among the families; and in Barra de la Cruz, two villages (Barra de la Cruz and Playa Grande) do the patrolling with supervision from the project's technical team.

In 2003 the project witnessed another important moment. After a long period of negotiation, mediated mainly by conservation-minded people in the Michoacán state government, the governors of Michoacán, Guerrero, and Oaxaca—the states that hold the major leatherback rookeries—gathered with federal authorities in Mexiquillo to sign the Tristate Agreement for the Recovery of the Leatherback Turtle. As part of this agreement, a commitment was made to actively involve the coastal communities in leatherback recovery efforts. This was the origin of the Community Network for the Recovery of the Leatherback in México (the Leatherback Network), which in 2004 brought together representatives from all the local groups that already were part of conservation projects in the three states with environmental authorities and scientists.

The first two meetings were a little awkward. It was the first time something like this had been done in the region, and the community representatives were expecting to be lectured about the relevance of protecting natural resources and then to receive some sort of economic support, whereas the authorities and scientists expected to be asked to give patronizing speeches about the importance of conservation, while other participants were probably just plain curious. None of them expected the meeting's outcome. For starters, no economic reward was ever given or promised to the communities for attending the meeting, and this was a surprise for people accustomed to being invited to political gatherings in which the lure was money. Instead, they were asked to give presentations about their protection activities and results, which meant speaking to a good-sized crowd of about fifty participants. The authorities and scientists were not asked to speak at all to the crowd; instead, they were asked to share the table with the representatives of the communities, to listen to their concerns and needs, and to share conservation experiences on an equal level. No academic degrees, no logos, no political parties—just a group of people concerned for the leatherback turtle.

After seven meetings, the Leatherback Network has turned around the odds for the Mexican Pacific leatherback population, from grim to hopeful. Thanks to the training component of the meetings, the community groups have a better understanding of the leatherback's life cycle and are able to identify the hazards that require attention and action. The groups from coastal towns now have a forum in which they know they can speak to the authorities, share their concerns, and be listened to, and the authorities have a forum to talk to the communities and divulge information. Everyone has a space to share experiences and obtain feedback. Groups that previously worked only with other sea turtle species are expanding their field seasons to protect leatherback clutches (fig. 9.2). Issues regarding the con-

FIGURE 9.2. Young people like Jonathan work in the community committees to protect leatherback clutches at the Mexican index beaches.

servation of this species are being considered in environmental policies of the three states. México's Comisión Nacional de Áreas Naturales Protegidas (CONANP, National Commission for Protected Areas) structured the national action program for the conservation of the leatherback, including input from the Leatherback Network. Most important, the groups from communities close to the nesting beaches are determined now more than ever to protect their turtles.

In the meantime, the Mexican leatherback population is not recovering. Nesting numbers fluctuate every season, but overall they have continued to decline (fig. 9.3). Yet despite this trend, the participants of the Leatherback Network have high hopes that they are making a substantial contribution to leatherback conservation. Proyecto Laúd and the Leatherback Network have been able to protect half of the clutches laid along the Mexican Pacific for the past few seasons, a tremendous feat considering the

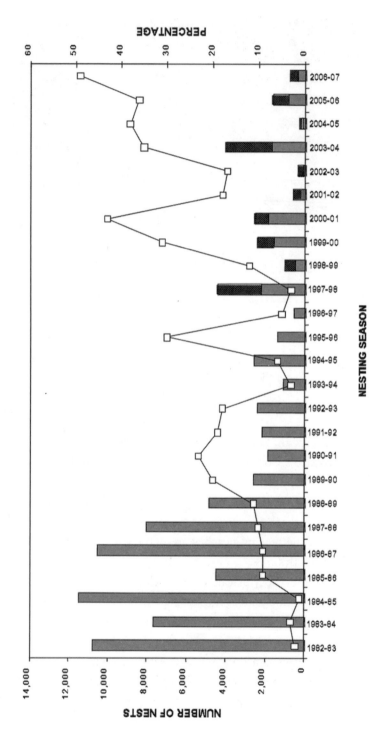

FIGURE 9.3. Historical results of protection efforts at the Mexican Pacific leatherback nesting beaches.

length of the coast (2,200 km) and the number of beaches in the region that register at least one leatherback nest each season (around 150 beaches). Since the nesting activities concentrate on the index beaches, efforts focus on these places, where more than 90 percent of the clutches are protected.

The problems surrounding the recovery of the leatherback turtle are complex. It is clear that the turtles face numerous threats throughout their oceanic life, which we are still in the early stages of understanding. But after their long travels, whenever the females return to Mexican beaches to reproduce, they will meet with an increasing number of people from various backgrounds concerned for their fate and engaged in their conservation.

The Olive Ridley Turtle

The olive ridley is by far the most abundant sea turtle species in the Mexican Pacific (plates 8 and 9). It is so common that in many coastal communities people refer to it just as *la tortuga* ("the turtle"), while all the other species receive different local names. México had several olive ridley arribada beaches, where the synchronous mass nesting events of this species occur periodically during the nesting season (plate 8). Mismaloya in Jalisco, Piedra de Tlalcoyunque in Guerrero, and Escobilla and Morro Ayuta in Oaxaca were considered arribada beaches prior to the 1960s, with nesting populations exceeding 50,000 females per season (fig. 9.4; Márquez et al. 1976). Such abundance highlighted the olive ridley as a plentiful source of protein for coastal communities and later sustained a structured commercial fishery. The records for this commercial fishery started in 1948, and, according to René Márquez, the first decade had "a low and stable capture of sea turtles," with a mere 500 metric tons per year, or roughly 11,500 turtles. The popularity of sea turtle products and the opening of the international market for turtle skin in 1960 with the ban of crocodile products pushed the capture levels to a maximum of 14,330 metric tons per year in 1968, which equates to an estimated 358,300 individual turtles (Márquez 1976b).

Most of the commercial capture occurred in front of the arribada beaches, where great numbers of turtles concentrated for the mass nesting events. Arribadas are peculiar phenomena (see chapter 13)—all kinds of predators wait eagerly around the beaches to take their share of eggs, hatchlings, and weakened nesting females, and humans were no exception. At the

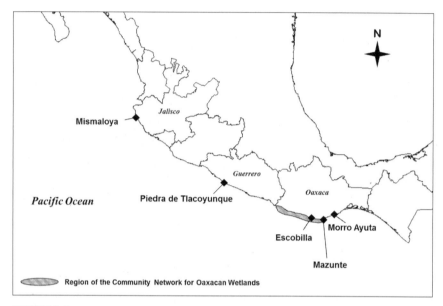

FIGURE 9.4. Places of interest regarding olive ridley conservation in the Mexican Pacific.

same time that the commercial fishery took thousands of females at sea, inhabitants of nearby towns came to the beaches to take eggs by the truckloads, despite a ban on consumption of sea turtle eggs dictated in the early 1920s (Secretaría de Agricultura y Fomento 1922). The same happened at beaches with olive ridley solitary nestings, where females and eggs were harvested extensively during the nesting season. In some small coastal towns, the turtles served as a bank of sorts—instead of keeping the money at home, when people needed cash they came to the beach for a few egg clutches or females to sell.

Not surprisingly, the exorbitant harvest could not be sustained for long. By 1970, the commercial capture plummeted to less than 4,000 metric tons per year. As a result, all sea turtle capture was banned from 1971 to 1973 in an attempt to reorganize the fishery. After 1973 the Mexican government implemented strict regulations and quotas, and the new system of franchises (*cooperativas*) maintained the legal capture around 100,000 adult turtles per year, while the fishermen groups were required to establish an egg protection program.

But the damage was done. The arribada populations in Mismaloya and Tlalcoyunque (fig. 9.4) had vanished and still have not recovered. All along the Mexican Pacific, nesting populations were severely depleted, and people noticed that it was harder and took longer to gather eggs or turtles than a few years earlier. By the mid 1980s some cooperativas in Michoacán were voluntarily giving up the franchise because they couldn't meet the quotas—there weren't enough turtles left to make business. The commercial fishery for olive ridleys in México ended in 1990, with the permanent ban on the commerce of sea turtle products (Secretaría de Pesca 1990).

The decline of the olive ridley directly affected small coastal communities along the Mexican Pacific; many cooperativas were formed by fishermen at small towns, such as San Agustinillo or Mazunte, close to Escobilla, who made a living at it. Even before the establishment of that system, they were turtle hunters for generations.

Without the possibility of exploiting turtles, what do the people do? The answer to this question has been quite diverse. Unfortunately, many turtle fishermen turned to the black market. In Oaxaca, nesting occurs at distant, isolated areas, and the strong cultural attachment of people to the consumption of turtle eggs drives poachers to all the beaches in the state. However, the level of egg and turtle harvest has been only a fraction of what it used to be. Today, fewer people risk capture by the *marinos* (Mexican Navy) and PROFEPA (Mexican Federal Attorney of Environmental Protection), who now guard most of the important beaches. Jail sentences are stricter than before: poachers face up to nine years in jail, with no bail if the turtle products come from a protected area, not counting the unpleasant experience of being captured and dragged to the closest federal police facility, often many miles away. Currently, much of the turtle harvest in Oaxaca occurs as small-scale family commerce, which is still illegal but considerably harder to stop.

Other communities tried alternative sources of income, but socio-economic conditions and lack of knowledge made that very difficult. To this day, many communities along the Mexican Pacific, especially in the South, remain uneducated, isolated from the rest of the country's economy, and poor by modern global standards. Government programs alleviate some of the needs but are not enough. A few people in some small communities started wondering what was going on with the turtles that were part of their world for so long, and if there was anything they could do about it.

While this happened, turtle biologists dealt with the problem of a declining olive ridley population. Several protection programs started all along the Mexican Pacific, run by the federal and state government, universities and NGOs. In Oaxaca, the old slaughterhouse and biological center where eggs were incubated was turned into Centro Mexicano de la Tortuga (Mexican Turtle Center), which along with environmental education programs now manages the conservation programs at the last arribada beaches. Groups of biology students or government technicians started camps at some nesting beaches during the nesting season, patrolling the beaches to protect the turtles and talking to the locals about their natural resources. A few people in some small communities started wondering what was going on with these "crazy" people that walked the beaches all night chasing turtles, competing with poachers and relocating clutches into hatcheries. For the people that lived at the coastal towns, it was unclear if anything good could come out of the biologists' activities.

It was just a matter of time before this mix developed into something that no one foresaw at the beginning. An extraordinary example is the Network for Oaxacan Wetlands (Red de los Humedales de la Costa de Oaxaca; fig. 9.4). This truly community-based conservation organization was born during several training workshops for the restoration of mangrove ecosystems promoted by biologists from Lagunas de Chacahua National Park in 2001. The workshops gathered representatives of the communities surrounding the Chacahua National Park to talk about the loss of natural resources and the actions that should be taken to reduce this situation. People realized that they had a lot to loose if they did not care about their surroundings and if they kept extracting their resources to exhaustion. With the technical advice from some NGOs and the Chacahua National Park personnel, the network was legally constituted in 2003 (Vásquez-Ruíz et al. 2009).

From the beginning, the Network for Oaxacan Wetlands had a different perspective than other conservation organizations; they worried about the welfare of their communities, and the welfare of nature was seen as an important part of that objective (fig. 9.5). They have five main lines of action: management and conservation of wildlife, improvement and diversification of community products, restoration of habitat, community development, and institutional strengthening. In 2010, the network comprised twenty-two small Oaxacan coastal communities, and their star projects are

FIGURE 9.5. The participants of PROTUMAR, the sea turtle program of the community-based Network for Oaxacan Wetlands, take great care in following the most accepted management procedures.

the Mangrove Conservation Program (PROMANGLAR) and the Sea Turtle Conservation Program (PROTUMAR). The latter program had six community camps, La Ventanilla, El Tomatal, Los Naranjos, La Vainilla, La Tuza, and Cerro Hermoso, with an interest to add more. All continue to be run by community committees, all have government permits, and all use standard field forms and methods. They mainly protect olive ridley clutches, given the abundance of the species in Oaxaca, but La Tuza is also a priority beach for leatherback nesting.

I have been involved with the development of PROTUMAR since 2004, first through Kutzari A.C. (a Mexican NGO dedicated to the conservation of sea turtles) and later as part of CONANP, as technical adviser for the program, participating in several training workshops at the communities in the past few years. The level of commitment of these communities to sea turtle conservation is impressive; they are former poachers or consumers of sea turtle products, but all have "quit the industry" and are now involved in sea turtle conservation. They speak to their neighbors and organize talks for children in their community schools; if things get especially difficult in one of the communities, they get together and as a group deal with the issue. They personify the old saying that "union is strength," which is very important since they are swimming against the tide. More than once somebody in the network has received death threats; the community restaurant in Escobilla was lost to arson, and the perpetrators were never caught.

They have another aspect that is quite unusual for coastal communities: they give considerable importance to scientific knowledge as a guide to their conservation activities. With no disregard of their traditional knowledge of sea turtles, the groups of PROTUMAR have a great thirst for learning about biology and the reasons for management procedures—they are not happy to learn only *that* turtle eggs should be relocated immediately; they want to know *why*. Hence, training workshops with them can be quite challenging. They take great pride in knowing by heart the scientific names of sea turtles, they get worried if they feel there is a chance that the information they give in talks at schools might be inaccurate, and they get really worried if their hatching percentages are below the average for the species—outstanding for people whose average level of education does not exceed elementary school.

The network's enthusiasm for protecting their nesting beaches and their own lifestyles is very contagious and is attracting many other communities that want to join the organization. They provide advice to many of them, but in an attempt to prevent the network getting too large and out of control, the board of directors has decided (wisely) to limit the entry to the network to those communities with proven organizational skills and commitment to protect nature. As with any conservation organization, financial resources are very limited, and aside from their conservation projects, they need to fund the community production projects and the goal of improved livelihoods for their families. Some signs of success are evident: La Ventanilla is a well-known ecotourism center where visitors enjoy a motor-free trip to the lagoon for crocodile and bird watching. El Tomatal produces organic peanuts, which are turned into peanut butter and candy by the local cooperative run by women. More recently, La Tuza started the production of honey. All these projects are done under a strict philosophy of environmental awareness advised by CONANP and various NGOs.

The community of Escobilla, adjacent to the arribada beach, is part of the network but does not participate as a PROTUMAR camp. The beach was decreed a sanctuary for the olive ridley turtle in 2002 and is operated by Centro Mexicano de la Tortuga. Since an arribada beach cannot be managed as a "traditional" turtle camp, Centro Mexicano de la Tortuga is figuring out ways to ensure the participation of the local community, by including them in the monitoring of arribadas and supporting their ecotourism project. Thanks to an integral conservation approach that includes beach protection, community participation, and regulations to reduce mortality at sea (fig. 9.5), the olive ridley population at La Escobilla Sanctuary has had a spectacular recovery over the last fifteen years, with more than a million estimated nestings per year.

The Network for Oaxacan Wetlands has turned into a conservation icon for the state. They have been invited to give talks to several meetings countrywide; they presented their results at the International Sea Turtle Symposium in Loreto, Baja California Sur, México, in 2008 and are part of the Technical Advisor Council for the Sea Turtles of Oaxaca, which invited them to participate as the voice of the coastal communities related to sea turtles in Oaxaca.

The beaches at which PROTUMAR operates camps are not the most important for the olive ridley or other species regarding nesting numbers, but their value is heightened by the dedication and commitment of the people that work there, and for being an example to other coastal communities in the country. With any luck, others will learn and follow.

Conclusions

The stories of sea turtles and people are intimately bound at the nesting beaches. The two examples reviewed above teach us that even in aspects that could seem cold and distant, like government policies or resource management, in the end everything in turtle conservation is powered by the up close and personal experiences that people have at a nesting beach, whether they are biologists, fishers, peasants, or volunteers (plates 15 and 16).

Also, these two stories are just a token from hundreds of sea turtle conservation stories in México. Other noteworthy stories include the activities of the Pomaro communities in Michoacán for the conservation of the black turtle in Colola and Maruata, and the recent Latin American network for the conservation of the Pacific hawksbill, one of the most endangered sea turtle populations in the world. If these populations have a chance for recovery, it is because of the commitment of many people to make such a recovery happen.

Nesting beach conservation along the Mexican Pacific is complicated, to say the least. The turtles are still doing what they have done for millions of years: they come to isolated beaches to lay their eggs. It is the human factor that grew in complexity, with new threats appearing while we struggle to stop the old ones. But as long as humans value their relationship with sea turtles, these ancient reptiles still have a chance. Sea turtles remain part of the culture and traditions of the Mexican Pacific coastal towns, but now their images decorate restaurants, hotels, and jewelry, instead of ancient pottery and rituals.

Now it is the responsibility of conservation projects to keep the balance and serve as bridge between the interests of the turtles and the people, and time has shown that no single conservation project will succeed in the task if it remains isolated. It is no accident that the two stories reviewed in this chapter are about networks. More and more people in México realize that reaching out and sharing information and experience is the key to face the challenges and to succeed. As the people from the Leatherback Network like to say, referring to the recovery of their beloved turtles, "together we'll do it."

ACKNOWLEDGMENTS

I'd like to express my huge gratitude to Laura Sarti, Ninel García, and Carlos López for sharing their experiences and stories and for letting me be part of the wonderful Proyecto Laúd for so many years. Thank you also to Hugh Wheir from Animal Alliance for believing in us, to the crew of *LightHawk* for the aerial surveys in 1996 and 1997, and to Sandy Lanham for her invaluable help in the aerial surveys from 1998 to 2010—the story of the Proyecto Laúd wouldn't be the same without them.

Thanks to all the people from Red de Humedales de la Costa de Oaxaca and from Red de Comunidades por la Recuperación de la Tortuga Laúd for their outstanding example of commitment in the conservation of the sea turtles in the Mexican Pacific.

REFERENCES

León-Portilla, M. 1972. De Teotihuacan a los Aztecas: Antología de fuentes e interpretaciones históricas. Serie Lecturas Universitarias. Universidad Nacional Autonoma de Mexico, México DF.

Márquez, R. 1976a. Reservas naturales para la conservación de las tortugas marinas en México. Instituto Nacional de la Pesca, Subsecretaria de Pesca. Serie Información, INP/SI:i83, México DF.

Márquez, R. 1976b. Estado actual de la pesquería de tortugas marinas en México. Instituto Nacional de la Pesca. Serie Información, INP/SI:i46, México DF.

Márquez, R., A. Villanueva, and C. Peñaflores. 1976. Sinopsis de datos biológicos sobre la tortuga golfina *Lepidochelys olivacea* (Eschscholtz, 1829) en México. Instituto Nacional de la Pesca (INP), Sinopsis sobre la Pesca No. 2. INP, México DF.

Márquez, R., A. Villanueva, and C. Peñaflores. 1981. Anidación de la tortuga laúd Dermochelys coriacea schlegelli en el Pacífico mexicano. Ciencia Pesquera 1(1), 45–52.

Peckham, S. H., D. Maldonado-Díaz, V. Koch, A. Mancini, A. Gaos, M. T. Tinker, and W. J. Nichols. 2008. High mortality of loggerhead turtles due to bycatch, human consumption and strandings at Baja California Sur, Mexico, 2003–2007. Endangered Species Research 5, 171–183.

Pritchard, P. C. H. 1971. The leatherback or leathery turtle. IUCN Marine Turtle Series Monograph No. 1. International Union for Conservation of Nature and Natural Resources, Morges, Switzerland.

Pritchard, P. C. H. 1982. Nesting of the leatherback turtle, *Dermochelys coriacea* in Pacific Mexico, with a new estimate of the world population status. Copeia 1982, 741–747.

Sarti, L., A. R. Barragán, D. García, N. García, P. Huerta, and F. Vargas. 2007. Conservation and biology of the leatherback turtle in the Mexican Pacific. Chelonian Conservation and Biology 6:70–78.

Secretaría de Agricultura y Fomento. 1922. Acuerdo fijando las disposiciones reglamentarias a que se sujetará la explotación de tortugas en aguas federales. Diario Oficial de la Federación, 22 de abril de 1922. México DF.

Secretaría de Pesca. 1990. Acuerdo por el que se establece veda para las especies y subespecies de tortuga marina enaguas de jurisdicción Federal del Golfo de México y Mar Caribe, así como en las del Océano Pacífico, incluyendo el Golfo de California. Diario Oficial de la Federación, 31 de mayo de 1990. México DF.

Vásquez-Ruiz, F., A. Reyes-Sánchez, and A. R. Barragán. 2009. PROTUMAR: 6 years of community participation on sea turtle conservation in the coast of Oaxaca, México. Poster presented at the 29th Symposium on Sea Turtle Biology and Conservation, Brisbane, Australia, February 2009.

10

Saving the Eastern Pacific Hawksbill from Extinction

Last Chance or Chance Lost?

ALEXANDER R. GAOS AND INGRID L. YAÑEZ

Summary

Only a few years ago, most of the science and conservation community thought hawksbill turtles had essentially been wiped out in the eastern Pacific Ocean. The simultaneous discoveries of several significant foraging and nesting aggregations for the species has catalyzed an international effort to recover the population. Key to these findings has been heeding the advice of and collaborating closely with coastal community members, particularly fishermen and egg "poachers." Several conservation projects have now been established, and new hawksbill sightings and habitat are rapidly being documented (see plates 6 and 7), a fact that bodes well for the future outlook of the population. These efforts are facilitated by a passionate group of individuals and organizations cohesively spearheading work throughout the region. A unique life-history paradigm has also been revealed for the species, which has led to the implementation of innovative research and conservation techniques and may help explain how this population evaded detection for so long. The vibrant and growing movement to learn about and conserve hawksbills in the eastern Pacific is both timely and essential, as it may very well be now or never for recovery of this severely depleted population.

Introduction

The initial question was whether there were enough hawksbills (*Eretmochelys imbricata*) remaining in the eastern Pacific Ocean that someone could actually go out and find them. Many local fishermen in Baja California Sur, México, insisted there were but that encountering them would be difficult. Furthermore, the only way we would do so was by looking in habitats distinct from those used by their well-studied and often-captured green turtle (*Chelonia mydas*) counterparts. While these accounts were motivating, we were not convinced of their validity until a small hawksbill caused the buoy line of our tangle net to dip under the surface of the water. Indeed, hawksbills were present, and we could predictably find and capture them in particular areas by heeding the advice of fishermen.

As one of the world's three Critically Endangered sea turtle species (Mortimer and Donnelly 2008), hawksbills are threatened with extirpation in the eastern Pacific (Nichols 2003; Seminoff et al. 2003; Gaos et al. 2010) and are considered one of the most threatened sea turtle populations on the planet (Wallace et al. 2011). Only a few years ago much of the sea turtle conservation community considered hawksbills ecologically extinct in the eastern Pacific and believed efforts to recover the population were essentially hopeless.

It is not difficult to conceive how such a notion came about. When reviewing the scientific and "gray" literature at the time, it quickly became apparent that almost no information was available for hawksbills in this region of the world. Once common in neritic habitats from México to Ecuador (Cliffton et al. 1982), the species reportedly became rare to absent in most localities in the eastern Pacific by the late 1960s (Caldwell 1962; Cliffton et al. 1982; Mortimer and Donnelly 2008). Even anecdotal reports of nesting or foraging were rare (e.g., Gaos et al. 2006). Considering the numerous sea turtle conservation projects established along the approximately 15,000 km of eastern Pacific coastline that is suitable for hawksbills (extending from northern México to northern Perú), one would assume that if hawksbills were out there, they would have been reported years ago. Since that was not the case, people assumed essentially none were left.

With our first in-water capture of a hawksbill in Baja California Sur, and observation of nine more individuals of the species in the region during the next two months, as well as the discovery of several nesting sites along the Pacific coast of the Americas during that time, the ability to observe and conserve eastern Pacific hawksbills suddenly became a real possibility. The findings redirected the conservation outlook for this perilous population, sparking optimism for its recovery and opening new opportunities for research.

Still, information continues to be extremely limited and represents one of the principal barriers to conservation. The International Union for Conservation of Nature's Marine Turtle Specialist Group recognized the lack of data on the population as a serious obstacle for developing appropriate and effective management strategies and considers studies to generate information a globally critical conservation need (Mast et al. 2004).

But how does one investigate a species teetering on the brink of extinction about which almost nothing is known? In the case of sea turtles, seeking out the help and knowledge of coastal community members, particularly fishers who depend on the ocean for their existence, has proven to be an extremely successful method (see plates 15 and 16). Archie Carr, arguably the first sea turtle conservationist, was the first to harness the power of local knowledge to locate Tortuguero, Costa Rica (Carr 1956), one of the most important sea turtle nesting sites in the Caribbean. Around that same time, another sea turtle pioneer, Henry Hildebrand, confirmed the existence of a previously undiscovered nesting rookery of more than 40,000 Kemp's ridley turtles (*Lepidochelys kempii*) at Rancho Nuevo, México, by engaging coastal community members (Hildebrand 1963).

Threats to Eastern Pacific Hawksbills

As is the case with all sea turtles, both adult and juvenile eastern Pacific hawksbills continue to face an amalgam of threats, including the collection of their eggs and meat for consumption, the alteration and destruction of nesting and foraging habitats, and incidental capture in industrial and artisanal (i.e., small and midscale) fishery operations (see plates 11 and 12). Industrial fisheries are particularly worrisome because they have been reported to kill hundreds of thousands of sea turtles each year (Lewison et al. 2004), with older age classes being particularly affected (Lewison and Crowder 2007). Artisanal fisheries have also been found to affect a much higher number of sea turtles than originally suspected, rivaling effects of industrial-scale fisheries (Peckham et al. 2007).

Because most artisanal fisheries operate in near-shore waters, they pose particular threats to hawksbills, which are also known to inhabit neritic habitats (Cliffton et al. 1982; Witzell 1983; Grant et al. 1997; Seminoff et al. 2003; Gaos et al. 2011). This presents somewhat of a paradox: while local fishers represent one of the greatest resources for researchers trying to locate and conserve hawksbills, they also represent one of the gravest threats to the species' survival. In our experience, local fishermen are some of the most respectable, kind, and honest people on Earth, often simply struggling to provide the most basic commodities for their families. Finding a way to conserve hawksbills and other sea turtles while not infringing on the ability of these people to survive is one of the most challenging endeavors we face.

Another major threat and one unique to the species is the trade centered around hawksbills' intricately patterned and highly valued carapace scutes—known as tortoiseshell, *penca* (Spanish), or *bekko* (Japanese)—which are used to produce artisanal crafts such as combs, pennants, hairpins, and other novelties (Chacón and Arauz 2002; Mortimer and Donnelly 2008; see plate 12). Entire hawksbills are also collected, dried, stuffed, and sold as curios to hang on the walls of restaurants, homes, and businesses (King 1979) throughout the world (fig. 10.1). The lucrative fishery and trade industry centered on hawksbills continues to this day and has played a major role in the species' listing in Appendix I of the Convention on International Trade in Endangered Species of Wild Fauna and Flora (CITES).

Additional threats to eastern Pacific hawksbills include pollution, egg predation, and large-scale changes in oceanographic conditions and nutrient availability (Lewison and Crowder 2007), all of which present unique challenges we must confront and minimize if we are to save the population.

Discovering Hawksbills in the Eastern Pacific

Having encountered eastern Pacific hawksbills on a handful of occasions years earlier while working in Costa Rica, we began doubting the widely accepted idea that hawksbills were past the point of no return. During seasonal stints working with the ProCaguama loggerhead (*Caretta caretta*) sea turtle project in northwest México, we had the opportunity to speak extensively with local fishermen, many of whom insisted hawksbills could still be found in the country's Gulf of California, also known as the Sea of Cortez. The information coincided well with the fact that of the limited number of reports of the species in the eastern Pacific at the time, the

FIGURE 10.1. Juvenile hawksbill cadavers being dried at a fish camp in the Gulf of California, México (© Alexander Gaos). They will later be stuffed and sold as curios.

majority focused on the gulf and wider Baja California peninsula region of México (Aschmann 1966; Felger and Moser 1985; Nichols 2003; Seminoff et al. 2003; Saenz-Arroyo et al. 2006).

Innately and irresistibly attracted by the challenge and high-threat scenario presented by hawksbills in the eastern Pacific, we began conceptualizing a project in northwest México to seek out what remained of the population. After patching together a few thousand dollars, we bought a derelict 1993 Ford F-150 pickup and drove 2,500 km (~1,500 miles) around the gulf over the course of two and a half months. During this initial survey we visited virtually every community along the coast to spend time with fishermen and their families and glean information on hawksbills. We were calling our effort the Eastern Pacific Hawksbill Initiative. Our objectives were fairly straightforward at the time: compile existing information, raise awareness, and attempt to observe individuals of the species. While our fieldwork was based principally out of México, our goal was to eventually develop an international effort geared toward investigation and conservation of the population throughout the eastern Pacific.

During our survey trip we compiled a wealth of knowledge on the species, including observations of recently caught specimens and information on the marine habitats and potential capture sites where the species might still be found. While visiting the small town of Cabo Pulmo on the eastern cape of the Baja California peninsula, we were told that hawksbills could be seen fairly easily just offshore. Figuring the likelihood of actually seeing one was slim, we decided to snorkel around and take a look anyway. After only ten minutes we encountered a juvenile hawksbill casually eating algae off a rock. Struck by the relative ease with which we found the turtle, we wondered whether it was a stroke of pure luck or a sign that more hawksbills persisted in the eastern Pacific than originally believed.

A few months after completing our survey trip, we returned with the specific objective of encountering hawksbills at the sites fishermen had indicated as having the most potential. Since we were working in a strictly foraging habitat, we deployed tangle nets, specially modified fishing nets, in an effort to capture them. Unlike at nesting beaches, where sea turtles emerge from the ocean and can easily be observed and quantified, in marine habitats turtles can be much more difficult to detect, and tangle nets have proven to be an invaluable tool for researchers. It is no surprise that tangle nets are a product of local fisher knowledge. In México, as well as many other countries around the world, sea turtles once made up a large portion (and in some cases still do) of the local diet, and tangle nets were often the primary tool used for their capture. Essentially large gill nets, tangle nets float on the surface of the water but are modified to function with less weight so that when a turtle becomes entangled it is able to surface and breathe. For fishermen it means a fresh catch, while for researchers it represents a live turtle.

Juan de la Cruz is a fisherman and expert ex-turtle hunter who lives near Loreto, midway down the gulf side of the Baja California peninsula.

FIGURE 10.2. Expert fishermen and ex–turtle hunter Juan de la Cruz indicates areas to set tangle nets for hawksbill turtles near Loreto, Baja California Sur, México (© Alexander Gaos). Directions from local fishermen are invaluable for locating the species.

We had visited Juan during our survey trip, and he claimed hawksbills were definitely still around and that he would help us find them. He had helped Jeffrey Seminoff and Wallace J. Nichols, the pioneers of sea turtle conservation in northwest México, catch their first green turtle in the region some fifteen years earlier, so we were optimistic about his comments and opted to visit him for one of our first monitoring sessions.

Juan advised us on where and how to tend our two tangle nets (fig. 10.2), and we did so using a small Zodiac inflatable boat. At 1 a.m. on our sixth night of monitoring, we located the first of the two nets and lifted the near end out of the water, when something quickly dipped into the water halfway down its length. A couple of minutes later the first hawksbill of the project was aboard, a lively juvenile measuring 48 cm (curved carapace length, CCL), which we dubbed Joalin. After the capture we relocated both nets and called it a night, ecstatic to have success.

Upon pulling up to the first net the next morning, the tail end of another large turtle projected out of the water. Automatically assuming it to be a green turtle due to its size and the prevalence of the species in the region, we were flabbergasted when a large (73 cm CCL) hawksbill lifted its head out of the water. With the hawksbill aboard, we made way over to the second net, where another large hawksbill had been ensnared and was resting at the water's surface.

Thus it was that, by following Juan's leads, in two nights we captured three hawksbills belonging to a population that was virtually unknown and considered by most to be past the point of no return. Over the course of the next several weeks we observed an additional seven hawksbills in the wild and several more held in captivity that had been caught previously by fishermen. Coinciding with previous reports, all the hawksbills were juveniles. The number of young hawksbills in the gulf, coupled with the fact that at the time almost no hawksbill nesting had been documented anywhere along the Pacific coast of the Americas, led us to question—where were all these little turtles coming from?

Shifting the Panorama

The twenty-eighth annual International Sea Turtle Symposium (ISTS) was held in Loreto, Baja California Sur, México, and provided us with an opportunity to share our preliminary findings and begin featuring eastern Pacific hawksbills in the global sea turtle conservation spotlight. We had arranged a small breakfast meeting with interested individuals working in the eastern Pacific to discuss the situation facing hawksbills and potential opportunities for investigation.

Mike Liles and Mauricio Vasquez attended the meeting and had brought a poster to share at the ISTS displaying the results from a sea turtle census they had carried out with Carlos Hasbún in El Salvador. Having completed only half of their survey, consisting of interviews with fishers and coastal community members, they had already documented more than seventy hawksbill nesting events at two principal sites known as Bahía Jiquilisco and Los Cóbanos. Upon further discussion, it became evident that they were not even remotely aware of the incredible significance of their data. In fact, Mike would later tell that upon arriving at the ISTS they were somewhat embarrassed to show their poster because they thought the results were meager and unimportant. Exciting discussions ensued regarding the situation facing eastern Pacific hawksbills, our efforts with the Eastern Pacific Hawksbill Initiative, and the fact that they had just documented more nesting by hawksbills than all previous reports in the eastern Pacific combined.

As a leading sea turtle researcher in the eastern Pacific, Jeff Seminoff was excited about the hawksbill information being discovered and shortly after the ISTS suggested we hold a more formal regional workshop in El Salvador to compile existing information on the population and help raise awareness of the country's recently discovered importance for nesting by the species. Within weeks the meeting was confirmed and planning had begun.

During the planning process, Jeff unexpectedly received an email from Ecuadorian researcher Andrés Baquero. Andrés, along with Juan Pablo Muñoz and Felipe Vallejo, had begun monitoring sea turtle activity at a 1-km beach called La Playita, within Machalilla National Park. Andrés was seeking turtle flipper tags for the project and mentioned to Jeff that the tags would be important for marking the nesting hawksbills they were recording. Instantly overwhelmed by the serendipitous timing of the information, Jeff excitedly began explaining the upcoming workshop and subsequently called us with the new revelation.

The First Workshop of the Hawksbill Turtle in the Eastern Pacific was held in Los Cóbanos, El Salvador, in July 2008 and brought together this and other critical information on hawksbills. The meeting was attended by sea turtle experts from the United States, México, Guatemala, El Salvador, Nicaragua, Costa Rica, Colombia, and Ecuador. Most of the participants were acquainted with each other before the workshop and unselfishly shared information on the species, leading to the first ever pair of region-wide maps featuring the aggregated knowledge (at the time) of hawksbills in the eastern Pacific. The group also created a resolution outlining the principal threats to the population and calling for international efforts to investigate and protect the population (Gaos et al. 2010).

Among the most significant achievements of the workshop was the constitution of an international eastern Pacific hawksbill working group and the commitment of its members to continue sharing information on the species. Although the Eastern Pacific Hawksbill Initiative was originally how we referred to our project in northwest México, because of the regional connotation of the name it was adopted by the working group.

Inspired by the workshop, attendees returned home and began focusing monitoring and conservation efforts at the newly discovered sites or began actively seeking out hawksbills in other areas. Under Mike's leadership, permanent nest conservation projects were established at the two sites in El Salvador, which together receive 200–350 nests annually (Liles et al. 2011). Andrés and his team solidified a similar project in Ecuador, which hosts considerably less nesting but remains the only verified hawksbill nesting site in South America.

Accompanied by our recently born son, Joaquinn, we took another long road trip in search of hawksbills, this time covering several thousand kilometers between the United States and Panamá. In Nicaragua we connected with José Urteaga to follow up on the twenty hawksbill nesting events he had reported at the workshop. We found strong physical and anecdotal evidence for hawksbill nesting in Estero Padre Ramos and together initiated a monitoring program, which proved the area to be the single densest nesting site yet identified in the entire eastern Pacific, receiving 150–300 nests annually.

Even in Costa Rica, the country hosting the largest number of sea turtle research projects in the eastern Pacific and where few sea turtle nesting beaches remain unexplored, new information began emerging on nesting hawksbills. Didiher Chacón Chaverri conducted a pilot study on the isolated coast of Corcovado National Park, located on the Osa Peninsula, and documented twenty-four hawksbill nests during less than a four-week span.

Sporadic hawksbill nesting was also documented at several beaches along the Pacific coasts of southern Nicaragua and central México, Honduras, and Panamá. However, it wasn't just nesting, as marine sightings of hawksbills also began emerging across the region. In Colombia, Diego Amorocho began observing dozens of juvenile hawksbills on the coral reefs around the island of Gorgona and established an ongoing in-water monitoring program at the site. Aarón Esliman, Hoyt Peckham and their team began consistently encountering juvenile hawksbills at El Pardito, Cabo Pulmo and other sites in northwest México.

On our trip, we also teamed up with local lobster "hookah" fishers and hand captured various juvenile and adult hawksbills along central Pacific México. In Costa Rica, we assisted Randall Arauz and Erick Lopez in setting the first ever tangle nets along the country's Pacific coast, at a site near the Playa Celetas-Ario National Wildlife Refuge. Working with local fishermen, we caught a hawksbill in less than an hour on our first set. Randall, Erick, and Javier Carrión subsequently began routinely catching hawksbills at the site and established a permanent in-water monitoring program. Other important foraging areas have also been identified in El Salvador, Honduras, Nicaragua, Costa Rica, Panamá, and Ecuador.

What Took So Long?

The fact that hawksbill turtles have been present in relatively substantial numbers in the eastern Pacific but went virtually undetected for decades is somewhat difficult to fathom. The research thus far offers some possible explanations. Hawksbills appear to be using habitat for both nesting and foraging that is distinct from that used by the species in other parts of the world (Gaos et al. 2011).

In both Bahía Jiquilisco and Estero Padre Ramos, hawksbills are literally coming into estuaries to nest, where they use small, secluded, vegetation-choked patches of shoreline that resemble riverbanks more than ocean beaches (fig. 10.3). The estuarine habitat is a far cry from the fluffy expanses of sandy beach representative of most of today's sea turtle nesting beaches. Hawksbill nests are also interspersed throughout the estuaries in very low numbers (e.g., one nest per night along more than 30 km of fragmented shorelines), so locating and quantifying nests is extremely challenging. Creating projects in such areas, and hence documenting and reporting activity, is difficult because logistical aspects are daunting and traditional "nightly beach patrols" are not feasible.

Several postnesting female hawksbills equipped with satellite tags in El Salvador, Nicaragua, and Ecuador have shown that they spend an extraordinary amount of time in these same mangrove and estuarine habitats for foraging (Gaos et al. 2011). In fact, some of the turtles never leave the estuaries after nesting. These findings contrast sharply with the reputation of adults of the species migrating between nesting and feeding areas and

FIGURE 10.3. A typical hawksbill nesting beach within Bahía Jiquilisco, El Salvador (© Bryan Wallace). It is unusual nesting and foraging habitat such as this that scientists believe might have contributed to hawksbills going undetected in the eastern Pacific for so long.

being associated almost exclusively with coral reef habitats in other parts of the world (Carr et al. 1966; van Dam and Diez 1996; Gaos et al. 2011).

Eastern Pacific hawksbills may also be using habitat that differs from that used by other sea turtle species, and this peculiar niche has likely led to decreased reports. Although in-water monitoring has occurred in northwest México for more than a decade, hawksbill captures have been limited (e.g., Seminoff et al. 2003). Nonetheless, directed efforts to locate the species, as described earlier in this chapter, resulted in the observation of ten individuals during a mere two-month time span. These captures occurred by setting tangle nets directly over shallow rocky shoals, often near the entrance to mangrove inlets, whereas prior monitoring efforts aimed at green turtles tended to consist of nets being set farther offshore. Unless specifically asked where to find hawksbill turtles, fishers will understandably indicate sites where the greatest number of sea turtles can be found—

which are generally farther-from-shore green or olive ridley (*Lepidochelys olivacea*) turtle foraging areas. Additionally, nets set over rocky shoals can easily become entangled and damaged, requiring extensive labor for removal and reparation, so setting in these areas would otherwise not occur.

While these factors might have contributed to why hawksbills were undetected for so long in the eastern Pacific, the relatively small population in the region likely also played a role. The eastern Pacific has fewer coral reefs than many of the world's other oceans, which generally represents the typical habitat for the species, so there simply may not be enough appropriate habitat to support large numbers of the species. While it may be that the eastern Pacific hawksbill population has adapted to using mangrove estuaries (Gaos et al. 2011), these habitats have encountered serious decline over the past thirty years (Field et al. 1998; Polidoro et al. 2010). Additionally, the eastern Pacific Ocean is resource limited compared with the world's other oceans, which likely lessens the growth rate, survival, and reproduction of leatherback turtles in these regions (see chapter 2). If this is true, it is not hard to conceive that similar large-scale resource-limiting factors might be negatively affecting hawksbill population size in the region.

In summary, it appears that eastern Pacific hawksbills are using unique and previously undocumented habitats for both nesting and foraging, in locations where researchers simply had not looked. Coupled with the reduced number of individuals surviving in the region (albeit more than previously thought), one can begin to understand why hawksbills went undetected for so long.

Developing a Regional Perspective

With sightings of eastern Pacific hawksbills infrequent and dispersed across such large swaths of ocean and coastline, individual reports might at first seem insignificant, particularly to researchers working in the field who observe at most a few individuals each season. Given the imperiled state of the population, however, it is evident that every event represents an important piece of the puzzle in understanding and protecting the population. Discovering areas where hawksbills continue to aggregate was the first step in determining that hope remains for the population and was made possible only by sharing what many individuals initially considered to be insignificant information. The Second Workshop of the Hawksbill Turtle in the Eastern Pacific was held in Estero Padre Ramos, Nicaragua, in July 2010 and was attended by more than fifty researchers. The growth of the vibrant Eastern Pacific Hawksbill Initiative network and the continued sharing of information and experiences have and will continue to be fundamental to effective conservation and recovery efforts.

As with sea turtles everywhere, eastern Pacific hawksbills cannot be saved by focusing on a single nesting beach or by protecting a single life stage (Carr 1967). Likewise, because hawksbills use multiple habitats that often span multinational jurisdictions, they cannot be saved by individual organizations or political institutions. Carefully planned and executed multinational conservation strategies involving members from private, public, and governmental sectors, such as that being implemented by Eastern Pacific Hawksbill Initiative, may be the most effective way to save eastern Pacific hawksbills.

Current and future projects to protect the population must incorporate extensive monitoring, as conservation strategies and management plans require reliable long-term quantitative and qualitative information to be effective (Taylor 1995; Brook et al. 1997). We must continue to gather biological information via nesting beach and in-water projects that will enable us to better understand the population and determine such things as demography, population trends, genetic stocks, and conservation priorities. Once available, the information will better enable scientists, managers, and politicians to invoke the laws and strategic planning necessary for recovery of the population. For example, El Salvador—largely motivated by the documentation by local nongovernmental organizations of hawksbill nesting and subsequent egg collection by humans, and catalyzed by the Eastern Pacific Hawksbill Initiative workshop—officially prohibited the collection, consumption, and sale of sea turtle eggs and products (Executive Orders 343 and 74, 4 February 2009), after decades of nonaction. The move represents one of the most significant advancements for hawksbill conservation in the eastern Pacific to date and was made possible only through detailed data capture and international collaborative efforts.

Equally important to leadership by governmental and nongovernmental institutions is leadership by local community members. The future survival of hawksbills may ultimately lie in the hands of these stakeholders, especially fishermen, and ensuring that they are convinced the species is worth saving is a task not easily achieved. Doing so depends greatly on field researchers, who must engage local inhabitants on a very human level, approaching them as partners in trying to achieve the same end, a better future for hawksbill turtles and marine resources in general (plates 15 and 16).

The Road to Recovery

The essential parameter to survivability of wildlife populations is their size (O'Grady et al. 2004), so the fundamental need of our efforts to save eastern Pacific hawksbills is to increase the number of individuals. One way to achieve this is by augmenting hatchling production by establishing nest protection programs. The recently established projects mentioned in this chapter are essential in this respect. However, it is crucial that new nesting areas be identified and similar projects be developed in other parts of the eastern Pacific.

Equally important to the protection of nests, eggs, and hatchlings is the protection of the reproductively active adults. As several previous studies have pointed out (e.g., Crowder et al. 1994; Heppell et al. 1996; Heppell 1998), sea turtle population models illustrate that survival of late-stage individuals (e.g., adult and subadult turtles) has a much larger impact on the annual growth rate of a population than the survival of early-stage individuals (e.g., small juveniles). In-water monitoring is a key tool in our ability to identify aggregation areas and migration corridors where we can subsequently attempt to provide protection and mitigate direct threats to different life stages.

In reality, we must find ways to protect all life stages of the population if we expect it to have a real chance to recover. In the case of eastern Pacific hawksbills, a small and immediately threatened population, we must do so rapidly in order to initiate recovery as quickly as possible. Interviews with fishermen and coastal community members are effective methods for identifying priority conservation areas by helping to determine exactly where to search for these elusive turtles. Interviews are also an effective means of raising awareness of eastern Pacific hawksbills among these stakeholders and thus also tackle one of the major threats to the population.

Next Steps

Among the many urgent needs, we must identify hawksbill aggregation areas both in water and on land and initiate conservation projects at those sites. Pacific Panamá lacks investigation, and our preliminary surveys of the western coast yielded evidence for both foraging and nesting by the species. The Darién Gap region shared between Panamá and its neighbor Colombia remains a totally unexplored frontier and one that, in addition to offering fascinating adventure and discovery, might also prove vital for the survival of eastern Pacific hawksbills. Although these underexplored areas might seem to be the most likely places to find new hawksbill hotspots, if the findings to date have taught us anything, it is that the species actually might be existing literally "under our noses."

The origin of the juvenile hawksbills in the Gulf of California continues to go unexplained. The ocean currents moving northward off coastal El Salvador (North Equatorial Current) and Ecuador (Humboldt Current) both shift in westerly directions before reaching the Gulf of California, making these sites unlikely sources. Although it is possible that hawksbill hatchlings originating from El Salvador or Nicaragua disperse out of the North Equatorial Current before it shifts west, more plausible seems the existence of a yet to be discovered nearby nesting rookery. A tantalizing possibility is a group of islands called the Islas Tres Marías, located about 90 km west of the Mexican state of Nayarit. The islands historically hosted important nesting aggregations of the hawksbills (Zweifel 1960; Woodes 1711, cited in Saenz-Arroyo et al. 2006) but is the location of a federal penal colony established in the early 1900s, so entrance has been and continues to be prohibited. The island's waters and beaches have never been surveyed by sea turtle scientists and are rumored to still harbor important numbers of nesting and foraging hawksbills (R. Briseño, pers. commun., July 2008).

There is no doubt that eastern Pacific hawksbills continue to be one of the most mysterious and endangered sea turtle populations on the planet, and much work still needs to be done for us to understand and effectively conserve the species in this region of the world. Fortunately, talking about their recovery is no longer an abstract idea but rather a tangible goal. It is critical that we act immediately and take advantage of what is likely our last chance to save the population, rather than risk future references to this as a time when that chance was lost.

ACKNOWLEDGMENTS

We gratefully acknowledge the numerous groups and individuals that have helped make our research possible. Special thanks to the many individual fishermen, community groups, and organizations in northwest México; to Juan de la Cruz and his wife, Norma, as well as the fisher families of El Pardito, who helped us capture our first hawksbills; to Wallace "J" Nichols for brainstorming early conceptions of our efforts in northwest México and being the source of many ideas for hawksbill work throughout the region; to Kama Dean and Chris Pesenti for their key organizational support during our fledgling efforts; and to the editor and our friend Jeffrey Seminoff for putting this together and providing unyielding support for our endeavors, as well as catalyzing much of the hawksbill research currently taking place in the eastern Pacific. We also gratefully recognize the initial seed funds provided by the National Fish and Wildlife Foundation, People's Trust for Endangered Species, and the U.S. Fish and Wildlife Service. Finally, we thank our families for sharing our enthusiasm and our son, Joaquinn, for being such a good sport during the many adventures.

REFERENCES

Aschmann, H. 1966. The Natural and Human History of Baja California—from Manuscripts by Jesuit Missionaries in the Decade 1752–1762. Publication No. 7. Dawson's Book Shop, Los Angeles, CA.

Brook, B. W., L. Lim, R. Harden, and R. Franklin. 1997. Does population viability analysis software predict the behavior of real populations? A retrospective study on the Lord Howe Island woodhen *Tricholimnas syvestris*. Biological Conservation 82:119–128.

Caldwell, D. K. 1962. Sea turtles in Baja California waters (with special reference to those of the Gulf of California), and the description of a new subspecies of northeastern Pacific green turtle. Contributions in Science Los Angeles City Museum 61:1–31.

Carr, A. 1956. The Windward Road. Florida State University Press, Tallahassee.

Carr, A. 1967. So Excellent a Fishe. Natural History Press, NY.

Carr, A., H. Hirth, and L. Ogren. 1966, The ecology and migration of sea turtles, 6. The hawksbill turtle in the Caribbean Sea. American Museum Novitates 2248:1–29.

Chacón, D., and R. Arauz. 2002. Diagnóstico sobre el comercio de las tortugas marinas y sus derivados en el istmo centroamericano. Red Regional para la Conservación de las Tortugas Marinas en Centroamérica, San José, Costa Rica.

Cliffton, K., D. O. Cornejo, and R. S. Felger. 1982. Sea turtles of the Pacific coast of Mexico. In: Bjorndal, K. A. (Ed.), Biology and Conservation of Sea Turtles. Smithsonian Institution Press, Washington, DC. Pp. 199–209.

Crowder, L. B., D. T. Crouse, S. S. Heppell, and T. H. Martin 1994. Predicting the impact of turtle excluder devices on loggerhead sea turtle populations. Ecological Applications 4:437–445

Felger, R. S., and M. B. Moser. 1985. People of the Desert and Sea: Ethnobotany of the Seri Indians. University of Arizona Press, Tucson.

Field, C. B., J. G. Osborn, L. L. Hoffmann, J. F. Polsenberg, D. D. Ackerly, J. A. Berry, O. Bjorkman, Z. Held, P. A. Matson, and H. A. Mooney. 1998. Mangrove biodiversity and ecosystem function. Global Ecology and Biogeography Letters 7:3–14.

Gaos, A. R., F. A. Abreu-Grobois, J. Alfaro-Shigueto, D. Amorocho, R. Arauz, A. Baquero, R. Briseño, D. Chacón, C. Dueñas, C. Hasbún, M. Liles, G. Mariona, C. Muccio, J. P. Muñoz, W. J. Nichols, M. Peña, J. A. Seminoff, M. Vásquez, J. Urteaga, B. P. Wallace, I. Yañez, and P. Zárate. 2010. Signs of hope in the eastern Pacific: international collaboration reveals encouraging status for severely depleted population of hawksbill turtles. Oryx 44: 595–601.

Gaos, A. R., R. Arauz, and I. L. Yañez. 2006. Hawksbill turtles on the Pacific coast of Costa Rica. Marine Turtle Newsletter 112:14.

Gaos, A. R., R. L. Lewison, M. Liles, W. J. Nichols, A. Baquero, C. R. Hasbún, M. J. Vasquez, J. Urteaga, and J. A. Seminoff. 2011. Shifting the life-history paradigm: discovery of novel habitat use by hawksbill turtles. Biology Letters; doi:10.1098/rsbl .2011.0603.

Grant, G. S., P. Craig, and G. H. Balazs. 1997. Notes on juvenile hawksbill and green turtles in American Samoa. Pacific Science 51:48–53.

Heppell, S. S. 1998. Application of life-history theory and population model analysis to turtle conservation. Copeia 2:367–375.

Heppell, S. S., D. T. Crouse, and L. B. Crowder 1996. A model evaluation of headstarting as a management tool for long-lived turtles. Ecological Applications 6:556–565.

Hildebrand, H. 1963. Discovery of the nesting site of the ridley sea turtle on the west coast of the Gulf of Mexico. Ciencia 22:105–112.

King, F. W. 1979. Historical review of the decline of the green turtle and the hawksbill. In: Bjorndal, K. A. (Ed.), Biology and Conservation of Sea Turtles. Smithsonian Institution Press, Washington, DC. Pp. 183–188.

Lewison, R. L., and L. B. Crowder. 2007. Putting longline bycatch of sea turtles into perspective. Conservation Biology 21:79–86.

Lewison, R. L., S. A. Freeman, and L. B. Crowder. 2004. Quantifying the effects of fisheries on threatened species: the impact of pelagic longlines on loggerhead and leatherback sea turtles. Ecology Letters 7:221–231.

Liles, M. J., M. V. Jandres, W. A. López, G. I. Mariona, C. R. Hasbún, and J. A. Seminoff. 2011. Hawksbill turtles (*Eretmochelys imbricata*) in El Salvador: nesting distribution and mortality at the largest remaining nesting aggregation in the eastern Pacific Ocean. Endangered Species Research 14:23–30.

Mast, R. B., B. J. Hutchinson, and N. J. Pilcher. 2004. IUCN SSC Marine Turtle Specialist Group news first quarter 2004. Marine Turtle Newsletter 104:21–22.

Mortimer, J. A., and M. Donnelly. 2008. Eretmochelys imbricata. In: IUCN Red List of Threatened Species, vol. 2010.1. Available: www.iucnredlist.org [accessed 17 April 2009].

Nichols, W. J. 2003. Biology and conservation of sea turtles in Baja California, Mexico. Ph.D. dissertation, University of Arizona, Tucson.

O'Grady, J. J., D. H. Reed, B. W. Brook, and R. Frankham. 2004. What are the best correlates of predicted extinction risk? Biological Conservation 118:513–520.

Peckham, S. H., D. D. Maldonado, A. Walli, G. Ruiz, L. B. Crowder, and W. J. Nichols. 2007. Small-scale fisheries bycatch jeopardizes endangered Pacific loggerhead turtles. PLoS One 10:e1041.

Polidoro, B. A., K. E. Carpenter, L. Collins, N. C. Duke, A. M. Ellison, J. C. Ellison, E. J. Farnsworth, E. S. Fernando, K. Kathiresan, N. E. Koedam, et al. 2010. The loss of species: mangrove extinction risk and geographic areas of global concern. PLoS One 5:e10095; doi:10.1371/journal.pone.0010095.

Saenz-Arroyo, A., C. M. Roberts, J. Torre, M. Cariño-Olvera, and J. P. Hawkins. 2006. The value of evidence about past abundance: marine fauna of the Gulf of California through the eyes of 16th to 19th century travelers. Fish and Fisheries 7:128–146.

Seminoff, J. A., W. J. Nichols, A. Resendiz, and L. Brooks. 2003. Occurrence of hawksbill turtles, Eretmochelys imbricata, near Baja California. Pacific Science 57:9–16.

Taylor, B. L. 1995. The reliability of using population viability analysis for risk classification of species. Conservation Biology 9:551–558.

van Dam, R. P., and C. E. Diez. 1996. Diving behavior of immature hawksbills (Eretmochelys imbricata) in a Caribbean cliff wall habitat. Marine Biology 127:171–178.

Wallace, B. P., A. D. DiMatteo, A. B. Bolten, M. Y. Chaloupka, B. J. Hutchinson, F. A. Abreu-Grobois, J. A. Mortimer, J. A. Seminoff, D. Amorocho, K. A. Bjorndal, J. Bourjea, B. W. Bowen, R. Briseño-Dueñas, P. Casale, B. C. Choudhury, A. Costa, P. H. Dutton, A. Fallabrino, E. M. Finkbeiner, A. Girard, M. Girondot, M. Hamann, B. J. Hurley, M. López-Mendilaharsu, M. A. Marcovaldi, J. A. Musick, R. Nel, N. J. Pilcher, S. Troëng, B. Witherington, R. B. Mast. 2011. Global Conservation Priorities for Marine Turtles. PLoS One 6(9):e24510.

Witzell, W. 1983. Synopsis of biological data on the hawksbill turtle, Eretmochelys imbricata (Linnaeus, 1766). Food and Agriculture Organization Fisheries Synopsis 137.

Zweifel, R. G. 1960. Herpetology of the Tres Marías Islands. In: Results of the Puritan-American Museum of Natural History Expedition to Western Mexico. Bulletin of the American Museum of Natural History 119(2).

11

Current Conservation Status of the Black Sea Turtle in Michoacán, México

CARLOS DELGADO TREJO AND
JAVIER ALVARADO DÍAZ

Summary

The coast of the state of Michoacán, México, collectively hosts one of the largest nesting populations of the black turtle (*Chelonia mydas agassizii*) in the eastern Pacific Ocean (plate 5). Historical records indicate that around 25,000 black turtle females nested every year at Colola, the main black turtle rookery in Michoacán. However, population numbers were severely reduced during the 1960s and 1970s as a result of the combined effect of harvest of breeding adults and poaching of nests; by the 1980s the population was near local extinction. In 2000, almost thirty years after black turtle conservation efforts began in Michoacán, the first signs of population recovery were recorded. Between that year and 2007, black turtle nesting numbers ranged between 1,500 and 2,000 females. The upward trend in black turtle nesting females recorded at Colola in recent years is undoubtedly a result of the conservation efforts of the local Nahua communities, which resulted in the protection of more than 45,000 nests and the production and release of more than 2.5 million hatchlings. However, while higher numbers of females nesting at Colola during these seven years represent early signs of population recovery, they are still at least an order of magnitude lower than the numbers of females nesting at Colola in the mid-1960s. Moreover, high rates of black turtle mortality registered in foraging grounds of Baja California and Sonora jeopardize the continued recovery of the Michoacán black turtle breeding population. Reduction of mortality in both breeding and feeding areas is critical to sustain and augment the recuperation of the Michoacán black turtle population.

Biology of the Black Turtle

Over geologic time, green turtles were able to mix across the Pacific and Atlantic ocean basins, because of a gap between the landmasses of North America and South America (Bowen and Karl 1997). However, the uplift of the Central American isthmus approximately four million years ago created an impassable barrier between these populations—the water around the southern ends of Africa and South America apparently is too cold for these tropical animals to pass through. Consequently, green turtles in the eastern Pacific have diverged at least morphologically—and perhaps in other ways—from other green turtle populations.

Most striking among these differences is the darker coloration pattern characteristic of the eastern Pacific green turtle, also known as the black turtle (see plate 4). The carapace and dorsal surfaces of head and flippers of the adult black turtle are olive-green to dark gray or black, and the plastron varies from whitish-gray to bluish or olive-gray. Hatchlings are black to dark gray above and white below, with a border around the dorsal edge of the carapace and flippers. Juveniles are brightly colored with a mottled or radiating carapacial pattern of light and dark brown, olive, and yellow (Caldwell 1963).

In addition to their distinctive coloration, the carapace of an adult black turtle is narrower, more strongly vaulted, and more indented over the rear flippers than that of the green turtle (Cornelius 1986). The black turtle is also markedly smaller and lighter than the green turtle, measuring between 70 and 90 cm in curved carapace length (CCL), compared with more than 90 cm typically seen in green turtles in other parts of the world (Seminoff 2007).

The black turtle is found along the Pacific coast of the Americas, from San Diego Bay, California, USA, and Baja California in the north to southern Perú, including such islands as Islas Revillagigedos and the Galápagos Islands (see chapter 3). In the Pacific coastal waters of México and Central America, the black turtle is the second most common sea turtle after the olive ridley (*Lepidochelys olivacea*). The main black turtle nesting sites are located in the state of Michoacán, México, and in the Galápagos Islands, Ecuador, with several other notable nesting sites along the American Pacific coastline (Seminoff 2007; plate 5).

Black turtles occupy three habitat types: high-energy beaches, convergence zones in the pelagic habitat, and benthic feeding grounds in relatively shallow, protected waters. Hatchlings leave the beach and apparently—as has been reported for the green turtle—move into convergence zones in the open ocean (Carr 1986). When they reach approximately 40 cm CCL, they leave the pelagic habitat and enter benthic feeding grounds (Seminoff et al. 2002b).

Adult black turtles are primarily herbivorous, feeding on seagrasses and algae. However, in some areas the black turtle feeds on a variety of animals in addition to plants. In Perú the black turtle's diet has been recorded to include algae (*Macrocystis* sp., *Rhodymenia* sp., *Gigartina* sp.), mollusks (*Nassarius* sp., *Mytilus* sp., *Semele* sp.), polychaetes, jellyfish, amphipods, and fish (sardine and anchovy) (Hays-Brown and Brown 1982). In the Galápagos Islands, Pritchard (1971) registered algae (*Caulerpa* sp., *Ulva* sp.) and mangrove leaves (*Rhizophora mangle*) as part of the black turtle's diet.

Black turtle nesting seasons vary with location, as typical mainland rookeries (and Galápagos) host nesting from late in the year through the early months of the next year: March–July at the islands of Socorro and Clarión (Márquez 1990), December–May on the Galápagos Islands (Green and Ortiz-Crespo 1982), October–March at Playa Naranjo, Costa Rica (Cornelius 1986), and August–January in Michoacán.

Black Turtles in Michoacán, México

Michoacán nesting beaches support more than one-third of the black turtle population in the eastern Pacific (Cliffton et al. 1982; Seminoff 2007). In Michoacán, the black turtle nests on twenty-three beaches situated within Nahua Indian lands, between the Río Nexpa estuary and the Faro de Bucerías Bay (fig. 11.1, table 11.1). Maruata Bay and the beach of Colola are the main rookeries. Known feeding areas for the Michoacán breeding population are located in México and Central America, mainly the west coast of the Península de Baja California (Laguna Ojo de Liebre, Laguna San Ignacio, and Estero Coyote), Bahía Magdalena and Canal de Infiernillo in the Gulf of California, Laguna Inferior and Lagunas Superior in Oaxaca, and the Golfo de Fonseca (Honduras, El Salvador, Nicaragua) and

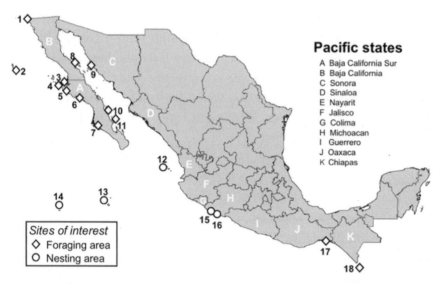

Pacific states

A Baja California Sur
B Baja California
C Sonora
D Sinaloa
E Nayarit
F Jalisco
G Colima
H Michoacan
I Guerrero
J Oaxaca
K Chiapas

Sites of interest
◇ Foraging area
○ Nesting area

FIGURE 11.1. Map of México showing foraging and nesting areas of interest for the east Pacific green turtle (Adapted from Seminoff 2007). See Table 11.1 for names of numbered sites.

Poza del Nance (Guatemala). As with black turtles generally, individuals from the Michoacán population are mainly herbivorous as adults.

In Michoacán, nesting occurs between August and January, with a peak in October–November (Márquez 1990). At Colola and Maruata, black turtle nesting occurs after the nesting peak of the olive ridley and before that of the leatherback. Black turtle females in Michoacán typically nest in three-year cycles, although four- and five-year cycles are common, and they lay from one to seven clutches (averaging of three) per season, at about twelve-day intervals. Clutch size in Michoacán averages 69.3 eggs and ranges from 1 to 130 eggs.

In most studied populations of sea turtles, occurrence of mating decreases once nesting has commenced, for example, green turtle (*Chelonia mydas*) in Australia (Booth and Peters 1972; Limpus 2008) and Hawaii (Balazs 1980). In contrast, black turtle mating in Michoacán apparently occurs both before and between nesting events (Alvarado and Figueroa 1991), although the significance of this behavior, apparently unique to the black turtle, is unknown. Considering the growth rate of juvenile black turtles in Baja California reported by Seminoff et al. (2002b), we estimate that black

TABLE 11.1. Summary of sites of interest for green turtles along the Pacific coast of México. See figure 11.1 for locations.

1. San Diego Bay (USA)	10. Loreto Bay National Park
2. Isla Guadalupe	11. Canal de San Jose
3. Laguna Ojo de Liebre	12. Tres Marías Islands
4. Bahía Tortugas	13. Socorro (island), Revillagigedos
5. Estero Coyote	14. Clarión (island), Revillagigedos
6. Laguna San Ignacio	15. Colola Beach, Michoacán
7. Bahía Magdalena	16. Maruata Beach, Michoacán
8. Bahía de los Angeles	17. Laguna Mar Muerta
9. Canal del Infiernillo	18. Poza de Nance (Guatemala)

turtles breeding in Michoacán reach sexual maturity at between twenty-four and thirty-eight years of age.

According to tag-recovery data, black turtle migrations occur between the northern and southern extremes of their range. Recoveries of nesting females tagged in Michoacán have been documented from Central America (El Salvador, Guatemala, Nicaragua, Costa Rica, and Colombia) and from Mexican waters, primarily from the Gulf of California but also from the coast of Oaxaca. Of ninety-four documented recaptures during 1989–2000, forty-four were incidental catches, mainly from shrimp trawlers, mostly in coastal waters. The longest recorded distance covered by a black turtle was 3,160 km by a turtle tagged at Michoacán and recovered at Charambira, Colombia. Other adult female black turtles have traveled more than 1,500 km between Michoacán breeding areas to marine feeding areas (Byles et al. 1995; Seminoff et al. 2002a).

Human Use of Black Turtles

As with other sea turtle species, black turtles have been markedly affected by human exploitation, specifically due to consumption of meat and eggs (plate 12). The vast majority of black turtle nesting in the eastern

Pacific is restricted to a few major rookeries, all of which have supported extensive subsistence use and facilitated commercial land-based extraction of nests and nesting females. Moreover, the concentration of different segments of the population in localized feeding and/or developmental grounds and the occurrence of regular seasonal migrations have also facilitated a subsistence and commercial ocean-based extraction of adult and juvenile black turtles in the region.

Subsistence use of sea turtles and their eggs has occurred at many coastal places since precivilization times and has continued to modern times (Campbell 2007). Eggs are nutritious, easy to transport, and frequently in demand beyond the immediate extraction area. Over time, commercial extraction developed at some locations, expanding on or supplanting subsistence use (Nietschmann 1982). The black turtle's concentrated nesting and breeding in a few coastal areas in the eastern Pacific facilitated efficient extraction and transportation of eggs and turtles to regional and national markets and contributed to industrial-scale harvest of eggs on several rookeries.

The consumption of eggs and turtles in the eastern Pacific predates the arrival of Europeans (Nabhan 2003), and Márquez et al. (1990) noted the importance that black turtles played in the culture and diet of Mexican indigenous groups since pre-Columbian times, including the Comcáac (Seri) of the Sonora coast, the Nahuas in Michoacán, and the Huaves in Oaxaca. The Comcáac exploited black turtles in the Gulf of California, and turtle meat was one of the main components in their diet at least until the end of the 1800s (Caldwell 1963). The Comcáac also used turtle shells for housing material, and flipper skin was used as footwear. For the Nahua Indians of Michoacán, black turtle eggs were an important dietary staple (Alvarado and Figueroa 1991). In Oaxaca, Huave Indians traditionally consumed the meat of juvenile and adult black turtles. However, with the introduction of commercial demand for sea turtle products, the traditional subsistence use of the black turtle was replaced by a rapidly expanding commercial exploitation.

In northern Mexican feeding grounds, black turtles were first fished heavily at the turn of the century, when an estimated 1,000 black turtles per month were shipped from the Pacific side of Península de Baja California (Bahía Magdalena, Bahia Tortugas, Laguna Ojo de Liebre, and Laguna de San Ignacio) and the Gulf of California to San Diego and San Francisco in

FIGURE 11.2. During the days of legal harvest, black turtles were hunted and harvested from foraging areas in northwest México, in such places as Sonora, México (Top), and from waters adjacent to the nesting beaches of Michoacán (Bottom) (photos courtesy of Kim Cliffton). In the photo on the bottom, notice the preponderance of male (i.e., long-tailed) black turtles: in 1980, the Mexican government allowed a legal take of 250 male black turtles per month between September and December.

California (Cliffton et al. 1982). By the 1930s, the market for sea turtle meat had decreased in the United States, while in México, especially in the states of Baja California Norte, Baja California Sur, Sonora, and Sinaloa, the demand for turtle meat grew steadily (fig. 11.2). From 1956 to 1963, black turtles harvested in the northern Mexican feeding grounds were the most important component of the Mexican turtle fishery, with a total live weight of 3,430 metric tons (Groombridge and Luxmoore 1989).

While commercial exploitation of black turtles in feeding areas had been occurring for decades, exploitation of black turtles in nesting areas in Michoacán began much later. As late as the 1940s, coastal nesting sites were relatively undisturbed as the Michoacán coastline was virtually uninhabited and inaccessible. The Nahua people made sporadic trips from their pueblos in the sierra back to the black turtle nesting beaches to collect eggs, which were then transported by mule back to the villages. Eggs were eaten fresh or hard-boiled and dried for storage to supplement their diet. For many decades, Nahua people obtained their limited egg extraction without significantly affecting the black turtle population. However, during the 1950s, coastal areas were increasingly cleared, and the settlements of Maruata and Colola were established. During the 1960s, commercial markets for sea turtle products, including leather, eggs, and meat, expanded, and an intense exploitation was under way along the Mexican Pacific coast. When a market for sea turtle products was introduced to the Nahua in Michoacán in the early 1970s, the settlements of Colola and Maruata grew rapidly. In the early 1970s approximately 70,000 eggs were extracted each night during the nesting season at Colola beach, and an additionally 10,000–20,000 from Bahía de Maruata (Cliffton et al. 1982).

Until the early 1960s, sea turtle leather processing in México had occurred on a local, artisan level (Groombridge and Luxmoore 1989). From the mid-1960s to mid-1970s, black turtles in Michoacán were exploited mainly for their skin. During the mid-1970s breeding seasons, local fishermen were capturing forty to eighty black turtles each day in Maruata Bay— approximately 7,000–15,000 individuals per year—using shark gill nets in front of the nesting beaches (fig. 11.2; Cliffton et al. 1982). At about the same time, a coastal highway was completed that passed by Colola and Maruata, which provided the means to increase commerce in black turtle products between the nesting beaches in Michoacán and other states within México.

Beginning in the 1970s, the Mexican government attempted to intervene and manage the black turtle exploitation. In 1979, it declared a closed season for the nesting beaches, to gain control of the black turtle fishery. Nonetheless, approximately 3,000 black turtles were illegally taken that year (Cliffton et al. 1982). In 1980, the government allowed a legal take of 250 male black turtles per month between September and December. In 1990, all marine turtle exploitation in México was banned (DOF 1990).

However, an illegal black turtle fishery has continued along the Pacific coast of México, especially in Baja California Sur, Baja California Norte, Sinaloa, and Sonora. For example, a recent study estimated that roughly 35,000 turtles (mainly black turtles) were killed yearly either by illegal hunting or as bycatch, just along the Península de Baja California (Nichols 2003).

Black Turtle Population Trends

Before commercial extraction, the black turtle was abundant in the region. Historically high numbers of black turtles used the feeding grounds within the Gulf of California and along the Pacific coast of the Península de Baja California (Cliffton et al. 1982). An indication of its previous abundance is given in the report of the visit of the vessel *Albatross* to Bahia Tortugas on the Pacific coast of the Península de Baja California in 1889, when 200 black turtles were captured in a single haul by a 200-m seine (Parsons 1962). As late as the 1960s the black turtle was still abundant in the nesting area of Michoacán. During the late 1960s, an estimated 500–1,000 females nested nightly in Colola beach during peak season. Cliffton et al. (1982) extrapolated that about 25,000 black turtle females nested annually at Colola at that time. Earlier observations corroborate this figure; Peters (1956) reports tracks of some 250 turtles on a 0.8-km stretch of beach at Bahía de Maruata in August 1950, two months before peak nesting. Cliffton et al. (1982) estimated that about 900 turtles must have nested at Bahía de Maruata within several days of Peters's (1956) observations.

As a result of the intense exploitation throughout its range, black turtle numbers had been significantly reduced in Michoacán nesting grounds by the late 1970s. By the time conservation actions were begun at Colola and Maruata in 1981, the number of nesting females at Colola had been reduced by more than an order of magnitude relative to the late 1960s. In the 1980s, an average of 400 black turtle females per season nested at Colola (from a minimum of 93 turtles in 1988 to 1,090 turtles in 1982). Beginning in 2000 an upward trend in black turtle females nesting at Colola has been detected. From 2001 to 2007 numbers of nesting females per season averaged nearly 1,500 females per season, ranging from 1,154 in 2006 to 2,500 in 2001. The upward trend registered in recent years suggests the beginning of the recovery of the black turtle population in Michoacán (fig. 11.3). Although this

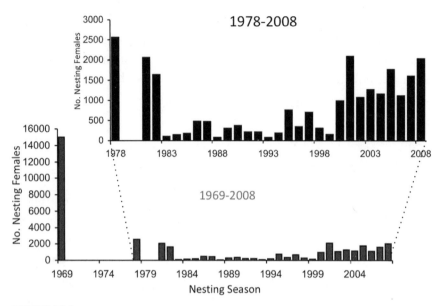

FIGURE 11.3. Annual nesting trends for black turtles at Colola, the primary nesting beach along the Pacific coast of México. Annual trends are displayed for two intervals: 1969–2008 (bottom), to provide context of historic declines, and 1978–2008 (top), to highlight the recent increases in annual nesting.

trend is encouraging, the recovery of the population to historical levels (i.e., pre-1960s) is still far from being achieved.

What explains the recent upward trend registered for the black turtles in Michoacán? Limpus and Nicholls (1988) report that numbers of nesting females in *Chelonia mydas* may be related to the oceanographic phenomena known as El Niño–Southern Oscillation, which affect conditions on turtle feeding grounds and thus affect turtle reproduction (see chapter 2). However, a study evaluating the possible relation between El Niño occurrences in the eastern tropical Pacific and nesting numbers of black turtles in Michoacán during 1985–2003 indicated no correlation (Calvillo 2007). Although further research is needed to uncover possible relationships between oceanographic conditions and patterns of black turtle reproduction, the recent increase in black turtle nesting populations appears to be due to something other than environmental factors.

Black Turtle Conservation in Michoacán

In 1979, Kim Cliffton, then of the Arizona-Sonora Desert Museum in Tucson, Arizona, began a conservation project for the black turtle in Michoacán. In 1982, the responsibility for the project passed to the Universidad de Michoacán, with support of the Pomaro fishing cooperative from Maruata, the turtle group from Colola, the U.S. Fish and Wildlife Service, and the World Wildlife Fund-USA. Since 2004, full responsibility of the black turtle conservation project in Michoacán has belonged to local groups from Maruata and Colola, with the support of the Universidad de Michoacán and the U.S. Fish and Wildlife Service.

Cliffton began the conservation project in 1978, and by 1980 the Mexican government had established a legal annual (September–December) quota of 250 adult black turtle males per indigenous fishing cooperative along the Mexican Pacific coast. However, at that point the Maruata fishing cooperative Pomaro voluntarily declined its right to the quota as a measure to prevent the disappearance of Maruata Bay's black turtles (Cliffton et al. 1982). This attitude of the fishing community of Maruata toward sea turtle conservation has remained to the present.

The black turtle conservation project in Michoacán was designed with the dual goals of first stopping the decline of black turtle population and then recovering the population to historical numbers. Strategic objectives were (1) to incorporate local communities in the conservation activities and eventually pass full responsibility of the project to them, (2) to generate scientific information about key aspects of black turtle natural history and population dynamics to delineate sound management and conservation strategies, (3) to protect at least 90 percent of the females nesting at Colola and Maruata and the same proportion of males in nearshore waters, and (4) to protect at least 80 percent of eggs deposited in Colola and Maruata and the same proportion of hatchlings.

Direct conservation actions in Colola from 1981 to 2007 have resulted in the protection of 45,000 nests and more than 3 million eggs, including egg relocation to beach hatcheries and surveillance of natural nest areas. These efforts have resulted in the production and subsequent release of approximately 2.5 million hatchlings. Most of these activities have been

FIGURE 11.4. Local villager Don Gabino Valencia in conservation activities in Colola hatchery.

carried out by local people, including children, who were economically compensated for each nest translated to the hatcheries (fig. 11.4). Many of the adults in charge of the project today participated as children decades ago. In large part, it is because of the involvement of these locals that the project and management of eggs have prospered. Hatchery results have been particularly good at Colola. Elapsed time between oviposition and hatchery relocation has been less than 45 min, a relatively short time considering Colola's size (4.8 km long and ~95 m wide). Furthermore, emergence success from hatchery nests at Colola has ranged between 76 and 83 percent. Through a coordinated effort of local people, Universidad de Michoacán biologists, and volunteers, approximately 15,000 of black turtle nesting females at Colola have been tagged throughout the years, and population numbers have been monitored for twenty-seven consecutive years.

An important component of any conservation project is evaluation and assessment. This is particularly challenging for sea turtle conservation projects, as the time lag between implementation of management actions and observable results is on the scale of decades. The Michoacán black turtle project has collected more than twenty years of data, which helps assess whether the black turtle project is achieving its conservation goals in Michoacán. To do this, Alvarado-Díaz et al. (2001) used a threat reduction assessment (Margoluis and Salafsky 1998), which indexes success by how much threats are reduced (as a percentage of total potential threat reduction). Considering reduction on main threats as identified at the beginning of the project, such as illegal and legal extraction of adult turtles (commercial and subsistence), illegal egg harvest (commercial and subsistence), and destruction and degradation of nesting habitat, the threat reduction assessment index was 82.4 percent. This score suggests a relatively high level of success for the project in Michoacán. However, if this index were extended to consider the entire geographic distribution of the black turtle, the score likely would be lower, considering as yet unmitigated threats in other parts of the region.

If we consider the black turtle as a metapopulation that uses a series of habitats at different life stages, the nesting beaches of Bahía de Maruata and Colola in Michoacán may very well represent the major habitat source, assuring a level of recruitment into the breeding ranks. Therefore, the main achievement of the project is providing relatively safe conditions in Michoacán for breeding, nesting, and hatchling production. Unfortunately, at least some of the feeding areas represent habitat sinks. For instance, along the Península de Baja California, perhaps the most important feeding area for black turtles in the region, the mortality rate (due to direct and indirect take of adults and juveniles) has been estimated to exceed 10,000 turtles annually (Nichols 2000).

Nonetheless, considering the time elapsed since the beginning of conservation activities at Colola and Maruata in 1979 and the beginning of an increase in the annual nesting population in 2000, the adult females nesting in recent years may well be recruits from hatchlings produced at the start of the project back in the 1970s. Before the start of the nesting beach projects, hatchling production was close to zero because of decades of intense egg extraction.

However, despite the increase in nesting females at these beaches, and in males engaging in courtship and mating activities in the offshore at Colola, the situation in the rest of the historical black turtle nesting areas in Michoacán is somewhat less encouraging. Cliffton et al. (1982) reported significant black turtle nesting on seventeen Michoacán beaches in 1978. In 1982, 42.8 percent of black turtle nesting in Michoacán occurred at Colola and Maruata (the only protected beaches), while the remainder occurred on the other fifteen unprotected beaches. By 2005, the proportion of black turtle nesting on unprotected beaches had fallen to 7.7 percent, suggesting that lack of protection resulted in an almost complete extirpation of the populations nesting at these secondary beaches. Therefore, it is critical to extend the conservation strategies from Colola and Maruata to all secondary nesting beaches and to continue multiannual surveys of all nesting beaches in Michoacán.

Considering the black turtle as a metapopulation using a series of habitats at different life stages, recovery to historical population levels will not be achieved unless protection is extended to all critical habitats throughout its geographic range. Despite the ban that took effect in 1990 on all marine turtle exploitation in México (DOF 1990), an illegal black turtle fishery has continued along the Pacific coast of México, especially in Baja California, Sinaloa, and Sonora. Fortunately, several conservation projects are under way in these areas, working with local fishing communities to reduce sea turtle bycatch. As demonstrated by the Michoacán black turtle conservation project, if these efforts are carried out over the long term, they will contribute to further population recovery of black turtles in the eastern Pacific.

REFERENCES

Alvarado, J., and A. Figueroa. 1991. Comportamiento reproductivo de la Tortuga negra *Chelonia agassizii*. Ciencia y Desarrollo 17:43–49.

Alvarado-Díaz, J., C. Delgado-Trejo, and I. Suazo-Ortuño. 2001. Evaluation of the black turtle project in Michoacán, México. Marine Turtle Newsletter 92:4–7.

Balazs, G. H. 1980. Synopsis of Biological Data on the Green Turtle in the Hawaiian Islands. NOAA Tech. Memo. NMFS-SWFC-7. U.S. Department of Commerce, U.S. National Marine Fisheries Service, Miami, FL.

Booth, J., and J. Peters. 1972. Behavioral studies on the green turtle (*Cheonia mydas*) in the sea. Animal Behavior 20:808–812.

Bowen, B. W., and S. A. Karl. 1997. Population genetics, phylogeography, and molecular evolution. In: Lutz, P. L., and J. A. Muisck (Eds.), The Biology of Sea Turtles. CRC Press, Boca Raton, FL. Pp. 29–50.

Byles, R., J. Alvarado, and D. Rostal. 1995. Preliminary analysis of post-nesting movements of the black turtle (*Chelonia agassizi*) from Michoacán, México. In: Proceedings of the Twelfth Annual Workshop on Sea Turtle Biology and Conservation. NOAA Tech. Memo. NMFS-SEFCS-361. U.S. Department of Commerce, U.S. National Marine Fisheries Service, Miami, FL. Pp. 12–13.

Caldwell, D. K. 1963. The sea turtle fishery of Baja California, México. California Fish and Game 49(3):140–151.

Calvillo, G. Y. 2007. Comportamiento de la anidación de tortuga negra (*Chelonia agassizii*) en la playa de Colola, Michoacán en relación a aspectos climáticos locales y globales. Tesis Profesional, Facultad de Biología, Universidad Michoacana de San Nicolás de Hidalgo, Morelia, Michoacán, México.

Campbell, L. 2007. Understanding human use of olive ridleys: implications for conservation. In: In: Plotkin, P. T. (Ed.), Biology and Conservation of Ridley Sea Turtles. Johns Hopkins University Press, Baltimore, MD. Pp. 23–44.

Carr, A. 1986. New Perspectives on the Pelagic Stage of Sea Turtle Development. NOAA Tech. Memo. SEFSC-190. U.S. National Marine Fisheries Service, Miami, FL.

Cliffton, K., D. O. Cornejo, and R. S. Felger. 1982. Sea turtles of the Pacific coast of México. In: Bjorndal, K. A. (Ed.), Biology and Conservation of Sea Turtles. Smithsonian Institution Press, Washington, DC. Pp. 199–209.

Cornelius, S. E. 1986. The Sea Turtles of Santa Rosa National Park. Fundación de Parques Nacionales, San José, Costa Rica.

DOF. 1990. Acuerdo por el que se establece veda total para todas las especies y subespecies de tortugas marinas en aguas de jurisdicción nacional de los litorales del Océano Pacífico, Golfo de México y Mar Caribe. Diario Oficial de la Federación, México, 31 Mayo, pp. 21–22.

Green, D., and F. Ortiz-Crespo. 1982. Status of sea turtle populations in the central eastern Pacific. In: Bjorndal, K. A. (Ed.), Biology and Conservation of Sea Turtles. Smithsonian Institution Press, Washington, DC. Pp. 221–223.

Groombridge, B., and R. Luxmoore. 1989. The green turtle and hawksbill (Reptilia: Cheloniidae): world status, exploitation and trade. Secretariat of the Convention on International Trade in Endangered Species of Wild Fauna and Flora, Lausanne, Switzerland.

Hays-Brown, C., and W. M. Brown. 1982. Status of sea turtles in the southeastern Pacific: emphasis on Peru. In: Bjorndal, K. A. (Ed.), Biology and Conservation of Sea Turtles. Smithsonian Institution Press, Washington, DC. Pp. 235–240.

Limpus, C. J. 2008. A biological review of Australian marine turtles: green turtle *Chelonia mydas* (Linnaeus). Queensland Environmental Protection Agency, Brisbane, Australia.

Limpus, C. J., and N. Nicholls. 1988. The Southern Oscillation regulates the annual numbers of green turtles (*Chelonia mydas*) breeding around northern Australia. Australian Wildlife Research 15:157–161.

Margoluis, R., and N. Salafsky. 1998. Measures of Success: Designing, Managing, and Monitoring Conservation and Development Projects. Island Press, Washington, DC.

Márquez, M. R. 1990. Sea Turtles of the World. An Annotated and Illustrated Catalogue of Sea Turtle Species Known to Date. FAO Species Catalogue, FAO Fisheries Synopsis 11(125). Food and Agriculture Organization of the United Nations, Rome.

Márquez, R., J. Vasconcelos, and C. Peñaflores. 1990. XXV años de investigación, conservación y protección de la tortuga marina. Instituto Nacional de la Pesca, Secretaría de Pesca, Mexico City, México.

Nabhan, G. 2003. Singing the Turtles to Sea. The Comcáac (Seri) Art and Science of Reptiles. University of California Press, Berkeley.

Nichols, W. J. 2000. Biology and conservation of sea turtles in Baja California, México. Doctoral dissertation, University of Arizona, Tucson.

Nichols, W. J. 2003. Sinks, sewers, and speed bumps: the impact of marine development on sea turtles in Baja California, México. In: Seminoff, J. A. (Comp.), Proceedings of the Twenty-Second Annual Symposium on Sea Turtle Biology and Conservation. NOAA Tech. Memo. NMFS-SEFC-503. U.S. Department of Commerce, U.S. National Marine Fisheries Service, Miami, FL. Pp. 17–18.

Nietschmann, B. 1982. The cultural context of sea turtle subsistence hunting in the Caribbean and problems caused by commercial exploitation. In: Bjorndal, K. A. (Ed.), Biology and Conservation of Sea Turtles. Smithsonian Institution Press, Washington, DC. Pp. 439–455.

Parsons, J. J. 1962. The Green Turtle and Man. University of Florida Press, Gainesville.

Peters, J. 1956. The eggs (turtle) and I. Biologist (39):21–24.

Pritchard, P. C. H. 1971. Galápagos sea turtles-preliminary findings. Journal of Herpetology 5:1–9.

Seminoff, J. A. 2007. Green Sea Turtle (*Chelonia mydas*) 5-Year Review: Summary and Evaluation. U.S. National Marine Fisheries Service, Silver Spring, MD.

Seminoff, J. A., J. Alvarado, C. Delgado, J. L. López, and G. Hoeffer. 2002a. First direct evidence of migration by an East Pacific green sea turtle from Michoacán, México to a feeding ground on the Sonoran coast of the Gulf of California. Southwestern Naturalist 47(2):314–316.

Seminoff, J. A., A. Reséndiz, W. J. Nichols, and T. T. Jones. 2002b. Growth rates of wild green turtle *Chelonia mydas* at a temperate foraging area in the Gulf of California, México. Copeia 3:610–617.

12

Empowering Small-Scale Fishermen to Be Conservation Heroes

A Trinational Fishermen's Exchange to Protect Loggerhead Turtles

S. HOYT PECKHAM AND
DAVID MALDONADO DÍAZ

Summary

Small-scale fisheries are ubiquitous in the world's oceans, and by-catch in these fisheries jeopardizes both fishermen's livelihoods and endangered species as much as any other fishing sector (plate 11). Because management of such fisheries is limited, conservation can depend on small-scale fishermen's direct participation. We empowered key fishermen from Japan, Hawaii, and México through a trinational exchange that offered them first-hand experience of the Pacific-wide impacts of their local bycatch. One of several successful outcomes came to fruition in Honolulu in August 2007 when a Mexican small-scale fisherman made a commitment that would spare thousands of endangered Japanese loggerhead sea turtles. In the embattled realm of marine conservation, in which the work of hundreds of people and millions of dollars usually saves just handsful of endangered animals at a time, Efraín de la Paz Regalado's move was historic. De la Paz's vision and resulting sacrifice distinguish him as a conservation hero, and the story of his transformation from callous to conscientious fisherman is both instructive and inspiring.

Introduction

Speaking before Japanese, Hawaiian, and Mexican fishermen, scientists, and policy makers assembled at a plush Honolulu conference room in August 2007, Efraín de la Paz Regalado, sun-leathered and solemn, vowed that his small fleet would voluntarily cease fishing the bottom-set longlines that were accidentally lethal to hundreds of the endangered turtles each year. In 1985 de la Paz and the twenty or so fishermen established their fishing cooperative at Santa Rosa, a remote inlet of the Baja California peninsula that offered previously unfished stocks. They prospered fishing from their *pangas* (small open boats), hand-hauling traps, tangle nets, and shrimp trawls. But in 2004, as their nearshore catches were waning, de la Paz sent his boats farther offshore with bottom-set longlines, the string of hooks branching off of a long mother line featured in Sebastian Junger's bestseller *The Perfect Storm*. Venturing 35–70 km offshore to fish in the peninsula's legendarily rich shelf waters, de la Paz and his partners soaked their longlines overnight and began catching tons of valuable demersal fish, but they also accidentally killed hundreds of endangered loggerhead turtles (*Caretta caretta*). As recently as 2005 de la Paz and his partners did not realize that their bycatch in México could drastically affect loggerheads back on their nesting grounds in Japan.

Endangered Migrators

Nesting exclusively in Japan, the North Pacific loggerhead has declined dramatically over the past few generations to no more than a couple of thousand turtles nesting per year, qualifying it as Endangered (Kamezaki et al. 2003). The thirty to forty years loggerheads need to reach maturity predispose them to endangerment. The fact that they spend those three to four decades traversing the North Pacific doesn't help; in addition to the natural gauntlet of predators and disease, juvenile loggerheads are exposed to lethal fishing gear throughout their trans-Pacific range. Fish weirs and pound nets around the coast of Japan, pelagic longlines and pirate drift nets throughout the open Pacific, and nets and longlines fished from small-scale vessels off the coast of México combine to kill thousands of loggerheads each year (plate 11).

The Wealth of Baja California Sur

The highly productive waters of Baja California Sur, México, have been revered for centuries for the abundance and diversity of large creatures they attract (Steinbeck 1941; Saenz-Arroyo et al. 2006). Persistent upwelling, fronts, and eddies result in unusually high production of phytoplankton and zooplankton. This richness fuels a perennial abundance of prey, including pelagic red crabs, sardines, and squid, which in turn consistently draw such predators as sea birds, sharks, billfish, tuna, dolphins, whales, sea turtles, and, of course, fishermen. Juvenile loggerheads gather for decades off the peninsula to forage their way to maturity on the region's abundant pelagic red crabs (Nichols 2003). But the unfortunate overlap between fishermen and foraging loggerheads results in one of the highest known rates of turtle bycatch and strandings in the world (Peckham et al. 2008).

The Lighthouse Keeper

Victor de la Toba commutes along a lonely 45-km beach in a dune buggy between his home in the fishing village of Puerto López Mateos, Baja California Sur, and the lighthouse his grandfather helped build in the 1940s at Cabo San Lázaro. The steady flow of the cold, nutrient-rich California Current runs down the peninsula uninterrupted until it slams into Cabo San Lázaro, making it a hazard for ships and a repository of flotsam. Driven by the prevailing wind and current, anything floating off the central coast of the peninsula is likely to wash ashore on Playa San Lázaro.

Lighthouse keeper de la Toba finds countless toothbrushes, plastic bottles, and cigarette lighters along with carcasses of the ocean's big animals, including whales, dolphins, sea lions, seabirds and sea turtles, along Playa San Lázaro each day. In the late 1990s de la Toba noticed an alarming increase in the number of loggerhead carcasses on the beach (fig. 12.1). Rather than speeding down the low-tide beach flats to work at the lighthouse, he had to slow down and dodge the increasing number of turtle carcasses. In 1997 de la Toba contacted Wallace J. Nichols, who at the time was a graduate student at the University of Arizona helping to pioneer sea turtle conservation across northwestern México. Nichols surveyed the beach with de la Toba various times over several years, and they found dozens of loggerhead carcasses each trip, leading Nichols to conclude in his doctoral dissertation that local fisheries bycatch was likely having a substantial impact on the loggerhead population (Nichols 2003).

In 2001 Nichols invited one of the authors (S.H.P.) to research the loggerhead bycatch problem. The first step was to formalize the beach sur-

FIGURE 12.1. Lighthouse keeper Victor de la Toba prepares to measure and process one of thousands of carcasses we have encountered along the 45-km Playa San Lázaro since 2003 (© S. H. Peckham/ProCaguama). Cause of death in this case is obvious—the animal had swallowed a hook and become entangled in the leader and associated gear.

vey. Daily surveys of Playa San Lázaro we have run on all-terrain vehicles since 2003 revealed that 400–500 loggerhead carcasses stranded along the 43-km beach each year, with most stranding in summer fishing months, amounting to one of the highest sea turtle stranding rates reported worldwide (Peckham et al. 2008).

Fishermen openly attributed the strandings to bycatch in local nets and longlines during interviews we conducted at Puerto López Mateos, Baja

California Sur, in 2003, reporting an average of four loggerheads caught per boat per week. Yet despite the relatively high frequency of bycatch they reported, fishermen generally believed that the impact of their bycatch mortality was insignificant. Typical of most of his peers, one fisherman asked us, "How can loggerheads be endangered when I caught forty in my nets just this morning?"

Proyecto Caguama

In 2003 we launched Proyecto Caguama (Project Loggerhead) to assess and mitigate loggerhead bycatch at Baja California Sur. Extending the conservation mosaic model developed by Nichols and the Grupo Tortuguero, northwestern México's regional sea turtle conservation network, we launched a bycatch awareness and mitigation program with the goal of empowering local fishermen to find their own answers to the question above and to act upon their new perspectives. The conservation mosaic consists of strategic integration of three approaches to achieving conservation: (1) building a conservation *network* of fishermen, students, teachers, activists, researchers, managers, and other coastal people; (2) drawing on these partnerships to derive new *knowledge* to develop locally practical bycatch solutions and alternatives; and (3) *communication* of this knowledge in resonant and appropriate ways to avoid bycatch and foster a sustainable sea ethic. Our methodologies together with detailed accounts of the partnerships we developed with net fishers of Baja California Sur can be found in two recent publications (Hall et al. 2007; Peckham et al. 2007).

Unknown Mortality

By 2005 we had made strong progress in partnership with local net fishers in reducing their bycatch, but it was clear from the strandings data that there were other sources of mortality in the region. Late in the summer of 2005, while conducting at-sea surveys, we ventured 100 km north of Puerto López Mateos, our principal study site. We found plenty of turtles, but not as we had hoped. Instead of countless loggerheads basking at the surface in the midday sun, we found a slew of fresh-dead carcasses bobbing askew at the surface, and then we came across several hooked on a longline.

We followed the string of longline buoys, counting carcasses, and eventually found two fishermen in their *panga*, checking and rebaiting their hooks. As we approached, the captain waved amiably while the bowman cut into the throat of a loggerhead carcass to retrieve a hook. The fishermen explained unapologetically that that they had been catching a dozen or so loggerheads a day all week, as had the other five boats in the Santa Rosa cooperative, and that they were frustrated catching so many loggerheads instead of their target sharks. In all we counted seventeen dead loggerheads that day, and we sighted only two live ones. We had stumbled upon a staggering source of loggerhead mortality.

That same afternoon we mobilized our team to drive up to Santa Rosa in order to reach out to the Santa Rosa fishermen and investigate their bycatch. To get to Santa Rosa we drove a couple of hours through the desert from the small village of Puerto López Mateos to the isolated fish camp of Santa Rosa where a dozen tar-paper houses and a chapel huddle at the top of a sand dune overlooking an estuary protected from the open Pacific by a narrow barrier island. When we arrived in the evening, fishermen were loading their day's catch into a freezer truck, and we were struck by the utility of the site, chosen by de la Paz because it offered quick access to an otherwise untapped expanse of the rich waters of Baja California Sur.

The fishermen welcomed us into their camp and their lives, and we set about systematically sharing our knowledge of turtles and documenting their bycatch. For a full week we accompanied the fishermen offshore to learn of the fishery and document bycatch. During seven day-trips that we observed, the Santa Rosa fleet caught twenty-six loggerheads, all but two of which were dead (fig. 12.2). Our sample size was admittedly small, but the fishermen were actually surprised at how few turtles they had caught, indicating that they usually catch more turtles per trip. Extrapolating the bycatch rate across the whole fleet and season, we realized the Santa Rosa fleet was killing more turtles per year than *any* other known source in the Pacific. But like the net fishers from Puerto López Mateos, based on their high bycatch rates the fishermen had always assumed loggerheads were abundant.

Nursery Habitat

If fishermen all over western México were catching loggerheads at rates similar to the Santa Rosa fleet, the conclusion that loggerheads were abundant might be true. But our data suggested otherwise. To elucidate the

FIGURE 12.2. Fishermen from de la Paz's Santa Rosa fleet handle a bycaught loggerhead (© S. H. Peckham/ProCaguama). In 2005 and 2006 the Santa Rosa fleet was catching an average of six loggerheads per day per boat, adding up to an estimated 1,800 turtles for the fleet each year, 90 percent of which were dead.

ecology of loggerheads in the northeast Pacific, we had continued a satellite tracking program that Nichols and colleagues had started in 1997. Because the first couple of animals tagged migrated to and toward Japan, we were amazed to find that the loggerheads we tracked subsequently spent most of their time in a very concentrated area off the coast of Baja California Sur. By 2005 it was clear that juvenile loggerheads forage in a very limited habitat at Baja California Sur, aggregating at extraordinarily high densities in a very small high-use area, or hotspot, whose center lies just off Santa Rosa and Puerto López Mateos (fig. 12.3; Peckham et al. 2007).

Exhibition at Santa Rosa

In Puerto López Mateos we had enjoyed the luxury of having time to develop trusting relationships over several seasons. But in Santa Rosa we had to act faster to address the staggering bycatch rate. We had lived and

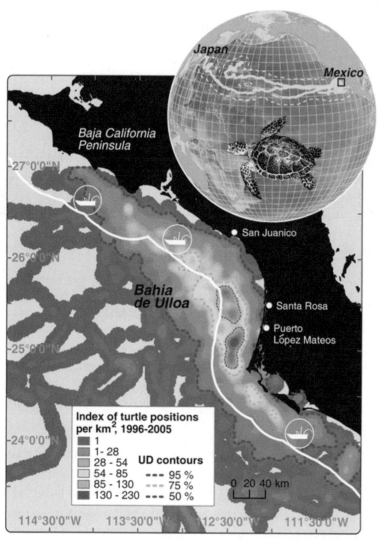

FIGURE 12.3. Loggerhead turtle habitat use in the North Pacific. Tracked loggerheads moved across the North Pacific (inset) but spent most of their time in very small area along the Baja California Sur coast. They spent the most time in a high-use area off Santa Rosa and Puerto López Mateos (the two dark gray clusters shown approximately 50 km offshore of the two ports), well within the 55-km range of the two ports' fishing fleets (white line). Reproduced from Peckham et al. (2007).

fished and eaten and laughed with the fishermen and families of Santa Rosa over the week, but we needed to perform a bycatch intervention.

Our strategy was to create an outdoor exhibition right in Santa Rosa, a spontaneous celebration of turtles and fishing. We rinsed off the side of a freezer truck, set up a projector and sound system off a couple of car batteries, and waited for sundown. We opened with a stirring slide show of images of Santa Rosa and the fishermen we photographed all week set to a couple of popular *Norteña* songs. Next, fisherman and then the coordinator of Grupo Tortuguero, Rodrigo Rangel, took over to share the strandings and tracking data, carefully pointing out the Santa Rosa fleet's overlap with the loggerhead hotspot. Then Rangel detailed the results of our bycatch observations at Santa Rosa and clearly enumerated the extrapolations and potential population impacts, concluding that the future of North Pacific loggerheads might rest largely in the hands of the Santa Rosa cooperative. Finally, we showed a short film we had produced in 2003 titled *Caguamas del Pacifico: In Whose Hands?* that conveys the importance of responsible fishing as told by regional fisher leaders whom the Santa Rosa *cooperativistas* personally recognized.

Before the show we had been worried that the fishermen might get angry and even deny their bycatch. Instead, in a show of the trust we had developed over the week, the fishermen called us out on the data, pointing out that our observations underestimated their bycatch and playfully protesting that we had cut their faces out of all the bycatch photos. Afterward, there was a growing perception among the fishermen that they needed to find a way to reduce their bycatch.

The Patrón

Our outreach was successful but didn't amount to the rapid action for which we had hoped. We had carefully built awareness among the fishermen, hoping each would choose to reduce his bycatch in light of his new perspectives. But we hadn't realized that the fleet was essentially controlled by one man, and he had taken no part in our outreach. Efraín de la Paz Regalado arrived to Santa Rosa each afternoon to take stock of the catch and to give orders for the next day's fishing before quickly returning to business elsewhere.

We came across him one evening on the 20-km sandy track that leads out to the Santa Rosa fish camp. Smoking continuously, gold rings and white teeth glinting in the long desert light, he gave us no more than a couple of minutes to explain our cause and case to him. When he heard our story, he laughed off our concern and summarily dismissed us: "We fish. We don't have the luxury to save a few turtles."

In 2006 we were unable to observe the Santa Rosa fleet until late September. That day our colleague Alex Gaos counted a staggering twenty-one dead loggerheads caught on one *panga*'s 200 hooks in a single day. The other fishermen complained of catching the same number on each of their boats that day, and all that week. No further observations were possible in 2006, because the fleet was making their seasonal switch to shrimp fishing the next week.

Small-Scale Fisheries and Their Unexpected Conservation Leverage

The high bycatch rates of both the Puerto López Mateos and Santa Rosa fleets led us to reflect on the impacts of small-scale fisheries elsewhere. Small-scale fisheries, including artisanal, traditional, and subsistence fisheries, are ubiquitous to coastal waters worldwide, employing more than 99 percent of the world's 51 million fishers (Berkes et al. 2001; Lewison and Crowder 2007). Surprisingly, the overall bycatch mortality caused by the two fleets we observed rivaled that of industrial-scale fleets that span the entire North Pacific. We were alarmed to conclude that our findings with the Santa Rosa and Puerto López Mateos fleets may be representative of effects worldwide wherever small-scale fisheries and the high-use areas of sea turtles and other megafauna overlap; small-scale fisheries may be among the greatest current threats to nontarget megafauna (Peckham et al. 2007).

But we also realized that the high bycatch per unit of effort of small-scale fisheries could provide an unexpected conservation advantage. Because many more turtles are caught per unit of effort (e.g., hook or day fished), for each unit of small-scale fishing effort modified to reduce bycatch, a much higher benefit accrues for the megafauna than might be expected for industrial-scale fisheries. For this reason, by helping to change the way de la Paz's fleet of six small boats fish at Baja California Sur, we could accomplish more than we would by modifying the fishing practices of entire industrial fleets.

But mitigating the bycatch of small-scale fleets like de la Paz's presents a daunting conservation challenge. Because small-scale fisheries occur primarily in developing nations, documentation and management of their bycatch have been limited or nonexistent (Panayotou 1982; Pauly 2006). Furthermore, the command–and–control approaches such as fisheries closures that have been used in industrial fleets are increasingly recognized as impractical and inadvisable, particularly in developing nations (Berkes et al. 2001; Hilborn et al. 2005; McClanahan et al. 2006). Realizing that bycatch solutions would depend on fishers' direct involvement and support in developing new social norms and economic alternatives (McClanahan et al. 2006; Jackson 2007), especially at Baja California Sur, we focused our efforts with Proyecto Caguama on identifying and empowering small-scale fishermen like de la Paz with the inspiration, knowledge, and resources to reduce their bycatch.

De la Paz's Proposition

Late in the fall of 2006, after many attempts, we finally tracked down de la Paz receiving a load of shrimp at the landing in Puerto López Mateos. After a few minutes of hearing us emphasize the importance of his fleet's bycatch, de la Paz cheekily offered to sell us all his longline gear for US $100,000. We replied, first, that we had nowhere near such funds and, second, that we were practically and philosophically opposed to buyouts. But we asked what his gear was actually worth, what it would cost to replace, and how much fishing profit he might lose in selling it. He said he thought that around $100,000 was the replacement cost of his gear and that he wanted to switch to cleaner fishing techniques, and then he made it clear that he had to get back to work. We agreed to keep in touch and left enthused by his openness to bycatch mitigation but wondering how we could ever make something work for him and his prosperous fleet.

Trinational Fishermen's Exchange

Back in 2005 enough net fishermen had committed to reducing their bycatch that we were able to ask them about their motivations. Through conversations and interviews, we had learned that many fishermen were motivated to take conservation action once they understood turtles' transpacific migrations and consequential vulnerability (Hall et al. 2007). Reflecting on the fact that turtles and fishermen face bycatch issues throughout the loggerheads' migratory range, spanning from Japan through Hawaii to México, we proposed a Trinational Fishermen's Exchange. With the help of Kama Dean, then codirector of the San Diego–based nonprofit Pro Peninsula, Proyecto Caguama's parent nonprofit, we won support from the World Wildlife Fund, Japan Foundation, U.S. National Marine Fisheries Service, and Western Pacific Fisheries Management Council to bring fishermen together from all three countries to share their bycatch challenges and develop international solutions.

For the first stage of the exchange, in November 2006 we assembled a delegation of fisher leaders, conservation scientists, and government representatives from each country to witness the decline of nesting in Japan firsthand and to brainstorm about bycatch solutions. Fishermen were selected to participate for being (1) influential and inspiring communicators recognized within their communities, (2) distinguished among their peers, and (3) committed to long-term conservation of sea turtles and natural resources in general.

Generously hosted by Naoki Kamezaki, Yoshimasa Matsuzawa, and the Sea Turtle Association of Japan, our first trinational exchange was successful in impressing upon all of us that declines in nesting have been steep and ongoing for decades and that the loggerheads we all work with represent a remnant population (Kamezaki et al. 2003). We also confirmed that the bycatch rates we observed in México are the highest known in the Pacific in terms of numbers of turtles caught per day per boat or per product. The Mexican fishermen took this information to heart and shared it widely upon their return among families, friends, and colleagues and throughout the Grupo Tortuguero network.

An Unlikely Delegate

Back in Baja California Sur in the spring of 2007, Efraín de la Paz Regalado accosted us while we prepared for our summer field season. "Oye, why didn't you take me to Japan with all the others?" We carefully explained the criteria for selecting the fisher delegates, emphasizing that each had to have demonstrated long-term commitment to conservation, of sea turtles in particular. We also reminded him that he was personally responsible for killing more turtles than anyone else we knew, clearly not the message we wanted to share with others. "For precisely that reason," de la Paz immediately answered, "I'm your best candidate: I can save more turtles than anyone in the Pacific."

For the second and third stages of the trinational exchange, we were set to visit México and Hawaii. After some debate we decided to invite de la Paz for exactly the reason he offered. We started by asking him to host the Japanese, Hawaiian, and Mexican delegations at his Santa Rosa camp in early August, the season when his fleet usually began fishing their bottom-set longlines that were so lethal to loggerheads. He generously agreed, and we hoped the timing and experience might change the way he fished.

Fiesta at Santa Rosa

The 20-km track leading from the main road to the Santa Rosa camp is narrow and dusty. So our Japanese and Hawaiian colleagues, jet-lagged, hot, and nauseous after weaving through all the cacti aboard the Grupo Tortuguero's turtle bus, were overjoyed to step into the ocean breeze and estuarine vistas of Santa Rosa. De la Paz and all the members of the Santa Rosa cooperative were gracious hosts, putting on a fish fry, a trinational volleyball game, and tours of their chapel, homes, boats, and fishing gear.

The Japanese, Hawaiian, and even the other Mexican fishermen were impressed by the simplicity, efficiency, and beauty of the livelihoods of the Santa Rosa crew. The foreign fishers were amazed that the half dozen boats of the Santa Rosa fleet, each with no more than a couple hundred hooks, could kill so many loggerheads. The Santa Rosa fishermen assured them it was true, amazed for their part that fleets so massive with boats as huge as the Hawaiians' could catch so much tuna and swordfish while taking so few turtles. De la Paz and his partners expressed a genuine embarrassment and concern for their high bycatch rates. When pressed by members of the foreign delegations about their plans for longline fishing in the summer of 2007, the cooperative members and de la Paz said they were looking for alternatives—welcome news, exactly what we had hoped to achieve.

Basking at Laniakea

A week later de la Paz waded in the warm clear waters of Laniakea Beach on the north shore of Oahu, Hawaii, alongside his Mexican, Japanese, and Hawaiian counterparts. De la Paz laughed boyishly, frolicking with the others in the gentle waves, and then suddenly fell silent. He had caught sight of a large green turtle gliding toward him over the white sand in knee-deep water. The turtle swam past him to nibble on the algae of a nearby rock, and de la Paz gazed while the others took snapshots, counting seven other turtles all within sight in the shallows (fig. 12.4).

Back on the beach, de la Paz questioned Stacy Hargrove, our host from the U.S. National Marine Fisheries Service who had taken us to see the turtles. It was her boss, George Balazs, who had succeeded in protecting their north-shore basking beach. De la Paz wanted to know how many turtles visit the spot, during what seasons, and why, but most of all, whether they had always been there. He was especially pleased with Hargrove's answer that the turtles had begun frequenting Laniakea over the past decade as their population recovered: "When I first arrived in Baja California Sur," recollected de la Paz, "turtles gathered in numbers like these in the shallows of the estuaries. Now I see there's hope they could recover to be like that again."

An Unexpected Commitment

De la Paz's interest at Laniakea represented an unlikely transformation. This was the fisherman who months earlier had callously brushed off our concern for the hundreds of turtles he and his fleet were killing. The day before his visit to Laniakea, we arranged a workshop at the headquarters of the Western Pacific Fisheries Management Council in Honolulu. With the support of director Kitty Simonds, Irene Kinan convened a meeting of turtle bycatch experts ranging from fishermen to physiologists to share current work with our trinational delegation.

Early on the morning of the meeting, we met with de la Paz one final

FIGURE 12.4. Efraín de la Paz Regalado (right) among basking green turtles with trinational exchange members S. H. Peckham and Cesareo Castro at Laniakea Beach, Hawaii, all pretending to have lost fingers to turtle bites (© ProCaguama).

time to confirm that he was comfortable with sharing his bycatch data. Fresh off the plane from México, he seemed so unconcerned that we once again wondered if he understood the gravity of his bycatch impact. We hoped that the contrast between the representation of his own bycatch impact and those from Hawaii and Japan would inspire de la Paz to strive to reduce his bycatch. It turned out that we were severely underestimating him.

In the council's elegant conference room high above the streets of Honolulu, de la Paz was seated with thirty-five experts working to reduce sea turtle bycatch from Hawaii and around the Pacific. When his turn came to stand up and briefly introduce himself before the formal presentations, de la Paz launched into an extended monologue. Since the translator was not expecting anything more than his name, he was unprepared for de la Paz's torrent of words, and only those few with Spanish could understand him. He shared his majestic full name, of course, and then explained the origins and resources of the Santa Rosa cooperative. The assembled scientists, administrators, and fishermen had begun shifting in their seats or gazing out the window by the time he began detailing his fleet's typical summertime catch—after all, only a third of them had yet introduced themselves, the meeting had not yet started, and this Mexican fisherman was droning interminably in incomprehensible Spanish. Concerned for the group, two fellow delegates were signaling to de la Paz when he made his entirely unexpected declaration.

"So as you see, it's costing me lots of money to be away from my fleet in México. I didn't come to Hawaii for fun, but rather with a purpose in mind." De la Paz paused dramatically, and smiled, ignoring our signals for him to finish. "I came here to declare that my fleet will no longer fish the bottom-set longlines that kill so many turtles." De la Paz paused, then finished simply with "Gracias," and sat down.

Almost everyone looked relieved, undoubtedly because de la Paz had finally stopped speaking rather than for what he had said. It wasn't until our joint presentation in the afternoon, following numerous talks detailing the great progress the Hawaiian fleet has made eliminating their bycatch almost entirely, that the group realized the significance of de la Paz's declaration. The rest of those assembled were discussing or reporting admirable but incremental reductions in sea turtle bycatch. De la Paz had just committed to saving hundreds of loggerheads each year through one swift action.

In a word, we were flabbergasted. Though we had joked with him about buying out his hooks for some exorbitant sum, there had never been any discussion about his voluntarily retiring his fleet from its high bycatch activity. We had floated the idea of somehow incentivizing a change in fishing practice with several organizations, but at that time we had not yet shared the possibility with de la Paz. When recognition of the implications of de la Paz's declaration finally sunk in for all of the experts assembled, a spontaneous celebration took place, with Kitty Simonds, director of the Western Pacific Regional Fisheries Management Council, and de la Paz exchanging memorable gifts (fig. 12.5).

FIGURE 12.5. Efraín de la Paz Regalado was heartily congratulated for his commitment to eliminate his fleet's loggerhead bycatch by fisher delegates from Japan, Hawaii, and México and North Pacific bycatch experts at a meeting convened by Grupo Tortuguero, Pro Peninsula, and the Western Pacific Regional Fisheries Management Council, 20 August 2007 (© ProCaguama).

The Santa Rosa Declaration

During the rest of the Hawaiian stage of our exchange and over the next month, we worried that de la Paz might back away from the commitment he had made in Honolulu. There were too many reasons that he might be unable to retire his longlines—starting with the direct costs, opportunity costs, and the need to answer to his cooperative.

But de la Paz and the entire Santa Rosa cooperative defied our doubts a second time. On 25 September 2007 de la Paz solemnly signed the historic Santa Rosa Declaration, and he and his fishermen loaded their longlines into a Grupo Tortuguero truck (fig. 12.6). The declaration detailed their commitment to (1) cease fishing lethal bottom-set longline gear in the hotspot, (2) do everything possible to avoid bycatch in its other fishing activities, and (3) collaborate with the Grupo Tortuguero to identify and minimize bycatch by neighboring fleets.

FIGURE 12.6. On 25 September 2007 Efraín de la Paz Regalado (left) signed the landmark Santa Rosa Declaration and shook hands on it with S. H. Peckham (right) of the Grupo Tortuguero, permanently retiring the Santa Rosa fleet's longline gear and sparing hundreds or thousands of loggerheads each year since then (© ProCaguama).

In the meantime, Wallace J. Nichols and Werner Chabot of the Ocean Conservancy launched a fundraising campaign for the Santa Rosa cooperative. Thanks to their hard work and compelling appeal, the broad membership of the Ocean Conservancy donated $10,000 in a record amount of time. Soon after the Santa Rosa Declaration, we had the honor of passing these funds along to the Santa Rosa cooperative as a way to help them transition away from the lethal bottom-set longlines to less destructive practices, including hook and line and seine fishing.

Coming Full Circle

Each January more than 400 fishers, their families, and other concerned citizens gather in Loreto, Baja California Sur, for the annual meeting of the Grupo Tortuguero (see chapter 1). The January 2008 meeting marked the Grupo Tortuguero's tenth anniversary and drew a couple hundred turtle researchers and conservationists from around the globe. We prepared a slideshow for de la Paz to present, but he wasn't satisfied with showing a few dozen photos. De la Paz took the stage, showed the photos, and proceeded to explain in his own words to all those assembled, unscripted and from the heart, the story of how he and his fleet had come to save hundreds of loggerheads in 2007 alone.

De la Paz explained that the switch was relatively easy for his fleet. It was just a question of realizing how serious their bycatch impacts were. This became most clear, de la Paz explained, talking with the fishermen from Japan and Hawaii and seeing the green turtles of Laniakea basking in the shallows in numbers as great as had once frequented Baja California Sur. After the visit of the Japanese and Hawaiians to Santa Rosa and after his trip to Honolulu, neither de la Paz nor the members of his cooperative wanted to fish bottom-set longlines.

De la Paz and his fleet didn't earn quite as much in 2007 and 2008 as in past years, but as de la Paz explained, what they lost in terms of cash they more than doubled in terms of their pride and dignity. Now they stand tall among other fishermen around the Pacific Rim for taking responsibility for the full ecological impacts of their fishing.

Governmental Support

Though increasingly depleted, the legendarily rich waters of the Baja California peninsula still offer fishermen like de la Paz the luxury of being able to switch among gears, techniques, and even fisheries in response to dynamic conditions and ephemeral fish populations. Versatile and highly responsive, the small-scale fishermen of the region switch gear, techniques, and fisheries five or six times a year, and they often switch fisheries suddenly, sometimes from one day to the next. As a result, the ten or so established fleets that fish the Baja California Sur loggerhead hotspot have many alternatives to fishing the bottom-set gear that is so lethal to loggerheads. De la Paz and other members of the Santa Rosa and López Mateos fleets are reaching out to these neighboring fleets to encourage them to focus on alternative techniques and areas in order to reduce their bycatch.

But each summer several nomadic fleets from neighboring states set up camps along the uninhabited estuaries that border the loggerhead hotspot. These outlaw transient fishermen, fishing from unregistered boats without permits, have zero incentive for local conservation and appear to be small-scale versions of the roving bandits recognized to be the scourge of many industrial-scale fisheries (Berkes et al. 2006). Because they are fishing bottom-set gear in the loggerhead hotspot, these fleets are likely to be causing severe levels of loggerhead mortality. Furthermore, we have sighted and received many reports of international longline vessels fishing off the loggerhead hotspot. A handful of these vessels fishing their hundreds of hooks for even a few weeks a year in the hotspot could explain much of the loggerhead strandings at Playa San Lázaro.

De la Paz and many other Baja California Sur fishermen are intent on discouraging both transient small-scale and pirate industrial fishing in the loggerhead area because they affect local fisheries as well as loggerheads. Accordingly, a coalition of fishermen and conservationists from around Baja California Sur is working with state and federal authorities to establish a new fisheries management plan and special loggerhead refuge that would offer better management and enforcement throughout the region, including, of course, bycatch reduction.

Emerging Conservation Opportunities

Increased awareness and stringent management have led to innovation and resulting bycatch reduction in several industrial-scale fishing fleets. The Hawaii-based longline fleet, for instance, has reduced its bycatch significantly (Gilman et al. 2007). But in focusing conservation efforts primarily on industrial fleets, we all may have missed an unexpected conservation opportunity. Because of their exceedingly high bycatch rates, small-scale fleets such as de la Paz's may affect marine megafauna such as loggerheads more than we ever expected, offering unprecedented conservation leverage.

Our long-term community-based initiatives in Baja California Sur communities characterized by the conservation mosaic enabled us to identify and partner with de la Paz and his fleet. Our Trinational Fishermen's Exchange empowered de la Paz with a truly pan-Pacific perspective on the importance of his fleet's bycatch. De la Paz, in turn, acted on his new knowledge honorably and decisively, exhibiting exemplary leadership and sacrifice.

We are hopeful that a combination of these approaches could lead to the identification of high-conservation-leverage opportunities and empowerment of local leaders elsewhere. Similar efforts by the Sea Turtle Association of Japan in conjunction with the Trinational Fishermen's Exchange have contributed to promising relationships with Japanese fisher leaders. Though not formulaic, combining these approaches could enable important conservation gains. Based on our experiences, success will likely depend on identifying, inspiring, and empowering fisher leaders like Efraín de la Paz Regalado.

ACKNOWLEDGMENTS

We thank the participants in the exchange, both those who traveled (below) and also the many fishermen, their families, and others who so graciously hosted our delegations in Japan, Hawaii, and México. We also thank Wallace J. Nichols for entrusting us with the Baja California Sur bycatch issue and for his wise counsel throughout. We are deeply grateful to Kama Dean for helping to formalize, fund, and carry out the exchange. The teams of Proyecto Caguama, Pro Peninsula, Grupo Tortuguero, and the Sea Turtle Association of Japan all worked to make the Trinational Fishermen's Exchange a success, especially Y. Matsuzawa, T. Ishihara, M. Kojiro, translators H. Ida and T. Shimada, I. Kinan, G. Balazs, S. Hargrove, R. Briseño-Dueñas, J. Laudino-Santillán, C. Lucero-Romero, G. Ruiz, and D. Durazo. We finally acknowledge the invaluable work of the Proyecto Caguama team to conduct the underlying research and engage fishermen of Baja California Sur, in particular, V. de la Toba, J. Laudino, A. Gaos, R. Ochoa, N. Rossi, R. Donadi, R. Rangel, and E. Caballero-Aspe.

The Trinational Fishermen's Exchange was supported initially by the World Wildlife Fund and fully funded by the Japan Foundation's Center for Global Partnership, U.S. National Marine Fisheries Service, and the Western Pacific Fisheries Regional Management Council. We thank Kim Davis of the World Wildlife Fund for her early adoption of this initially farfetched fishermen's exchange. Our collaboration with the Sea Turtle Association of Japan was initiated with support from the Pacific Rim Research Program. Proyecto Caguama was funded during 2003–2008 by the U.S. National Marine Fisheries Service, the David and Lucille Packard Foundation, the Western Pacific Fisheries Management Council (NOAA grant FNA05NMF4411092), the PADI Foundation, Project AWARE, and University of California Institute for México and the United States. We received institutional support from Pro Peninsula, Grupo Tortuguero de las Californias A.C., Sea Turtle Association of Japan, Western Pacific Fisheries Management Council, and the Ocean Conservancy.

Lastly, we thank the entire Trinational Fishermen's Exchange team, including members from Japan (Akio Hashimoto, Shoji Yamashita, Akijiro Uta, Kunio Horiuchi, Kazuhiko Sasaki, Mitsuahru Kume, Daisuke Suda, Naoki Kamezaki, Yoshimasa Matsuzawa, Takashi Ishihara, Mizuno Kojiro), México (Jesus Lucero Romero, Israel Ritchie Sánchez, Juan Ignacio Romero Aguilar, Cesareo Castro, Efraín de la Paz Regalado, Johath Laudino-Santillán, David Maldonado Díaz, Raquel Briseño-Dueñas, Georgita Ruíz, Graciela Tiburcio Pintos), and the United States (Leland Oldenburg, Diane Oldenburg, Hoyt Peckham, Kama Dean, Irene Kinan-Kelly, Wallace J. Nichols, and members of the Hawaiian Longline Association).

REFERENCES

Berkes, F., T. P. Hughes, R. S. Steneck, J. A. Wilson, D. R. Bellwood, et al. 2006. Globalization, roving bandits, and marine resources. Science 311(5767):1557–1558.

Berkes, F., R. Mahon, P. McConney, R. Pollnac, and R. Pomeroy. 2001. Managing Small-Scale Fisheries: Alternative Directions and Methods. International Development Research Centre, Ottawa, Canada.

Gilman, E., D. R. Kobayashi, T. Swenarton, N. Brothers, P. Dalzell, et al. 2007. Reducing sea turtle interactions in the Hawaii-based longline swordfish fishery. Biological Conservation 139:19–28.

Hall, M. A., H. Nakano, S. Clarke, S. Thomas, J. Molloy, S. H. Peckham, J. Laudino-Santillán, W. J. Nichols, E. Gilman, E. Cook, et al. 2007. Working with fishers to reduce bycatches. In: Kennelly, S. J. (Ed.), Bycatch Reduction in the World's Fisheries. Springer, New York. Pp. 325–288.

Hilborn, R., J. M. Orensanz, and A. M. Parma. 2005. Institutions, incentives and the future of fisheries. Philosophical Transactions of the Royal Society of London, Series B 360: 47–57.

Jackson, J. B. C. 2007. Economic incentives, social norms, and the crisis of fisheries. Ecological Research 22:16–18.

Kamezaki, N., Y. Matsuzawa, O. Abe, et al. 2003. Loggerhead turtles nesting in Japan. In: Bolten, A. B., and B. Witherington (Eds.), Loggerhead Sea Turtles. Smithsonian Books, Washington, DC. Pp. 210–218.

Lewison, R. L., and L. B. Crowder. 2007. Putting longline bycatch of sea turtles into perspective. Conservation Biology 21:79–86.

McClanahan, T. R., M. J. Marnane, J. E. Cinner, and W. E. Kiene. 2006. A comparison of marine protected areas and alternative approaches to coral-reef management. Current Biology 16:1408–1413.

Nichols, W. J. 2003. Biology and conservation of sea turtles in Baja California, Mexico. Ph.D. dissertation, University of Arizona, Tucson.

Panayotou, T. 1982. Management Concepts for Small-Scale Fisheries: Economic and Social Aspects. Food and Agriculture Organization, Rome.

Pauly, D. 2006. Major trends in small-scale marine fisheries, with emphasis on developing countries, and some implications for the social sciences. Maritime Studies 4:7–22.

Peckham, S. H., D. Maldonado, A. Walli, G. Ruiz, W. J. Nichols, et al. 2007. Small-scale fisheries bycatch jeopardizes endangered Pacific loggerhead turtles. PLoS One 2(10): e1041.

Peckham, S. H., D. Maldonado-Díaz, V. Koch, A. Mancini, A. Gaos, et al. 2008. High mortality of loggerhead turtles due to bycatch, human consumption and strandings at Baja California Sur, Mexico, 2003–7. Endangered Species Research 5:171–183.

Saenz-Arroyo, A., C. M. Roberts, J. Torre, M. Carino-Olvera, and J. P. Hawkins. 2006. The value of evidence about past abundance: marine fauna of the Gulf of California through the eyes of 16th to 19th century travelers. Fish and Fisheries 7:128–146.

Steinbeck, J. 1941. The Log from the Sea of Cortez. Viking Press, New York.

13

Interpreting Signs of Olive Ridley Recovery in the Eastern Pacific

PAMELA T. PLOTKIN, RAQUEL BRISEÑO-DUEÑAS, AND F. ALBERTO ABREU-GROBOIS

Summary

The most abundant sea turtle in the world, the olive ridley (*Lepidochelys olivacea*), which for countless generations provided food and medicine and utilitarian and spiritual values to coastal communities, became the target of large-scale industrial fisheries in the eastern Pacific that peaked in the 1960s and 1970s and supplied a global luxury leather market. The take of adult females and males in their nearshore breeding areas landed several million turtles during its operation and compounded the persistent egg exploitation along the coasts of México and Central America that reduced recruitment into the population. After two decades of intensive and scarcely regulated exploitation, even some of the massive arribada olive ridley populations collapsed at Mexican nesting beaches where the females once gathered to nest *en masse* and where fishing intensity was greatest.

Since the 1990 closure of the turtle fishery and the implementation of protection programs for nests and breeding females on the most important nesting beaches, starting in the 1960s but particularly from the end of the 1970s, large-scale declines have been halted and population growth is discernible in much of the eastern Pacific. Along the Mexican coast, although most rookeries have not fully recovered, increases are widely reported and a few new rookeries have emerged. Similar trends have been noted in Costa Rica. These recent and sustained increases of olive ridleys nesting at key rookeries are encouraging signs, but we must be cautious and not overstate their significance. Our insufficient knowledge of olive ridley reproductive behavior and ecology limits our ability to fully comprehend the impact of past exploitation, present recovery trends, and future extinction risk from other threats and emerging risk factors of natural and anthropogenic origins.

Introduction

In the ancient cultures of tropical regions, the sea turtle was always an important component of the symbols with which man interpreted the mysteries of life and divinized them in his mythology. Primitive human manifestations use turtles as an allegory for the origin of Earth, as a custodian of time, dreams, and wisdom, but also as a symbol of fertility, sexual vigor, and abundance (Molina 1981). In the constant exchange between the past and the present in the imaginary of modern societies, these symbols permeate the different expressions, uses, and customs of local people and become a key reference for conservation science that addresses sea turtle research within an interdisciplinary approach.

The olive ridley (*Lepidochelys olivacea*; see plate 8) is without a doubt the most abundant sea turtle in the eastern Pacific Ocean. The turtle's use of both marine and terrestrial habitats and the extraordinary abundance at sea and at nesting beaches were widely appreciated by the coastal communities in the region. The presence of turtles on beaches during the breeding seasons in predictable cycles allowed the coastal tribes in México to alternate their maize harvests with the collection of eggs and meat from the sea turtles (Mountjoy and Claassen 2005).

The earliest chronicles of the European explorers who first navigated the waters of the Pacific faithfully reflect the fantastic proliferation of olive ridleys. As early as the turn of the seventeenth century, Kimsey-Owen (cited by Robertson 1979) described how the Mayos—a tribe still present in northern Sinaloa, México—had developed great skill in capturing turtles at sea with "fisgas" (a type of rustic harpoon) and reported on the methods used to dry and salt turtle meat and to collect eggs and fat from turtles. Carapaces were used to store salt and meats, as headrests to sleep on, as cradles for babies, and as shrouds for the dead. Traditional healers combined preparations derived from turtles in the herbal remedies they used for therapies in shamanic rituals. The centuries-long traditional harvests were probably sustainable. But despite its abundance, the olive ridley could not withstand decades of industrial fishing when it opened up in the middle of the twentieth century.

Unlike the collapse of several commercially important marine fish species, which seized the world stage in the 1990s and has remained in the spotlight for more than a decade, the collapse of the olive ridley in the eastern Pacific was merely a blip on the radar screen. Understandably, olive ridleys never did feed a massive international market at the level of the Atlantic cod (*Gadus morhua*), but the turtle fishery did supply a global luxury leather market. This directed fishery, the largest marine turtle fishery in the world, was actually two fisheries operating independently in the eastern Pacific, one located in México (Frazier 1980; Cliffton et al. 1982) and the other in Ecuador (Frazier 1980; Green and Ortiz-Crespo 1982). During their years of operation, these fisheries landed millions of turtles.

Collapse was most visible at Mexican beaches, where hundreds of thousands of olive ridleys once emerged to lay their eggs (Márquez et al. 1976; Cliffton et al. 1982). The species' unique reproductive behavior of synchronized mass nesting, known as the *arribada*, supplied fishermen with an easy, reliable, and predictable source of turtles offshore from nesting beaches, where males and females gathered in large numbers to copulate, and on nesting beaches, where females crawled ashore (plate 8). Most arribadas were reduced in size and frequency, and some reportedly disappeared (Cliffton et al. 1982).

Signs of recovery are now evident. At some beaches arribadas occur more frequently, and the number of olive ridleys nesting today may exceed the number nesting before the industrial turtle fishery (Márquez et al. 1996; Peñaflores-Salazar et al. 2000). How did this turtle, once hanging in the balance, begin to recover?

Our goal in this chapter is to review the collapse of olive ridley arribada rookeries and recent signs of recovery in the eastern Pacific Ocean. We focus on the effects of the turtle fishery and its subsequent closure, and the trend toward eliminating or at least controlling egg collection. We recognize that other factors influence olive ridley population growth (e.g., incidental capture in other fisheries), but it is not within the scope of this chapter to review these in great detail. To that end, we refer the reader to recent reviews of at-sea and land-based threats to olive ridleys (Cornelius et al. 2007; Frazier et al. 2007). The political and socioeconomic dimensions of the turtle fishery and egg collection are also not within scope for this chapter, as these topics have also been reviewed (Trinidad and Wilson 2000; Campbell 2007). In the first part of this chapter we describe the biological phenomenon that predisposes the olive ridley to overexploitation. We then review the turtle fishery statistics to reveal the magnitude of fishing pressure. Next we discuss past and present abundance of olive ridleys. Finally, we discuss the signs of recovery and future challenges.

The Biological Phenomenon

Olive ridleys are behaviorally one of the most intriguing of all sea turtles—and of all animals—because of their two very different nesting behaviors. While in some rookeries olive ridleys will emerge individually from the sea to lay their eggs in the sand like most other species, at other rookeries or under certain circumstances not entirely understood, they can come ashore en masse, with hundreds of thousands of turtles emerging synchronously from the ocean in a short time interval to nest in close proximity (Hirth 1980; Eckrich and Owens 1995; Kalb 1999; Plotkin and Bernardo 2003, 2007; see plate 8). This reproductive behavior of synchronized mass nesting, known as the arribada (Carr 1967; Pritchard 1969; Hughes and Richard 1974; plate 8), occurs at just a few specific beaches in the eastern Pacific, Indo-Pacific, and western Atlantic. Very little is known about this reproductive behavioral polymorphism and how it is maintained in populations (Bernardo and Plotkin 2007).

In the eastern Pacific, arribadas occur in México, Nicaragua, Costa Rica, and Panamá (table 13.1). It is unknown how long this phenomenon has been present in the region, but it was first reported in the scientific literature in 1961 (Pritchard 1969). In addition to large aggregations of females nesting on land, both genders occur annually just offshore from arribada beaches, gathering in large numbers to copulate in the weeks preceding an arribada (Márquez et al. 1976; Plotkin et al. 1997).

The timing of arribadas in the eastern Pacific coincides annually with the rainy season (Cornelius and Robinson 1986), and they generally occur monthly. However, there are instances when arribadas occur more frequently than monthly (Hirth 1980; Ballestero 1996) or less frequently

TABLE 13.1. Overview of history and current condition of *Lepidochelys olivacea* arribada rookeries in the eastern Pacific.

Location	History	Current conservation status and population trend	Current conditions
MEXICO			
Playón de Mismaloya, Jalisco	Former arribada rookery; collapsed from fishing pressure	Depleted, increasing	Federal Sea Turtle Sanctuary category. Some egg poaching, chronic levels of illegal extraction of animals.
Ixtapilla, Michoacán	No significant olive ridleys nesting at this rookery before 1994 (Sánchez and Reyes 1998)	Stable, possibly increasing	Lack of formal conservation programs. Incipient monitoring using standardized methodology. More than five arribadas per year with 15,000 to 45,000 nesters/arribada (E. Albavera-Padilla, pers. commun.).
Piedra de Tlalcoyunque, Guerrero	Former arribada rookery; collapsed from fishing pressure	Depleted, stable	Federal Sea Turtle Sanctuary category. Little known, inconsistent monitoring. High levels of egg poaching, turtle extraction.
Chacahua, Oaxaca	Former arribada rookery; collapsed from fishing pressure	Depleted, irregular population trend	Federal Sea Turtle Sanctuary category. Reports of sporadic arribadas, e.g., 1997 (Aguilar–Reyes 2007)
Escobilla, Oaxaca	Arribada rookery; declined from fishing pressure	Increasing in abundance, number of arribadas/year, and duration of individual arribadas (Márquez et al. 2007) until about 2006; appears to be currently stable (E. Albavera-Padilla, pers. commun.)	Federal Sea Turtle Sanctuary category. Largest arribada rookery in the world (Abreu-Grobois and Plotkin 2008): about eight arribadas/year. Currently affected by declining hatchling production due to predation by dung beetle (*Omorgus*), destruction of nests by nesting females, low hatching efficiencies (Albavera-Padilla 2009).
Morro Ayuta, Oaxaca	No historical information available	Stable or possibly increasing; irregular monitoring; about eight arribadas/year (E. Albavera-Padilla, pers. commun.)	Second largest olive ridley arribada in México. High levels of egg poaching. Irregular and difficult protection but exhibits high hatching efficiency (>80 percent; Albavera-Padilla 2009).

NICARAGUA			
Boquita	Former arribada rookery; collapsed from intense commercial egg collection (Nietschmann 1975)	?	Extinct as an arribada rookery.
Masachapa	Former arribada rookery; collapsed from intense commercial egg collection (Nietschmann 1975)	?	Extinct as an arribada rookery.
Pochomil	Former arribada rookery; collapsed from intense commercial egg collection (Nietschmann 1975)	?	Extinct as an arribada rookery.
Chacocente	Former(?)/present arribada rookery	Increasing	Controlled egg collection July–January; unlimited rest of the year (Hope 2002) Wildlife refuge status.
La Flor	Former(?)/present arribada rookery	Increasing	Controlled egg collection throughout arribada nesting season of <10 percent of eggs since early 1990s (Honarvar 2007). Steep increase in abundance from early 1990s (Urteaga and Motha 2007).

(continues)

TABLE 13.1. (*continued*)

Location	History	Current conservation status and population trend	Current conditions
COSTA RICA			
Nancite	Former arribada rookery; collapsed	Depleted but stable	Wildlife refuge. Geographic isolation has provided natural protection. Little affected by exploitation. Decline probably due to natural causes (Valverde et al. 1998; Fonseca et al. 2009).
Ostional	Relatively recent arribada rookery	Increasing or stable	Wildlife refuge. Largest arribada rookery in Costa Rica. Controlled commercial harvest of eggs by locals.
PANAMA			
Isla Cañas	Arribada are still observed though of much lower abundances; high predation of turtles and eggs in the 1980s	Declining; number of arribadas/year declined from five in the mid-1990s to three in 2003	National Wildlife Refuge (L. Vargas, Autoridad Nacional del Ambiente, pers. commun., 2007).

when the turtles miss a month (Plotkin et al. 1997). Sometimes an individual arribada lasts for a single day (Plotkin 1994) or for as long as thirty days (Ballestero 1996).

While olive ridley arribada behavior and beaches have been the focal point of research and management of the species, the nesting range for the species in the eastern Pacific extends far beyond these select beaches (Bernardo and Plotkin 2007; see plate 9). Indeed, there are many more non-arribada than arribada rookeries. Some individual olive ridleys display a mixed strategy, nesting solitarily sometimes and during arribadas other times. So while arribadas occur at select beaches in México, Nicaragua, Costa Rica, and Panamá, solitary olive ridleys are also emerging to nest along nearly the entire coastline from México to Ecuador on non-arribada beaches (Carr 1967; Pritchard 1969, 1979; Bernardo and Plotkin 2007). Because of the sheer abundance of turtles, arribada rookeries drive the population trends for the species in any given region where they occur. This is particularly true for the eastern Pacific, where so many arribadas exist. For this reason but also because data documenting past and present abundance of olive ridleys at non-arribada beaches are generally lacking (NMFS/USFWS 2007; Abreu-Grobois and Plotkin 2008), our review of the collapse and subsequent increase of olive ridleys in the eastern Pacific centers on arribada rookeries.

The evolutionary significance of arribada behavior has been debated and likely will continue to be for many years (Bernardo and Plotkin 2007). However, it is clear that this behavior is highly disadvantageous to the turtle when humans fish them and collect their eggs for uncontrolled commerce. The spatial and temporal predictability of this massive pulse of turtle resource, both at sea and on land, greatly facilitated olive ridley exploitation.

The Largest Turtle Fishery in the World

Sea turtle hunting dates back to prehistoric human communities that used these animals for meat, skin, oil, eggs, shell, and bones (Frazier 2003). On the Pacific coast of México, sea turtle hunting dates back to 5,500 years BP, and corollary reductions in local availability of sea turtles resulting from such use is now evident (Kennett et al. 2007; Smith et al. 2007).

FIGURE 13.1. Capture of sea turtles by pre-Colombian Mexican coastal communities illustrated in the Florentine Codex, from the sixteenth century (drawing by Blas Nayar from a copy of an illustration in Vilches-Alcázar 1980 from Sahagún 1975).

They continued to be extensively used by pre-Colombian cultures (fig. 13.1). In Oaxaca, México, the traditional extraction and chores associated with sea turtle exploitation were divided among ethnic groups and even differently between men and women. The Huaves and Matellanos, for example, hunted black and olive ridley juveniles in coastal lagoons, whereas Chontales and other coastal communities collected sea turtle eggs (Márquez et al. 2007), and while men fished, hunted turtles, or harvested their eggs, women processed the products and were in charge of their commerce (Becerra-Figueroa 2004, cited in Márquez et al. 2007).

During the last two centuries, however, human use of sea turtles escalated beyond subsistence hunting as human populations grew and migrated to new worlds (Nietschmann 1982; Sáenz-Arroyo et al. 2006). Fishing methods became increasingly sophisticated, and commercial markets in sea turtle parts and products developed. Consequently, contemporary sea turtle populations worldwide, particularly hawksbill (*Eretmochelys imbricata*), green (*Chelonia mydas*), and olive ridley, the three most commercialized species (Mack et al. 1982), declined to levels that bear little resemblance to their former abundance (Jackson 1997; Bjorndal and Jackson 2003; McClenachan et al. 2006).

México

It is unknown exactly when sea turtle hunting in the eastern Pacific shifted from subsistence use to a commercial fishery. According to Mexican government records, there was a small-scale turtle fishery during the 1940s that targeted turtles from the Gulf of México and Caribbean and, to a lesser extent, east Pacific green turtles (*Chelonia agassizii*) (fewer than 1,000 turtles annually; Montoya 1967; Márquez et al. 1976). But the turtle fishery evolved rapidly in the 1960s in response to a global shortage of crocodilian skin used for luxury goods (Mack et al. 1982; Trinidad and Wilson 2000; Frazier et al. 2007). International demand for exotic wildlife skin was soaring, and traders began to look elsewhere for potential substitutes. Sea turtle flipper skin was a worthy candidate. Because the olive ridley was abundant and easy to capture and its skin was appealing and durable when tanned into leather (Márquez et al. 1976), it was the leading alternative. The Mexican turtle fishery (fig. 13.2) promptly filled the open niche created by the shortfall of crocodilian skins and shifted its target species to the olive ridley. Annual turtle take increased from a few hundred turtles in the 1940s to hundreds of thousands by the late 1960s (fig. 13.3). Turtle skin initially was the focal point of this fishery, but the Mexican government later encouraged total use of turtles and commercialized turtle meat, shell, bone, cartilage, and oil (Cliffton et al. 1982; Mack et al. 1982).

Turtles were fished where the densest nesting aggregations once occurred, in front of Jalisco and Oaxaca, but also in concentrations at their foraging sites and migratory corridors in the northwestern region, off Baja California Sur and Sinaloa (Márquez et al. 1976; Mack et al. 1982). Female ridleys were captured by hand at their nesting beaches, but the largest take was of males and females captured by leaping on them at foraging sites or while they mated in the nearshore waters in front of the nesting beaches (Márquez et al. 1976). Use of nets was less desirable, because turtles often drowned before they were landed, and the turtle processing plant preferred to receive live animals (Márquez et al. 1976).

FIGURE 13.2. Accumulated olive ridley carapaces discarded by the San Agustinillo slaughterhouse in Oaxaca, México (Photo by P. Pritchard).

The largest and earliest historical landings were from fishing areas of Sinaloa and Baja California Sur. Just for Sinaloa, an equivalent of one million turtles were estimated to have been extracted during the 1967–1970 peak period, if official and unofficial statistics are combined (Briseño-Dueñas 2006a, 2006b). The huge harvests reported from Sinaloa and Baja California Sur would not be surprising given the rich shrimping grounds in northwestern Mexican waters where a well-developed fishery, including commercial and artisanal fishing fleets, was readily available for the rising turtle fishery (fig. 13.3). It was only later that the sea turtle fishery began moving southward, toward the coastal waters off Jalisco, Michoacán, and eventually Oaxaca. By the mid-1980s sea turtles had become so scarce off Sinaloa that their extraction became commercially unsound (Briseño-Dueñas 2006b). Aside from the obvious impacts on populations that nested in Sinaloan beaches, given that flipper tags from turtles tagged while nesting in Oaxaca were also extensively returned from the Sinaloan and Baja Californian fisheries (Márquez et al. 2007), it is clear that this huge fishery

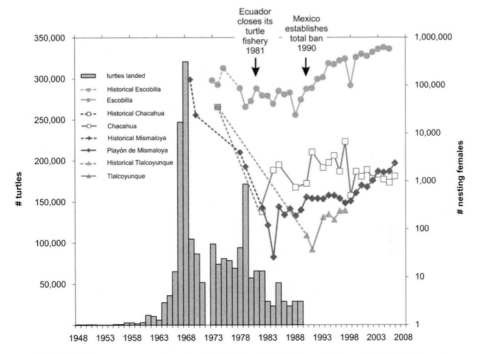

FIGURE 13.3. Estimates of turtles landed by the Mexican fishery (1948–1989) and population trends (1960s to 2008) for Mexican arribada rookeries that were affected by the exploitation. Landing data are converted to numbers of individuals from weight data in Márquez et al. (1976), Frazier (1980), Mack et al. (1982), and Peñaflores-Salazar et al. (2001). For trends, unless stated, census of nest information is converted to number of nesting females by dividing by female nest frequency of 2.5. Sources for population trend data are as follows. Historical Escobilla: census estimates of nesting females in Márquez et al. (1976). Escobilla: Dirección General de Vida Silvestre/SEMARNAT, México; C. Peñaflores, Centro Mexicano de la Tortuga. Historical Chacahua: census estimates of females in Márquez et al. (1976). Chacahua: from Aguilar-Reyes (2007). Historical Mismaloya: census estimates of nesting females in Montoya (1969), Márquez et al. (1976), Casas-Andreu (1978), and Villa Guzmán (1980). Mismaloya: Dirección General de Vida Silvestre/SEMARNAT, México; Márquez et al. (1998); A. Trejo, Universidad de Guadalajara, pers. commun. Historical Tlalcoyunque: census estimates of nesting females in Márquez et al. (1976). Tlalcoyunque: Márquez et al. (1998).

affected an entire swath of Mexican rookeries, certainly including all of the historic arribada rookeries of Jalisco, Guerrero, and Oaxaca.

The height of the turtle fishery took place from the mid-1960s through the late 1970s. It reached its peak in 1968, when an estimated 350,000 turtles were landed (Márquez et al. 1996). Steep declines in the number of turtles landed during the next two years prompted a full closure of the fishery from mid-1971 through 1972, with seasonal closures and quotas implemented thereafter to reduce fishing pressure (Márquez et al. 1976; Trinidad and Wilson 2000). Estimates of the number of turtles captured vary considerably by source and because data were reported in weight/tons of turtles and had to be converted (fig. 13.3). Mack et al. (1982) reported 500,000–1,400,000 turtles were captured in México from 1965 to 1977. These numbers were acquired from the Mexican fishery cooperatives (the fishermen) and from Antonio Suarez, owner of the turtle processing plant in México. Another estimate for this same time period, reported by the México National Fishery Institute, indicates that 1.3 million turtles were captured (Márquez et al. 1976). Based on official and unofficial reports, Frazier (1980) estimated that by 1979 more than 1.6 million turtles had been landed. Several unsupported and vastly larger estimates of the magnitude of the take have proliferated over the years, adding to the uncertainty of the number of turtles landed during the height of the fishery (Cliffton et al. 1982). The catch rate was relatively stable for a few years after the 1972 closure and then decreased again from 1980 through 1989 (Peñaflores-Salazar et al. 2000) as the demand for leather dropped and the turtle fishery became limited to Oaxaca. The cumulative turtle take by the fishery, using data reported by the México National Fishery Institute from 1945 through 1989 (fig. 13.3; Márquez et al. 1976; Peñaflores-Salazar et al. 2001), is estimated at just over 2 million turtles before the Mexican government closed the fishery in 1990 (Aridjis 1990).

Ecuador

By the 1970s, another industrial turtle fishery developed in the eastern Pacific in Ecuador (Green and Ortiz-Crespo 1982), with some assistance from Colombia (Hurtado 1982). The Ecuadorian turtle fishery captured primarily adult olive ridleys between 1970 and 1981 (Green and Ortiz-Crespo

1982). Turtles initially were fished for their meat, but meat became less important when commerce in turtle skin developed (Frazier 1980). Turtles were fished from pelagic waters offshore Ecuador in areas where olive ridleys concentrate to feed during the nonreproductive season. The origin of olive ridleys captured from these feeding grounds is believed to be México and Central America for several reasons. First, the mainland and islands belonging to Ecuador have no significant olive ridley nesting beaches that could account for the large number of turtles captured offshore. Second, there is direct evidence from mark-recapture studies that olive ridleys tagged while nesting in México and Costa Rica subsequently were captured by the Ecuadorian fishery (Márquez et al. 1976; Frazier 1980; Green and Ortiz-Crespo 1982; Cornelius and Robinson 1985, 1986). Finally, additional evidence linking turtles from the Ecuadorian feeding grounds to nesting beaches in the region comes from studies of postnesting olive ridleys tracked via satellite from Costa Rica to Ecuador (Plotkin 1994, 2010; Plotkin et al. 1995). Consequently, the Ecuadorian turtle fishery also contributed to the decline of olive ridleys at nesting beaches in México and Central America.

Estimates of the number of olive ridleys landed by the Ecuadorian turtle fishery vary widely by source and are complicated by the fact that most data are reported as weight of skins, number of pairs of flippers, or weight of meat (table 13.2). Frazier (1980) reported statistics from the Food and Agriculture Organization of the United Nations (FAO) for 1971 to 1978, data not recognized for their veracity, which indicated a total take of 85,800 turtles. Other estimates from Ecuadorian sources indicate the take was substantially larger. Several sources calculated the number of turtles fished using trade statistics reported by the Ecuador National Fishery Institute (i.e., the quantity or weight of turtle parts exported from Ecuador), divided by the average weight of turtle flippers, the average weight of meat per turtle, or both (Frazier 1980; Green and Furtando 1980; Green and Ortiz-Crespo 1982; Hurtado 1982). At its peak in 1979, the Ecuadorian fishery landed an estimated 133,233–148,036 turtles (Hurtado 1982). The cumulative take by the Ecuadorian turtle fishery is estimated at 421,894–468,059 turtles (Green and Ortiz-Crespo 1982; Hurtado 1982) before Ecuador closed the fishery in 1981 (Frazier and Salas 1982).

The cumulative legal take by both fisheries was approximately 2.5 million turtles in the thirty-year period from 1960 to 1990. The actual num-

TABLE 13.2. Estimates of turtles landed by the Ecuador fishery, 1970–1981.

Year	1	2 Min	2 Max	3 Mean	4 Min	4 Max	5	Combined[b] Min	Combined[b] Max
1970		352	528	440				352	528
1971	6,600	4,430	6,644	5,537				4,430	6,644
1972	4,400	6,626	9,939	8,283				6,626	9,939
1973	11,000	4,212	4,212	5,247				4,212	4,212
1974	11,000	16,351	16,351	20,737				16,351	16,351
1975	11,000	16,301	16,301	17,207				16,301	16,301
1976	11,000	24,778	24,778	27,502				24,778	24,778
1977	15,400	59,476	67,788	66,491				59,476	67,788
1978	15,400	80,535	89,483	85,009	80,535	89,483	86,916	80,535	89,483
1979		69,950[c]	77,722[c]	147,672	133,233[d]	148,036[d]	93,232	133,233[d]	148,036[d]
1980					18,749	20,832		18,749	20,832
1981					56,851	63,167		56,851	63,167
Total	85,800	283,011	313,746	384,152	289,368	321,518	180,148	421,894	468,059

[a]Sources: 1, FAO statistics reported in Frazier (1980); 2, Green and Ortiz–Crespo (1982); 3, Frazier (1980); 4, Hurtado (1982); 5, Green and Furtando (1980).
[b]Estimates from source 2 and 4 combined.
[c]Partial-year catch data.
[d]Full-year catch data.

ber of turtles removed by fisheries was likely far greater, however. The international demand that inspired a legal fishery for olive ridley skin also gave rise to a black market. Illegal turtle fishing was believed to be widespread in the eastern Pacific (Cornelius 1982; Mack et al. 1982; Trinidad and Wilson 2000). Fishing regulations imposed on the legal fishery in México also fueled the black market. In Oaxaca, the illegal catch was reportedly higher than the legal catch after quotas were implemented (Trinidad and Wilson 2000). The magnitude of uncontrolled turtle fishing is unknown, but it undoubtedly further affected turtles throughout the eastern Pacific.

Egg Exploitation

Along the coasts of México and Central America, human collection of sea turtle eggs occurred before, during, and after the turtle fisheries in the eastern Pacific had landed several million olive ridleys (Cliffton et al. 1982; Cornelius 1982; Hope 2002; see plate 12). The relative impact of egg exploitation on olive ridley population growth rate in the eastern Pacific is unknown; however, it did reduce potential recruitment before and during the time that mature adults were being removed by the turtle fisheries. Nonetheless, in accessible and unprotected non-arribada beaches, the typical extraction levels easily reach 100 percent of all nests.

Eggs from all sea turtle species nesting on east Pacific beaches were collected—some for subsistence use, others for small-scale markets, and some for large-scale commercial markets. Egg collection has been legal in Nicaragua and Costa Rica (Cornelius et al. 1991; Camacho and Cáceres 1995; Valle 1997; Hope 2002; Campbell 2007) and illegal in México since 1927, though egg poaching has occurred there despite the law (Trinidad and Wilson 2000). The massive overexploitation of eggs is believed to have contributed to the olive ridley decline throughout the eastern Pacific (Cliffton et al. 1982; Cornelius 1982). For example, intense commercial egg collection in Nicaragua reportedly led to the disappearance of arribadas at two beaches, Masachapa and Pochomil, in the 1970s (Nietschmann 1975). On the other hand, controlled egg collection at Ostional by the local residents has not resulted in a decline there even after twenty years of extraction (Chaves et al. 2005; Campbell 2007), but long-term protection of this resource by the local residents has occurred coincidental with the harvest.

On non-arribada beaches in the eastern Pacific, egg collection was also widespread and just as intense (Cornelius et al. 2007). Lagueux (1991) estimated that 100 percent of the olive ridley eggs have been collected on Honduran beaches adjacent to the Gulf of Fonseca since the 1940s, and perhaps even earlier.

The Rapid Decline

The intense fishing pressure in the eastern Pacific led to a rapid, large decline of olive ridley populations nesting in the region, particularly at Mexican beaches (fig. 13.3), where they were broadly reported as collapsed, crashed, or extinct (Frazier 1980; Cliffton et al. 1982; Peñaflores-Salazar et al. 2000). Undoubtedly, the turtle fisheries likely affected other arribada rookeries in the region, but the declines elsewhere were not as apparent as in México, for several reasons. First, and most important, no other arribada colonies were subjected to the same fishing intensity as México's. The Ecuadorian turtle fishery was intense, but it landed olive ridleys believed to have originated from multiple nesting beaches. Thus, no one rookery was directly targeted, and the impacts were likely spread among rookeries from a much larger geographic area. Second, turtles were fished intensely in México for a significantly longer duration (about three decades) than they were in Ecuador (about one decade). Finally, sea turtle conservation and protection programs in México developed coincidental with and in response to the turtle fishery (Márquez et al. 1998, 2007; Trinidad and Wilson 2000) and provided long-term abundance indicators. Such programs were slow to develop in other countries, and quantitative data of past abundance at some of these beaches are unavailable.

Decline in México

Historically, there were four large arribada rookeries in México where tens to hundreds of thousands of olive ridleys nested annually: Playón de Mismaloya, Piedra de Tlalcoyunque, Chacahua, and Escobilla (fig. 13.3) (Pritchard 1969; Cliffton et al. 1982; Márquez et al. 1996). A fifth rookery of significant magnitude nesting at Morro Ayuta in Oaxaca has received much less attention (Aguilar-Reyes 2007). The first signs of distress occurred in late 1960s to early 1970s, when the catch rate in the Mexican turtle fishery declined (Trinidad and Wilson 2000) and two arribada rookeries (Mismaloya and Tlalcoyunque) were depleted (Márquez et al. 1996; Peñaflores-Salazar et al. 2000). It is possible that these rookeries were first to decline because the intensity of the fishing pressure there was greatest (Trinidad and Wilson 2000), but equally feasible, the impact on these was greater since they never approached the size of those in Oaxaca. The Mexican government closed the fishery for eighteen months to reorganize in 1971–1972 (Trinidad and Wilson 2000). By 1973, the fishery reopened with new regulations—seasonal closures and quotas—and large numbers of turtles were again harvested that year. Turtle captures declined in Sinaloa, Jalisco, Michoacán, and Guerrero, and fishing pressure shifted south to Oaxaca (Trinidad and Wilson 2000; Briseño-Dueñas 2006a). Unfortunately, fishing regulations were disregarded, illegal turtle catch exceeded the legal catch, and olive ridley abundance continued to decrease (Trinidad and Wilson 2000). By the late 1970s, the arribadas at Mismaloya, Tlalcoyunque, and Chacahua had disappeared (Peñaflores-Salazar et al. 2000; Trinidad and Wilson 2000). The size and frequency of the last remaining arribadas at Escobilla had declined as well (Cliffton et al. 1982), reaching a low point in 1988 (figs. 13.3, 13.4) (Márquez et al. 1996). The turtle fishery took its toll along the entire Mexican coast and depleted both arribada and non-arribada nesting beaches (Abreu-Grobois and Plotkin 2008). Survival prospects of olive ridley indeed seemed grim, and the government closed the fishery in 1990 and banned trade in sea turtle products.

Decline at Other Arribada Beaches

Other arribada rookeries in the region also declined, but these declines have been weakly linked to the eastern Pacific turtle fisheries (Fonseca et al. 2009). Overharvest of eggs and incidental capture of ridleys by shrimp fisheries are also believed to have been primary threats to these rookeries (Cornelius 1982). Historically, there were several arribada beaches in Nicaragua: Boquita, Masachapa, and Pochomil (table 13.1; Cornelius 1982). Unfortunately, no quantitative data are available to document olive ridley past abundance on these beaches. But residents there reported fewer and smaller arribadas by the 1970s (Cornelius 1982), and two of the arrib-

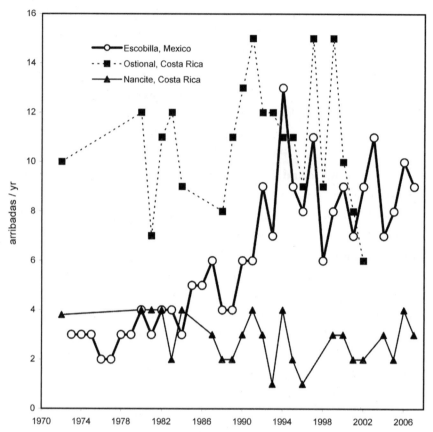

FIGURE 13.4. Number of arribadas per year registered at Escobilla, Ostional, and Nancite, 1972–2007. Data sources: Escobilla, Márquez et al. (2007) and C. Peñaflores, Centro Mexicano de la Tortuga, pers. commun.; Ostional, Chaves et al. (2005), G. Chaves, Proyecto de monitoreo de la tortuga lora (*Lepidochelys olivacea*) en el RNVS Ostional, Costa Rica, Escuela de Biología, Universidad de Costa Rica, pers. commun., and R. Valverde, pers. commun.; Nancite, Valverde et al. (1998) and Fonseca et al. (2009).

ada rookeries reportedly collapsed from unsustainable collection of eggs (Nietschmann 1975) used for food within Nicaragua and shipped elsewhere in Central America (Cornelius 1982).

In Costa Rica two arribada beaches, Nancite and Ostional, were discovered by the scientific community in 1970 (see tables 13.1, 13.3) (Richard and Hughes 1972; Hughes and Richard 1974). At Nancite, a significant decrease in both frequency of arribadas (fig. 13.4) and size has occurred since its discovery: the number of nesting turtles decreased by 42 percent between 1971 and 1984 (Valverde et al. 1998; Fonseca et al. 2009). Nancite is a very small, remote beach located in a protected area that was once a national park and now a conservation area. Egg collection did not occur at Nancite (other than by occasional egg poachers), nor were adult turtles fished there as they had been in México. Some turtles might have been taken from Nancite during the late 1960s and early 1970s when foreign fishing vessels illegally captured turtles from Costa Rican beaches (Cornelius 1982), presumably for the turtle skin trade. Nancite may have deteriorated naturally (Bernardo and Plotkin 2007) as a result of high levels of density-dependent egg mortality and an insufficient nest microenvironment (Honarvar et al. 2008; Fonseca et al. 2009).

Ostional residents reported that arribadas first appeared there in the early 1960s (Chaves et al. 2005). By 1971, an estimated 10,000 turtles nested there annually (Hughes and Richard 1974). Nesting beach censuses, begun in 1980, show an increase and stable trend throughout the 1980s and 1990s (Chaves et al. 2005). Arribadas ranged in size from 40,000 turtles to more than 100,000 turtles and occurred seven to sixteen times annually (Chaves et al. 2005). Ostional is also remote and nestled within a wildlife refuge, but it has never been completely protected like Nancite. A human community has populated the area since arribadas began and has harvested the eggs first for subsistence use and later for a national commercial market (Campbell 2007). For the last twenty years, the harvest has been controlled by the local community, and they have provided long-term protection of the eggs and hatchlings left in place. All evidence suggests that Ostional rookery is stable or increasing despite the egg extraction (table 13.1).

Past abundance estimates of olive ridleys at other areas in Central America are unavailable. Cornelius (1982) noted that at one time there were thirty beaches where the nesting aggregations were described as arribadas, but no other details are available.

Present Status

México

Since the turtle fishery closed in 1990, the three arribada rookeries that collapsed in México have remained depleted with respect to former abundances, though none has become extinct (table 13.1, fig. 13.3). Piedra

TABLE 13.3. Estimates of past and present olive ridley abundance at arribada beaches in the eastern Pacific.

Location	Data type	Past estimate		Present estimate	
		Years	Mean	Years	Mean
MEXICO					
Playón de Mismaloya	AE/PN	1969–1979	35,000–100,000 females	2001–2006	2,000–5,900 nests
Ixtapilla	AE	Before 1997	Sporadic to none	2008–2010	>150,000 nests
Piedra de Tlalcoyunque	AE/PN	1974	20,000–50,000 females	1993–1997	400–610 nests
Chacahua	AE/PN	1974	20,000–50,000 females	2000–2006	2,300–4,300 nests (16,650 from three arribadas in 1997)
Escobilla	AE/AE	1973–1975	180,000–401,000 females	2001–2008	700,000–1,193,000 nests
Morro Ayuta	AE	?	?	1997–2007	125,000–200,000 nests
NICARAGUA					
Chacocente	NN			2002–2004	42,541 nests
La Flor	NN	1993–1999	11,000–49,000 nests	2002–2006	65,000–190,000 nests
COSTA RICA					
Nancite	AE/AE	1970s	142,000–335,000 females	2007	17,876 females
Ostional	AE/AE	1971	10,000 females	2006	336,000 females
PANAMA					
Isla Cañas	AE/AE	1990s	15,000–60,000 nests	2006	About 9,000 females

All values are annual means unless otherwise stated.
[a]Codes for data type: AE, census estimate of nesting females during arribadas; NF, nesting females; NN, number of nests; PN, protected nests (excluding poached, predated, and otherwise lost).

References	
Past	Present
Montoya 1969; Márquez et al. 1976; Casas-Andreu 1978; Villa Guzmán 1980	Antonio Trejo, Universidad de Guadalajara, pers. commun.; Dirección General de Vida Silvestre, SEMARNAT, México
Sánchez and Reyes 1998	Centro Mexicano de la Tortuga, unpublished data; Albavera-Padilla 2009; Albavera-Padilla, pers. commun.
Márquez et al. 1976	Dirección General de Vida Silvestre, SEMARNAT México; Márquez et al. 1998; Peñaflores-Salazar et al. 2001
Márquez et al. 1976	Aguilar-Reyes 2007
Márquez et al. 1976 (August–October arribadas only)	Dirección General de Vida Silvestre, SEMARNAT, México; C. Peñaflores, Centro Mexicano de la Tortuga, unpublished data; Albavera-Padilla 2009; Albavera-Padilla, pers. commun.
No information	Albavera-Padilla 2009; Albavera-Padilla, pers. commun.
None	López-Carcache et al.(2008)
Ministerio del Ambiente y los Recursos Naturales, Nicaragua	Urteaga and Motha 2007
Valverde et al. 1998	Fonseca et al. 2009
Hughes and Richard 1974; Chaves et al. 2005	Solis et al. 2008
Evans and Vargas 1998	L. Vargas, Autoridad Nacional del Ambiente, pers. commun., 2007

de Tlalcoyunque appears stable, but few monitoring data are available. Chacahua exhibits high abundances in comparison to non-arribada beaches (1,000–1,500 nesting females/year) with an erratic trend, as well as a surprising single season in 1997 when four arribada events were reported (Aguilar-Reyes 2007) producing a total of more than 16,500 nests (equivalent to 6,600 nesting females). The rookery at Escobilla (plate 1), evidently the largest even before the turtle fishery, never did collapse. Within a few years after the closure, the number of turtles nesting at Escobilla increased significantly, and this increase has continued since (fig. 13.3). Nesting at Escobilla rebounded from a low point of approximately 55,000 nests in 1988 to more than 700,000 nests in 1994 (Márquez et al. 1996) and more than a million nests by 2000 (Márquez et al. 2005). Present abundance estimates indicate that more than a million olive ridleys nest on Escobilla annually. The annual frequency of arribadas also markedly increased after the fishery was closed (fig. 13.4), from an average of 3.6 arribadas (1973–1989) to 8.5 arribadas (1990–2007) (Peñaflores-Salazar et al. 2000; Albavera-Padilla 2009; E. Albavera-Padilla, pers. commun.), a value that has been sustained (Márquez et al. 2007). More significant, a new arribada rookery began in Ixtapilla, Michoacán, in the mid-1990s at a beach previously not known to host olive ridley nesting (Sánchez and Reyes 1998). Monitoring with adequate methodology is still incipient, but reports indicate abundances of 15,000 to 45,000 nests per arribada, with five or more occurring per year (Albavera-Padilla 2009; E. Albavera-Padilla, pers. commun.).

Nicaragua

At present there are two arribada beaches in Nicaragua, Chacocente and La Flor (plate 1), both believed to be increasing. Most recent estimates of abundance at Chacocente indicate 42,541 nests were laid in 2002–2003 (López-Carcache et al. 2008; J. Urteaga, pers. commun.) At La Flor, an estimated 65,000–190,000 nests were reported in 2002–2006 (Urteaga and Motha 2007), with an increasing trend of more than 10 percent per year.

Costa Rica

The Nancite rookery (plate 1) experienced a 90 percent decline between 1971 and 2007, and although arribadas still occur, they are smaller and less frequent (Fonseca et al. 2009). Nancite fits the model proposed by Bernardo and Plotkin (2007) where an ephemeral arribada rookery deteriorates over time. The sheer number of turtles nesting on this spatially limited beach resulted in high levels of density-dependent mortality of eggs, high concentrations of fungal and bacterial pathogens, and high nest predation (Bernardo and Plotkin 2007; Honarvar et al. 2008; Fonseca et al. 2009). Decades of nest protection did little to increase hatchling production and recruitment there (Cornelius et al. 2007; Fonseca et al. 2009).

In contrast, the Ostional rookery (plate 1), despite the egg extraction, has remained abundant (Chaves et al. 2005). Since censuses began in 1980, the size and frequency of arribadas have increased, and the nesting area used has increased significantly (Chaves et al. 2005). The average arribada size in the main nesting beach increased from 75,000 turtles in 1980 to 125,000 turtles in 2003 (Chaves et al. 2005). Discrepancies in the methods used for estimating arribada abundances have made it almost impossible to derive a clear understanding of the population trend. The most recent census data available using accepted survey methods (Valverde and Gates 1999) estimated 336,000 turtles nested at Ostional in 2006 (Solis et al. 2008).

Panamá

Olive ridley arribadas presently occur at Isla Cañas, a protected area within the Panamá National Wildlife Refuge system, as well as at Playa La Marinera (plate 1). In the late 1990s, 15,000–60,000 nests were laid at Isla Cañas annually during three to five arribadas (Evans and Vargas 1998). The rookery is believed to be in decline and was recently estimated at around 9,000 turtles (table 13.3). Preliminary information from the rookery at La Marinera indicates that it is larger; more than 10,000 females were estimated to have nested in arribadas there in 2008 (Abrego et al. 2010).

Interpreting the Signs

The overexploitation and collapse of olive ridley at multiple arribada beaches in México, followed by the impressive increase at Escobilla, and the new arribada rookery at Ixtapilla in less than twenty years, demonstrate a remarkable rate of rebound not previously reported in sea turtles. Other sea turtle species have exhibited some level of recovery after overexploitation and subsequent protection (Bjorndal et al. 1999; Balazs and Chaloupka 2004; Dutton et al. 2005; Troëng and Rankin 2005; Heppell et al. 2007), but none so rapidly as the olive ridley, at least in arribada rookeries.

Unfortunately, recovery is not so evident elsewhere in the eastern Pacific. Collapsed arribada rookeries in México, though stable and most showing signs of increase, host only solitary nesters and have not recovered regular arribadas. While the elimination of large-scale commercial exploitation of the olive ridley facilitated population growth at Escobilla, why is it that a similar rebound did not occur at the three former arribada rookeries in the same region under similar conservation circumstances? One plausible reason may be that these three rookeries, by being smaller, were critically affected and collapsed during the fishery while a much larger Escobilla proved more resilient to these impacts. We suggest that under some circumstances, the reduction of arribada rookeries below a critical density could, by itself, make it impossible or unlikely for an arribada to recover (i.e., a kind of Allee effect).

It must be stressed that recovery, in the sense of recuperating pre-1970s sizes, is also generally lacking at non-arribada rookeries in the eastern Pacific. These rookeries occur throughout this region, suffered analogous declines, but remain depleted (Abreu-Grobois and Plotkin 2008) even if many populations are growing in recent decades (e.g., Márquez et al. 1998). This is significant because most rookeries in the region are non-arribada rookeries, and (outside of México) most do not receive as much attention as arribada populations.

So, how do we interpret these signs exhibited by arribada rookeries? And is this recovery? When Márquez et al. (1996) first reported signs of recovery at Escobilla just a few years after permanent legal protection and closure of the fishery, the news was met with great skepticism. Ross (1996), and many others, believed the declaration was premature. Ross cautioned that the increase might reasonably be part of the natural annual variation observed at sea turtle nesting beaches worldwide, that a longer time series was needed, and that, at best, one could state that the decline had not continued. Pritchard (1997) was much more optimistic however. He noted that the increase in the number of turtles nesting at Escobilla was five times the typical level of annual nesting documented in recent years and double the level documented in the preceding twenty years.

As it turns out, Márquez et al. (1996) indeed was right. Recovery begun at Escobilla in the early 1990s has continued for more than eighteen years. The long time series of nesting data indicate the Escobilla rookery has increased 125 percent over two generations (about forty years) (Abreu-Grobois and Plotkin 2008). Márquez's (1996) suggestion that the number of arribadas per season can be used to gauge the status of arribada rookeries has also proven correct. It looks as if this value may be rookery specific (fig. 13.4) and Escobilla's—which dropped in the 1970s—reached an asymptotic maximum of eight to nine arribadas per year by the middle to late 1990s. The high growth rate of the Escobilla rookery has been ascribed to a combination of large cohorts of females produced through nest protection that began in the mid-1970s in the Pacific (assuming an average time to maturity of about thirteen years; Zug et al. 2006) and a significantly higher survival rate of adult females after the turtle fishery closed in 1990 (Pritchard 1997; Peñaflores-Salazar et al. 2000). Having before us a broad selection of long-term case histories provides us with privileged insight not only on how arribada rookeries fare under a diversity conservation regimes, but perhaps also on the inner workings of the mechanism for the establishment of arribadas.

In this vein, comparison of two arribada sites—one old and one new—represents the enigmatic nature of the arribada phenomenon. Reports of three arribadas in 1997 at Chacahua (fig. 13.3; Aguilar-Reyes 2007) that left three to five times the "normal" annual number of nests could provide rare insight into the inner workings of the process of arribada establishment. Assuming that it is not possible for the local population by itself to suddenly expand so rapidly and then wane back to normality, perhaps special or unusual circumstances led to abundant available breeders above a certain threshold that not only focuses mass nesting events of local individuals but also promotes the participation of nonlocals en route to their own nesting sites. Some sort of nesting site fidelity then would need to occur for the arribada to settle locally and repeat henceforth. The 1997 Chacahua arribadas appear to have been short-lived (despite formerly being an arribada site); thus, conditions did not remain adequate or repeat themselves. In contrast, in the case of Ixtapilla, the arribadas that began in the mid-1990s return year after year, which could explain how it became established where olive ridleys were not known to nest, that is, not by the transformation of the reproductive mode of a single rookery but from a mixture of breeders stemming from various origins, having been attracted by a rare event leading to a local overabundance of other breeders, and later fixing on the local nesting site.

Above all, it is clear that even the massive arribada rookeries are susceptible to overexploitation and collapse, even within a short period of time, at the hands of man, but given enough protection, and as long as populations do not fall below certain limits, they have the capacity to survive and recover. On the management front, many direct and indirect threats challenges lie ahead: directed take of adults, egg collection (Lagueux 1989; Campbell 2007; Cornelius et al. 2007), and bycatch in fisheries (Frazier et al. 2007). Though cases of extirpation of arribada rookeries from egg overexploitation exist (e.g., Boquita, Masachapa, Pochomil in Nicaragua—see table 13.1; Nietschmann 1975), all other cases of long-term egg extraction, as long as there is some degree of control, have allowed for stable or increasing arribada rookeries (e.g., Chacocente, La Flor, Ostional, Morro Ayuta; table 13.1). Fisheries bycatch is particularly worrisome given the increasing deterioration of the region's fisheries, a chronic overcapacity of the fleets, and more fishers going after ever diminishing stocks.

Recent, sustained increases in the abundance of olive ridleys at several key arribada rookeries in the east Pacific are indeed encouraging signs. The rapid increase at Escobilla is especially exciting because it clearly demonstrates that recovery is possible even after intense exploitation ceases and when long-term protection of other life stages occurs. While we are very optimistic that the large numbers of ridleys now nesting at Escobilla and Ostional are signs of recovery, we need to be cautious and not overstate the meaning of these signs. Our insufficient knowledge of olive ridley reproductive behavior and ecology limits our ability to fully comprehend the impact of past exploitation, present recovery trends, and future extinction risk from other threats and emerging risk factors of natural and anthropogenic origins. We understand so little about many key components: evolution, maintenance, and recovery of arribadas; the reproductive behavioral polymorphism unique to this species; the status and significance on nonarribada rookeries; basic life history traits and demography. It is our hope that this chapter stimulates further research in the eastern Pacific to help overcome these challenges and address remaining knowledge gaps.

ACKNOWLEDGMENTS

We thank the editors for their patience. We are grateful to the numerous individuals who provided us unpublished and in press data: Ernesto Albavera, Gerardo "Cachi" Cháves, Jack Frazier, Cuahutémoc Peñaflores, José Urteaga, and Roldán Valverde. María de los Angeles Herrera Vega helped source and correct the bibliography.

REFERENCES

Abrego, M., J. Rodríguez, and A. Ruiz. 2010. Datos de monitoreo de Playa La Marinera. Comunicación Personal. In: The State of the World's Sea Turtles. SWOT Report, vol. 5. Available: http://seaturtlestatus.org/report/view.

Abreu-Grobois, A., and P. Plotkin. 2008. *Lepidochelys olivacea*. In: 2008 IUCN Red List of Threatened Species. Available: www.iucnredlist.org.

Aguilar-Reyes, H. 2007. Las tortugas marinas en el Parque Nacional Lagunas de Chacahua. In: Fuentes, G., F. López, and M. A. Domínguez (Eds.), XXV Aniversario de conservación e investigación en tortuga marina. Tomo III. Un solo objetivo: las tortugas marinas. Universidad Autónoma "Benito Juárez" de Oaxaca, Oaxaca, México. Pp. 7–28.

Albavera-Padilla, E. 2009. Situación actual de la tortuga golfina (*Lepidochelys olivacea*) en playas de arribada del Pacífico mexicano. In: Sarti, L., A. Barragán, y C. Aguilar (Comp.), Memorias de la Reunión Nacional sobre Conservación de Tortugas Marinas. Veracruz, Ver., 25—28 de Noviembre de 2007. Comisión Nacional de Áreas Naturales Protegidas, SEMARNAT, Mexico City, México. Pp. 42–46.

Aridjis, H. 1990. Mexico proclaims total ban on harvest of turtles and eggs. Marine Turtle Newsletter 50:1–3.

Balazs, G. H., and M. Chaloupka. 2004. Thirty year recovery trend in the once depleted Hawaiian green sea turtle stock. Biological Conservation 117:491–498.

Ballestero, J. 1996. Weather changes and olive ridley nesting density in the Ostional Wildlife Refuge, Santa Cruz, Guanacaste, Costa Rica. In Keinath, J. A., D. E. Barnard, J. A. Musick, and B. A. Bell. (Comps.), Proceedings of the Fifteenth Annual Symposium on Sea Turtle Biology and Conservation. NOAA Tech. Memo. NMFS-SEFSC-387. U.S. Department of Commerce, U.S. National Marine Fisheries Service, Miami, FL. Pp. 26–30.

Bernardo, J., and P. T. Plotkin. 2007. An evolutionary perspective on the arribada phenomenon and reproductive behavioral polymorphism of olive ridley sea turtles (*Lepidochelys olivacea*). In: Plotkin, P. T. (Ed.), Biology and Conservation of Ridley Sea Turtles. Johns Hopkins University Press, Baltimore, MD. Pp. 59–88.

Bjorndal, K., and J. B. C. Jackson. 2003. Roles of sea turtles in marine ecosystems: reconstructing the past. In: Lutz, P. L., J. A. Musick, and J. Wyneken (Eds.), The Biology of Sea Turtles, Vol. 2. CRC Press, Boca Raton, FL. Pp. 259–273.

Bjorndal, K. A., J. A. Wetherall, A. B. Bolten, and J. A. Mortimer. 1999. Twenty-six years of green turtle nesting at Tortuguero, Costa Rica, an encouraging trend. Conservation Biology 12:126–134.

Briseño-Dueñas, R. 2006a. Tortugas Marinas: de recurso natural a especies en riesgo. In: Cifuentes Lemus, J. L., and J. Gaxiola López (Eds.), Atlas del Manejo y Conservación de la Biodiversidad y Ecosistemas de Sinaloa. El Colegio de Sinaloa, Culiacán, México. Pp. 121–130.

Briseño Dueñas, R. 2006b. ¿Cuándo las caguamas abundaron en nuestro mar? Red de Tortugueros de Sinaloa. Revista de Divulgación del Acuario Mazatlán 2(2):1–5.

Camacho, M. G., and J. G. Cáceres. 1995. Importancia de la protección y conservación de las tortugas marinas en Nicaragua y el aprovechamiento de huevos en el RVS Chacocente. In: Proceedings 1er Encuentro Ciclo de Conferencias Biológicas: Un Día con la Tortuga Marina. Parque Nacional Volcán Masaya, Septiembre 1994. Pp. 40–65.

Campbell, L. 2007. Understanding human use of olive ridleys: implications for conservation. In: Plotkin, P. T. (Ed.), Biology and Conservation of Ridley Sea Turtles. Johns Hopkins University Press, Baltimore, MD. Pp. 23–44.

Carr, A. 1967. So Excellent a Fishe. Garden City, NY: Natural History Press.

Casas-Andreu, G. 1978. Análisis de la anidación de las tortugas marinas del género Lepidochelys en México. Anales del Instituto de Ciencias del Mar y Limnologia 5(1):141–158.

Chaves, G., R. Morera, J. R. Aviles, J. C. Castro, and M. Alvarado. 2005. Trends of the nesting activity of the "arribadas" of the olive ridley (*Lepidochelys olivacea*, Eschscholtz 1829), in the Ostional National Wildlife Refuge (1971–2003). Proyecto de monitoreo de la tortuga lora (*Lepidochelys olivacea*) en el Refugio Nacional de Vida Silvestre Ostional, Costa Rica. Escuela de Biología, Universidad de Costa Rica, San José, Costa Rica. Available: www.mendeley.com/research/trends-of-the-nesting-activity-of-the-arribadas-of-the-olive-ridley-lepidochelys-olivacea-eschscholtz-1829-in-the-ostional-national-wildlife-refuge-19712003/.

Cliffton, K., D. O. Cornejo, and R. S. Felger. 1982. Sea turtles on the Pacific coast of Mexico. In: Bjorndal, K. (Ed.), Biology and Conservation of Sea Turtles. Smithsonian Institution Press, Washington, DC. Pp. 199–209.

Cornelius, S. E. 1982. Status of sea turtles along the Pacific coast of middle America. In: Bjorndal, K. A. (Ed.), Biology and Conservation of Sea Turtles. Smithsonian Institution Press, Washington, DC. Pp. 211–220.

Cornelius, S. E., M. A. Alvarado Ulloa, J. C. Castro, M. Malta De Valle, and D. C. Robinson. 1991. Management of olive ridley sea turtles (*Lepidochelys olivacea*) nesting at playas Nancite and Ostional, Costa Rica. In: Robinson, J. G., and Redford, K. H. (Eds.), Neotropical Wildlife Use and Conservation. University of Chicago Press, Chicago, IL. Pp. 111–135.

Cornelius, S. E., R. Arauz, J. Fretey, M. H. Godfrey, R. Márquez, and K. Shanker. 2007. Effect of land-based harvest of *Lepidochelys*. In: Plotkin, P. T. (Ed.), Biology and Conservation of Ridley Sea Turtles. Johns Hopkins University Press, Baltimore, MD. Pp. 231–252.

Cornelius, S. E., and D. C. Robinson. 1985. Abundance, Distribution and Movements of Olive Ridley Sea Turtles in Costa Rica. V. Final Report to the US Fish and Wildlife Service. Contract No. 14-16-0002-81-225. U.S. Fish and Wildlife Service, Albuquerque, NM.

Cornelius, S. E., and Robinson D. C. 1986. Post-nesting movements of female olive ridley turtles tagged in Costa Rica. Vida Silvestre Neotropical 1(11):12–23.

Dutton, D. L., P. H. Dutton, M. Chaloupka, and R. H. Boulon. 2005. Increase of a Caribbean leatherback turtle *Dermochelys coriacea* nesting population linked to long-term nest protection. Biological Conservation 126:186–194.

Eckrich, C. E., and D. W. Owens. 1995. Solitary versus arribada nesting in the olive ridley sea turtle (*Lepidochelys olivacea*): a test of the predator-satiation hypothesis. Herpetologica 51:349–354.

Evans, K. E., and A. R. Vargas. 1998. Sea turtle egg commercialization in Isla de Cañas, Panamá. In: Byles, R., and Y. Fernandez (Comps.), Proceedings of the Sixteenth Annual Symposium on Sea Turtle Biology and Conservation. NOAA Tech. Memo. NMFS-SEFSC-412. U.S. Department of Commerce, U.S. National Marine Fisheries Service, Miami, FL. P. 45.

Fonseca, L. G., G. A. Murillo, L. Guadamúz, R. M. Spínola, and R. A. Valverde. 2009. Downward but stable trend in abundance of arribada olive ridley (*Lepidochelys olivacea*) sea turtles at Nancite Beach, Costa Rica for the period 1971–2007. Chelonian Conservation and Biology 8(1):19–27.

Frazier, J. 1980. Marine Turtle Fisheries in Ecuador and Mexico: The Last of the Pacific Ridley? Unpublished paper, Smithsonian Institution, Washington, DC.

Frazier, J. 2003. Prehistoric and ancient historic interactions between humans and marine turtles. In: Lutz, P. L., J. A. Musick, and J. Wyneken (Eds.), The Biology of Sea Turtles, Vol. 2. CRC Press, Boca Raton, FL. Pp. 1–38.

Frazier, J., R. Arauz, J. Chevalier, A. Formia, J. Fretey, M. H. Godfrey, R. Márquez, B. Pandav, and K. Shanker. 2007. Human-turtle interactions at sea. In: Plotkin, P. T. (Ed.), Biology and Conservation of Ridley Sea Turtles. Johns Hopkins University Press, Baltimore, MD. Pp. 253–296.

Frazier, J., and S. Salas. 1982. Ecuador closes commercial turtle fishery. Marine Turtle Newsletter 20:5–6.

Green, D., and M. Furtando-G. 1980. Ridleys in Ecuador—a ray of hope? Marine Turtle Newsletter 16:1–5.

Green, D., and F. Ortiz-Crespo. 1982. Status of sea turtle populations in the central east Pacific. In: Bjorndal, K. A. (Ed.), Biology and Conservation of Sea Turtles. Smithsonian Institution Press, Washington, DC. Pp. 221–234.

Heppell, S. S., P. M. Burchfield, and L. J. Pena. 2007. Kemps ridley recovery. How far have we come, and where are we headed? In: Plotkin, P. T. (Ed.), Biology and Conservation of Ridley Sea Turtles. Johns Hopkins University Press, Baltimore, MD. Pp. 325–335.

Hirth, H. 1980. Nesting biology of sea turtles. American Zoologist 20:507–523.

Honarvar, S. 2007. Nesting ecology of olive ridley (*Lepidochelys olivacea*) turtles on arribada nesting beaches. Ph.D. dissertation, Drexel University, Philadelphia, PA.

Honarvar, S., M. P. O'Connor, and J. R. Spotila. 2008. Density-dependent effects on hatching success of the olive ridley turtle, *Lepidochelys olivacea*. Oecologia 157(2):221–230.

Hope, R. A. 2002. Wildlife harvesting, conservation and poverty: the economics of olive ridley egg exploitation. Environmental Conservation 29(3): 375–384.

Hughes, D. A., and J. D. Richard. 1974. Nesting of the Pacific ridley *Lepidochelys olivacea* on Playa Nancite, Costa Rica. Marine Biology 24:97–107.

Hurtado-G., M. 1982. The ban on the exportation of turtle skin from Ecuador. Marine Turtle Newsletter 20:1–4.

Jackson, J. B. C. 1997. Reefs since Columbus. Coral Reefs 16:S23–S32.

Kalb, H. J. 1999. Behavior and physiology of solitary and arribada nesting olive ridley sea turtles (*Lepidochelys olivacea*) during the internesting period. Ph.D. dissertation, Texas A&M University, College Station.

Kennett, D. J., B. Voorhies, T. A. Wake, and N. Martínez. 2007. Human impacts on marine ecosystems in Guerrero, Mexico. In: Rick, T. C., and J. M. Erlandson (Eds.), Human Impacts on Ancient Marine Ecosystems: A Global Perspective. University of California Press, Berkeley. Pp. 103–124.

Lagueux, C. J. 1989. Olive ridley (*Lepidochelys olivacea*) nesting in the Gulf of Fonseca and the commercialization of its eggs in Honduras. Master's thesis, University of Florida, Gainesville.

Lagueux, C. 1991. Economic analysis of sea turtle eggs in a coastal community on the Pacific coast of Honduras. In: Robinson, J. G., and K. H. Redford (Eds.), Neotropical Wildlife Use and Conservation. University of Chicago Press, Chicago, IL. Pp. 136–144.

López-Carcache, J., R. Vega, A. Carballo, M. Rodriguez, B. Cortez, S. Mota, M. Camacho, and J. Urteaga. 2008. Monitoring of isolated and arribada nests of olive ridley, *Lepidochelys olivacea*, in Chacocente Beach, Rio Escalante-Chacocente, Wildlife Refuge, Pacific Coast of Nicaragua (2002–2004). In: Proceedings of the Twenty-Fifth Annual Symposium on Sea Turtle Biology and Conservation. NOAA Tech. Memo. NMFS-SEFSC-582. U.S. Department of Commerce, U.S. National Marine Fisheries Service, Miami, FL. P. 135.

Mack, D., N. Duplaix, and S. Wells. 1982. Sea turtles, animals of divisible parts: international trade in sea turtle products. In: Bjorndal, K. A. (Ed.), Biology and Conservation of Sea Turtles. Smithsonian Institution Press, Washington, DC. Pp. 545–564.

Márquez, R. 1996. Las Tortugas Marina y Nuestro Tiempo. No. 144. Fondo de Cultura Económica–La Ciencia para Todos, México, DF.

Márquez, R., M. A. Carrasco, M. C. Jiménez, C. Peñaflores-S., and R. Bravo-G. 2005. Kemp's and olive ridley sea turtles population status. In: Coyne, M. S., and R. D. Clark (Comps.), Proceedings of the Twenty-First Annual Symposium on Sea Turtle Biology and Conservation. NOAA Tech. Memo. NMFS-SEFSC-528. U.S. Department of Commerce, U.S. National Marine Fisheries Service, Miami, FL. Pp. 237–239.

Márquez, R., M. C. Jiménez, M. A. Carrasco, and N. A. Villanueva. 1998. Comentarios acerca de las tendencias poblacionales de las tortugas marinas del género *Lepidochelys* después de la veda total de 1990. Oceánides 13(1):41–62.

Márquez, R., C. Peñaflores-S., and M. C. Jiménez-Q. 2007. Protección de la tortuga marina en la costa de Oaxaca por el Instituto Nacional de la Pesca. In: G. Fuentes-Mascorro, S. S. Martínez Blas, and F. A. López Rojas (Eds.), XXV aniversario de conservación e investigación en tortuga marina. Tomo II. Santuario "La Escobilla"—Un compromiso de conservación con la humanidad. Universidad Autónoma "Benito Juárez" de Oaxaca, Oaxaca, México. Pp. 8–29.

Márquez, R., C. Peñaflores, and J. C. Vasconcelos. 1996. Olive ridley turtles (*Lepidochelys olivacea*) show signs of recovery at La Escobilla, Oaxaca. Marine Turtle Newsletter 73:5–7.

Márquez, R., A. Villanueva-O., and C. Peñaflores-S. 1976. Sinopsis de datos biológicos sobre la Tortuga golfina *Lepidochelys olivacea* (Eschscholtz, 1829). INP/52 Sinopsis sobre las Pesca. Instituto Nacional de Pesca, Mexico City, México.

McClenachan, L., J. B. C. Jackson, and M. J. H. Newman. 2006. Conservation implications of historic nesting beach loss. Frontiers in Ecology and Environment 4(6):290–296.

Molina, S. 1981. Leyendo en la Tortuga (Recopilación). Martín Casillas Editores, México, DF.

Montoya, A. 1967. Recopilación de los datos del valor comercial y la captura anual de tortugas marinas en el periodo 1940–1965. Secretaría de Industria y Comercio, Instituto Nacional de Investigaciones Biológico-Pesqueras, Boletín del Programa Nacional de Marcado de Tortugas Marinas 1(10):1–10.

Montoya, A. E. 1969. Programas de investigación y conservación de las tortugas marinas en México. In: Proceedings of the Working Meeting of Marine Turtle Specialists organized by IUCN at Morges, Switzerland on 10–13 March. International Union for Conservation of Nature, Gland, Switzerland. Pp. 34–53.

Mountjoy, J., and C. Claassen. 2005. Middle formative diet and seasonality on the central coast of Nayarit, Mexico. In: Dillon, B. D., and M. A. Boxt (Eds.), Archaeology without Limits: Papers in Honor of Clement W. Meighan. Labyrinthos, Lancaster, CA. Pp. 267–282.

Nietschmann, B. 1975. Of turtles, arribadas, and people. Chelonia 2:6–9.

Nietschmann, B. 1982. The cultural context of sea turtle subsistence hunting in the Caribbean and problems caused by commercial exploitation. In: Bjorndal, K. A. (Ed.), Biology and Conservation of Sea Turtles. Smithsonian Institution Press, Washington, DC. Pp. 439–446.

NMFS/USFWS. 2007. Olive Ridley Sea Turtle (*Lepidochelys olivacea*) 5-Year Review: Summary and Evaluation. U.S. National Marine Fisheries Service and U.S. Fish and Wildlife Service, Miami, FL.

Peñaflores-Salazar, C., J. Vasconcelos-Pérez, E. Albavera-Padilla, and M. C. Jiménez Quiroz. 2001. Especies sujetas a protección especial. Tortuga golfina. In: Cisneros, M. A., L. F. Beléndez, E. Zárate, M. T. Gaspar, L. C. López, C. Saucedo, and J. Tovar (Eds.), Sustentabilidad y Pesca Responsable en México. Evaluación y Manejo. 1999–2000 [CD]. Instituto Nacional de la Pesca/SEMARNAT, México, DF. Pp. 1001–1021.

Peñaflores-Salazar, C., J. Vasconcelos Pérez, E. Albavera Padilla, and R. Márquez Millán. 2000. Twenty five years nesting of olive ridley sea turtle *Lepidochelys olivacea* in Escobilla Beach, Oaxaca, México. In: Abreu-Grobois, F. A., R. Briseño-Dueñas, R. Márquez, and L. Sarti (Comps.), Proceedings of the Eighteenth International Sea Turtle Symposium. NOAA Tech. Memo. NMFS-SEFSC-436. U.S. Department of Commerce, U.S. National Marine Fisheries Service, Miami, FL. Pp. 27–29.

Plotkin, P. T. 1994. Migratory and reproductive behavior of the olive ridley turtle *Lepidochelys olivacea* (Eschscholtz, 1829), in the eastern Pacific Ocean. Ph.D. dissertation, Texas A&M University, College Station.

Plotkin, P. T. 2010. Nomadic behavior of the highly migratory olive ridley sea turtle *Lepidochelys olivacea* in the eastern tropical Pacific Ocean. Endangered Species Research 13:33–40.

Plotkin, P., and J. Bernardo. 2003. Investigations into the basis of the reproductive behavioral polymorphism in olive ridley sea turtles. In: Seminoff, J. A. (Comp.), Proceedings of the Twenty-Second Annual Symposium on Sea Turtle Biology and Conservation. NOAA Tech. Memo. NMFS-SEFSC-503. U.S. Department of Commerce, U.S. National Marine Fisheries Service, Miami, FL. P. 29.

Plotkin, P. T., R. A. Byles, D. C. Rostal, and D. W. Owens. 1995. Independent vs. socially facilitated migrations of the olive ridley, *Lepidochelys olivacea*. Marine Biology 122:137–143.

Plotkin, P. T., D. C. Rostal, R. A. Byles, and D. W. Owens. 1997. Reproductive and developmental synchrony in female *Lepidochelys olivacea*. Journal of Herpetology 31:17–22.

Pritchard, P. C. H. 1969. Studies of the systematics and reproduction of the genus *Lepidochelys*. Ph.D. dissertation, University of Florida, Gainesville.

Pritchard, P. C. H. 1979. Encyclopedia of Turtles. T. F. H. Publications, Neptune, NJ.

Pritchard, P. C. H. 1997. A new interpretation of Mexican ridley population trends. Marine Turtle Newsletter 76:14–17.

Richard, J. D., and D. A. Hughes. 1972. Some observations of sea turtle nesting activity in Costa Rica. Marine Biology 16:297–309.

Robertson, T. A. 1979. Utopía del Sudoeste. Una colonia americana en México. Lorenzo Valdez, L. (Trans.). Originally published by Ward Ritchie Press, Los Angeles, CA, 1943.

Ross, J. P. 1996. Caution urged in the interpretation of trends at nesting beaches. Marine Turtle Newsletter 74:9–10.

Sáenz-Arroyo, A., C. Roberts, J. Torre, M. Cariño-Olvera, and J. P. Hawkins. 2006. The value of evidence about past abundance: marine fauna of the Gulf of California through the eyes of the 16th to 19th century travelers. Fish and Fisheries 7:128–146.

Sahagún, Fr. Bernardino de. 1975. Historia general de las cosas de Nueva España. Anotaciones y apéndices de Angel María Garibay. Editores Porrua, México, DF.

Sánchez, R., and G. Reyes. 1998. Ixtapilla, Michoacán: recolonization, wandering or a new olive Ridley nesting site? In: Book of Abstracts, 18th International Symposium on Sea Turtle Biology and Conservation, Mazatlán, México, 3–7 March.

Smith, C. B., D. J. Kennett, T. Wake, and B. Voorhies. 2007. Prehistoric sea turtle hunting on the Pacific coast of Mexico. Journal of Island and Coastal Archaeology 2:1–5.

Solis, D. S., C. M. Orrego, R. V. Blanco-Segura, M. R. Harfush-Melendez, E. O. Albavera-Padilla, and R. A. Valverde. 2008. Estimating arribada size: going global. In Rees, A. F., M. Frick, A. Panagopoulou, and K. Williams (Comps.), Proceedings of the 27th Annual Symposium on Sea Turtle Biology and Conservation. NOAA Tech. Memo. NMFS-SEFSC-569. U.S. Department of Commerce, U.S. National Marine Fisheries Service, Miami, FL. P. 249.

Trinidad, H., and J. Wilson. 2000. The bio-economics of sea turtle conservation and use in Mexico: history of exploitation and conservation policies for the olive ridley (*Lepidochelys olivacea*). In: Proceedings of the International Institute of Fisheries Economics and Trade Conference. Oregon State University, Corvallis, OR.

Troëng, S., and E. Rankin. 2005. Long-term conservation efforts contribute to positive green turtle *Chelonia mydas* nesting trend at Tortuguero, Costa Rica. Biological Conservation 121:111–116.

Urteaga, J., and S. Motha. 2007. Resultados de Monitoreo de Tortugas Marinas en al Paci-

fico de Nicaragua. Temporada 2006–2007. Unpublished report submitted to Ministerio del Ambiente y los Recursos Naturales, Managua, Nicaragua.

Valle, E. 1997. Marco jurídico de protección a las Tortugas marinas. In: Memoria del 1er
Taller Problemática de las Tortugas Marinas en Nicaragua, 18–19 Septiembre, El Cru-
cero, Managua, Nicaragua.

Valverde, R. A., S. E. Cornelius, and C. Mo. 1998. Decline of the olive ridley turtle (*Lepidochelys olivacea*) nesting assemblage at Playa Nancite, Santa Rosa National Park, Costa
Rica. Chelonian Conservation and Biology 3:58–63.

Valverde, R. A., and C. E. Gates. 1999. Population surveys on mass nesting beaches. In:
Eckert, K., K. Bjorndal, A. Abreu, and M. Donnelly (Eds.), Research and Management
Techniques for the Conservation of Sea Turtles. IUCN/SSC Marine Turtle Specialist
Group Pub. No. 4. International Union for Conservation of Nature Species Survival
Commission, Washington, DC. Pp. 56—60.

Vilches-Alcázar, R. 1980. Pesca prehispánica: artes, usos, costumbres. Banco Nacional Pesquero y Portuario, S.A., México, DF.

Villa Guzmán, J. 1980. Pesquería de Tortugas Marinas en el Estado de Jalisco. Tesis Profesional, Facultad de Ciencias, Universidad Nacional Autónoma de México, México,
DF.

Zug, G. R., M. Chaloupka, and G. H. Balazs. 2006. Age and growth in olive ridley sea turtles
(*Lepidochelys olivacea*) from the north-central Pacific: a skeletochronological analysis.
Marine Ecology 26:1–8.

Appendix

Citations for Data Used in Maps

State of the World's Sea Turtles: SWOT Guidelines for Data Use and Citation

The nesting data below correspond directly to the maps (see plates) and are organized by species and then alphabetically by country. These data are from the SWOT database via an official data request by the book's editors. Data originally come from a wide variety of sources and in many cases have not been previously published. To use data for research or publication, permission must be obtained from the data provider and must be cited to the original source as indicated in the "Data Source" field of each record.

In the records below, nesting data are reported from the most recent available year or nesting season or are reported as an annual average number of clutches based on the reported years of study. Count data are reported below and are displayed on the maps in generalized bins (e.g., 1–10 clutches, 11–100 clutches) for ease of interpretation. These ranges are generalized to encompass the wide variation in search effort, years monitored, and other factors and were determined on a species-specific basis, accounting for intraspecific variation in annual counts across all reported nesting sites.

Beaches for which count data were not available are listed as "unquantified." Additional metadata are available for many of these data records, including information on beach length, monitoring effort, and other comments and may be found in the SWOT online application on OBIS-SEAMAP (http://seamap.env.duke.edu/swot). For more information, contact the SWOT Database Manager at swotdata@gmail.com. For more information on the history, background, and aims of the SWOT program, and to access SWOT Reports and other products, please see the SWOT website (http://seaturtlestatus.org).

Plate 3. Leatherback, *Dermochelys coriacea*

COSTA RICA
Data Source: (1) Malaver, M., and Chacón, D. 2009. Informe Península de Osa Temporada 2008. Unpublished report. (2) Chacón-Chaverrí, D. 2008. Anidación de Tortugas Marinas en las Playas de Península DE Osa, Pacífico Sur, Costa Rica.
Nesting Beaches: Playa Carate, Rio Oro; Sirena and Corcovado
Year: 2008; Count: 1–10 per beach
SWOT Contact: Didiher Chacón Chaverri

Data Source: Gaos, A. R., Yañez, I. L., and Arauz, R. M. 2006. Sea Turtle Conservation and Research on the Pacific Coast of Costa Rica. Programa Restauración de Tortugas Marinas (PRETOMA). Technical Report.
Nesting Beaches: Costa de Oro, San Miguel
Year: 2004; Count: 1–10 per beach

SWOT Contact: Alexander Gaos

Data Source: Arauz, R., Viejobueno, M. S., Sunyer, S. P., and Naranjo, I. 2009. Conservación e investigación de tortugas marinas en el Pacífico de Costa Rica (Punta Banco, Refugio Nacional de Vida Silvestre Caeltas-Arío, San Miguel, Corozalito). Presentado a las autoridades del Area Conservación Tempisque (ACT) del Ministerio de Ambiente, Energía y Telecomunicaciones (MINAET) (Permiso de Investigación 2009).
Nesting Beach: Playa Caletas
Year: 2008; Count: 1–10

Data Source: Francia, G. 2008. Proyecto de Conservación Baulas del Pacífico de Junquillal. World Wildlife Fund.
Nesting Beaches: Playa Lagarto, Playa Frijolar, Playa Azul, Playa San Juanillo
Year: 2008; Count: 1–10 each beach

Data Source: Francia, G. 2008. Proyecto de Conservación Baulas del Pacífico de Junquillal. World Wildlife Fund.
Nesting Beach: Playa Junquillal
Year: 2008; Count 11–100

Data Source: Cháves, G., Morera, R., and Aviles, J. R. 2008. Seguimiento de la actividad anidatoria de las tortugas marinas (Cheloniidae y Dermochelyidae) en el Refugio Nacional de Vida Silvestre de Ostional, Santa Cruz, Guanacaste.
Nesting Beach: Playa Ostional
Year: 2008; Count: 11–100

Data Source: Piedra, R., and Vélez, E. 2005. Reporte de actividades de investigación y protección de la tortuga baula (*Dermochelys coriacea*), temporada de anidación 2004–2005, Playa Langosta. Unpublished manuscript, Proyecto de Conservación en Tortugas Marinas—Tortuga Baula, Parque Nacional Marino Las Baulas, Guanacaste, Costa Rica.
Nesting Beach: Playa Langosta
Year: 2005; Count: 101–500

Data Source: Leatherback Trust. Las Baulas Conservation Project—Archive 2004–2005 Field Report. www.leatherback.org/pages/project/report/report0405.htm.
Nesting Beach: Playa Grande and Playa Ventanas
Year: 2004; Count: >500 both beaches

ECUADOR
Data Source: (1) Barragán, M. J. 2006. Leatherback nesting in Ecuador: Personal communication. In: The State of the World's Sea Turtles Report, vol. 1. (2) Herrera, M., Coello, D., and Flores, C. 2009. Notas preliminares: Cabo San Lorenzo, su importancia como área de reproducción de tortugas marinas en el Ecuador.
Nesting Beach: San Lorezo
Year: 2006, 2009; Count: unquantified

EL SALVADOR
Data Source: Vasquez, M., Liles, M., Lopez, W., Mariona, G., and Segovia, J.
2008. Sea Turtle Research and Conservation, El Salvador. Technical Report.
FUNZEL, El Salvador.
Nesting Beaches: 7 beaches in Ahuachapan Department; 5 beaches in La Paz Department; 9 beaches in La Union Department; 1 beach in San Vincente Department; 9 beaches in Sonsonate Department; 6 beaches in Usulutan Department;
17 beaches in La Libertad Department
Year: 2008; Counts: 1–10 for all departments, except 11–100 for La Libertad
SWOT Contacts: Michael Liles, Mauricio Vasquez, Wilfredo Lopez, Georgina
Mariona, and Johanna Segovia

GUATEMALA
Data Source: Muccio, C., ARCAS. 2006. Leatherback nesting in the Hawaii area
of Guatemala. In: The State of the World's Sea Turtles Report, vol. 2 (2007).
Nesting Beaches: Hawaii, El Rosario, La Barrona, El Gariton
Year: 2006; Count: 1–10 per beach

Data Source: Pérez, J., Gómez, R., Estrada, C., Bran, A., and Alfaro, C. 2006.
Leatherback nesting in Guatemala: Personal communication. In: The State of
the World's Sea Turtles Report, vol. 1.
Nesting Beaches: Taxisco Beaches
Year: 2005; Count: 1–10

MÉXICO
Data Source: Amigos para la Conservacion de Cabo Pulmo and Grupo Tortuguero de las Californias. 2005. SWOT Database Online 2009.
Nesting Beach: Los Frailes
Year: 2004; Count: 11–100

Data Source: González, E., and Pinal, R. 2004. Informe final del programa de investigación y protección de la tortuga marina, y educación ambiental en el estado
de Baja California Sur. Temporada 2003–2004: ASUPMATOMA, A.C.
Nesting Beaches: Beaches between Agua Blanca and Todos Santos
Year: 2003; Count: 11–100

Data Source: Trejos, J. A., and Carretero, E. 2006. Leatherback nesting in
México. In: The State of the World's Sea Turtles Report, vol. 1.
Nesting Beaches: Santuario Playon, Playa del Coco
Year: 2003; Count: 11–100 per beach

Data Source: Vargas S. F., Vasconcelos, D., Ángeles, M. A., and Licea, M.
2004. Informe final de investigación de las actividades de conservación desarrollados en la Playa de Tierra Colorada durante la temporada 2003–2004. In: Sarti,
M., L., Barragán R., A. R., and J. A. Juárez. 2004. Conservación y evaluación de
la población de tortuga laúd Dermochelys coriacea en el Pacífico Mexicano,

temporada de anidación 2003–2004. DGVS-SEMARNAT-Kutzari, Asociación para el Estudio y Conservación de las Tortugas Marinas A.C.
Nesting Beaches: Mexiquillo, Barra de la Cruz, Tierra Colorada, Cahuitan
Year: 2004; Count: 101–500 per beach

NICARAGUA
Data Source: Urteaga, J. R. 2004. Conservación de tortugas tora, *Dermochelys coriacea*, en el Refugio de Vida Silvestre Río Escalante—Chacocente: Temporada 2003–2004, informe anual. Fauna and Flora International, Nicaragua.
Nesting Beaches: Playa El Coco, Playa La Flor, Tecomapa, Isla Juan Venado, Acayo-El Mogote, Refugio de la Vida Silvestre Rio Escalante
Year: 2003; Count: 1–10, 1–10, 11–100, 11–100, 11–100, respectively

Data Source: Torres, P. 2009. Informe Proyecto de Conservacion de Tortuga Tora (Dermochelys coriacea) en el Refugio de Vida Silvestre Rio Escalante-Chacocente, Nicaragua. Temporada 2008–2009.
Nesting Beaches: Veracruz de Acayo
Year: 2008; Count: 11–100

PANAMÁ
Data Source: Rodríguez, J., Ruíz, A., and Obrego, M. Comunicación personal, 2010.
Nesting Beach: Isla Santa Catalina
Year: 2009; Count: Unquantified

Data Source: Rodríguez, J., Ruíz, A., and Obrego, M. Comunicación personal, 2010.
Nesting Beach: La Barqueta
Year: 2004; Count: Unquantified

Data Source: Espino, D. Comunicación personal, 2010.
Nesting Beach: La Cuchilla, Morro de Puerco
Year: 2008; Count: Unquantified

Plate 5: Green Turtle, *Chelonia mydas*

COLOMBIA
Data Source: Barrientos-Muñoz, K. G., and Ramírez-Gallego, C. Personal communication, 2010; SWOT Database Online.
Site: El Valle, Choco
Year: 2008; Count: 1–10
SWOT Contact: Karla Barrientos-Muñoz

Data Source: Payan, L. F. 2009. Fortalecimiento del programa de monitoreo de tortugas marinas CIMAD—UAESPNN en el Parque Nacional Natural Gorgona. Unpublished report.

Nesting Beach: Playa Palmeras, Parque Nacional Natural Gorgona
Year: 2009; Count: 1–10
SWOT Contact: Diego Amorocho

COSTA RICA
Data Source: (1) Malaver, M., and Chacón, D. 2009. Informe Península de Osa
Temporada 2008. Unpublished report. (2) Chacón-Chaverrí, D. 2008. Anida-
ción de Tortugas Marinas en las Playas de Península de Osa, Pacífico Sur, Costa
Rica.
Nesting Beaches: Playa Carate, Rio Oro; Sirena; and Corcovado
Year: 2008; Count: 1–10, 1–10, 11–100, respectively
SWOT Contact: Didiher Chacón Chaverri

Data Source: Arauz, R., Viejobueno, M. S., Sunyer, S. P., and Naranjo, I. 2009.
Conservación e investigación de tortugas marinas en el Pacífico de Costa Rica
(Punta Banco, Refugio Nacional de Vida Silvestre Caeltas-Arío, San Miguel,
Corozalito). Presentado a las autoridades del Area Conservación Tempisque
(ACT) del Ministerio de Ambiente, Energía y Telecomunicaciones (MINAET)
(Permiso de Investigación 2009).
Nesting Beaches: Corozalito, Playa Caletas, Punta Blanco
Year: 2008; Count: 1–10 per beach

Data Source: Cháves, G., Morera, R., and Aviles, J. R. 2008. Seguimiento de la
actividad anidatoria de las tortugas marinas (Cheloniidae y Dermochelyidae) en
el Refugio Nacional de Vida Silvestre de Ostional, Santa Cruz, Guanacaste.
Nesting Beach: Playa Ostional
Year: 2008; Count: 11–100

Data Source: Seminoff, J. A. 2007. Green Sea Turtle (*Chelonia mydas*) 5-Year
Review: Summary and Evaluation. U.S. National Marine Fisheries Service, Sil-
ver Spring, MD, USA.
Nesting Beach: Playa Naranjo
Year: 2007; Count: 101–100

ECUADOR
Data Source: (1) Baquero, A., Muñoz, J. P., and Peña, M. 2010. Olive ridley
nesting in Ecuador. Personal communication. In: SWOT Report—The State of
the World's Sea Turtles, vol. 5. (2) Peña, M., Baquero, A., Muñoz, J., Puebla,
F., Macías, J., and Chalen, X. 2009. El Parque Nacional Machalilla: zona crítica
de anidación para la tortuga carey (*Eretmochelys imbricata*) y verde (*Chelonya
mydas*) en el Ecuador y el Pacífico Oriental. Temporadas 2007–2009. Memoria
del III Simposio Regional sobre Tortugas Marinas del Pacífico Suroriental.
Nesting Beaches: Bahía Drake, Puerto López, Tortuguita, Salango, La Playita
Year: 2008; Count: 1–10 per beach
SWOT Contact: Equilibrio Azul

Data Source: Herrera, M., Coello, D., and Flores, C. 2009. Notas Preliminares: Cabo San Lorenzo, Su Importancia Como Área de Reproducción de Tortugas Marinas en el Ecuador. Unpublished report.
Nesting Beaches: San Lorenzo
Year: 2007; Count: 1–10
SWOT Contacts: Daniel Rios, Dialhy Coello, and Marco Herrera

Data Source: Seminoff, J. A. 2007. Green sea turtle (*Chelonia mydas*) 5-year review: summary and evaluation. U.S. National Marine Fisheries Service, Silver Spring, MD, USA.
Nesting Beaches: Las Bachas, Las Salinas, Quinta Playa, Barahona, Galápagos Islands
Year: 2005; Counts: 101–1,000, 101–1,000, >1,000, 101–1,000, respectively

EL SALVADOR
Data Source: Vasquez, M., Liles, M., Lopez, W., Mariona, G., and Segovia, J. 2008. Sea Turtle Research and Conservation, El Salvador. Technical Report. FUNZEL, El Salvador.
Nesting Beaches: 7 beaches in Ahuachapan Department; 5 beaches in La Paz Department; 9 beaches in La Union Department; 1 beach in San Vincente Department; 9 beaches in Sonsonate Department; 6 beaches in Usulutan Department; 17 beaches in La Libertad Department
Year: 2008; Counts: 1–10 for Sonsonante, 101–1,000 for all others
SWOT Contacts: Michael Liles, Mauricio Vasquez, Wilfredo Lopez, Georgina Mariona, and Johanna Segovia

MÉXICO
Data Source: Seminoff, J. A. 2007. Green Sea Turtle (*Chelonia mydas*) 5-Year Review: Summary and Evaluation. U.S. National Marine Fisheries Service, Silver Spring, MD, USA.
Nesting Beaches: Isla Socorro, Isla Clarión, Revillagigedo Islands; Islas Tres Marías
Year: 2007; Counts: 11–100, Unquantified, Unquantified

Data Source: Seminoff, J. A. 2007. Green Sea Turtle (*Chelonia mydas*) 5-Year Review: Summary and Evaluation. U.S. National Marine Fisheries Service, Silver Spring, MD, USA.
Nesting Beaches: Bahía Maruata, Playa Colola, Michoacán
Year: 2007; Counts: >1,000 per beach

Data Source: Torres, P. 2009. Informe Proyecto de Conservacion de Tortuga Tora (Dermochelys coriacea) en el Refugio de Vida Silvestre Rio Escalante-Chacocente, Nicaragua. Temporada 2008–2009.
Nesting Beaches: Veracruz de Acayo
Year: 2008; Count: 1–10

PANAMÁ
Data Source: Rodríguez, J. Personal communication, 2010.
Nesting Beach: Ballena
Year: 1999; Count: Unquantified

PERÚ
Data Source: Wester, J. H., Kelez, S., Veléz-Zuazo, X. 2010. Tortugas marinas
en el norte de Perú: Nuevo limite sur de anidación de las tortuga verde *Chelonia
mydas* y golfina *Lepidochelys olivacea* en el Pacifico Este. Unpublished report.
Nesting Beach: Tres Cruces, Piura
Year: 2009; Count: 1–10

Plate 7. Hawksbill, Eretmochelys imbricata

COSTA RICA
Data Source: (1) Malaver, M., and Chacón, D. 2009. Informe Península de Osa
Temporada 2008. Unpublished report. (2) Chacón-Chaverrí, D. 2008. Anidación
de Tortugas Marinas en las Playas de Península de Osa, Pacífico Sur, Costa Rica.
(3) Chacón-Chaverrí, D. 2008. Hawksbill nesting in Costa Rica: Personal com-
munication. In: SWOT Report—The State of the World's Sea Turtles, vol. 3.
Nesting Beaches: Carate; Rio Oro; Playa Platanares; Sirena and Corcovado;
Manuel Antonio; Punta Rayo
Year: 2008; Count: 1–10 for Carate, 11–25 for the rest
SWOT Contact: Didiher Chacón Chaverri

Data Source: Arauz, R., Viejobueno, M. S., Sunyer, S. P., and Naranjo, I. 2009.
Conservación e investigación de tortugas marinas en el Pacífico de Costa Rica
(Punta Banco, Refugio Nacional de Vida Silvestre Caeltas-Arío, San Miguel,
Corozalito). Presentado a las autoridades del Area Conservación Tempisque
(ACT) del Ministerio de Ambiente, Energía y Telecomunicaciones (MINAET)
(Permiso de Investigación 2009).
Nesting Beaches: Playa Caletas, Camaronal, Punta Blanco
Year: 2008; Count: 1–10, 11–25, 11–25, respectively

Data Source: Piedra, R. 2008. Hawksbill nesting on Playa Langosta, Costa Rica:
Personal communication. In: SWOT Report—The State of the World's Sea
Turtles, vol. 3.
Nesting Beach: Playa Langosta
Year: 2005; Count: 1–10

ECUADOR
Data Source: (1) Baquero, A., Muñoz, J. P., and Peña, M. 2010. Olive ridley
nesting in Ecuador: Personal communication. In: SWOT Report—The State of
the World's Sea Turtles, vol. 5. (2) Peña, M., Baquero, A., Muñoz, J., Puebla,
F., Macías, J., and Chalen, X. 2009. El Parque Nacional Machalilla: zona crítica

de anidación para la tortuga carey (*Eretmochelys imbricata*) y verde (*Chelonia mydas*) en el Ecuador y el Pacífico Oriental. Temporadas 2007–2009. Memoria del III Simposio Regional sobre Tortugas Marinas del Pacífico Suroriental.
Nesting Beaches: Tortuguita, Salango, Frailes, Parque Nacional Machalilla
Year: 2008; Count: 1–10, 1–10, 1–10, 11–25, respectively
SWOT Contact: Equilibrio Azul

Data Source: Zárate, P. 2008. Hawksbill nesting in Ecuador: Personal communication. In: SWOT Report—The State of the World's Sea Turtles, vol. 3.
Nesting Beach: Las Tunas
Year: 2006; Count: Unquantified

Data Source: Herrera, M., Coello, D., and Flores, C. 2009. Notas Preliminares: Cabo San Lorenzo, Su Importancia Como Área de Reproducción de Tortugas Marinas en el Ecuador. Unpublished report.
Nesting Beaches: San Lorenzo
Year: 2009; Count: 1–10
SWOT Contacts: Daniel Rios, Dialhy Coello, and Marco Herrera

EL SALVADOR
Data Source: Vasquez, M., Liles, M., Lopez, W., Mariona, G., and Segovia, J. 2008. Sea Turtle Research and Conservation, El Salvador. Technical Report. FUNZEL, El Salvador.
Nesting Beaches: 17 beaches in La Libertad Department; 9 beaches in La Union Department; 6 beaches in Usulutan Department; and 9 beaches in Sonsonate Department.
Year: 2008; Counts: 1–10, 11–25, 11–25, 26–100, respectively
SWOT Contacts: Michael Liles, Mauricio Vasquez, Wilfredo Lopez, Georgina Mariona, and Johanna Segovia

Data Source: Vasquez, M., Liles, M., Lopez, W., Mariona, G., and Segovia, J. 2008. Sea Turtle Research and Conservation, El Salvador. Technical Report. FUNZEL, El Salvador.
Nesting Beaches: El Maculíz; Los Cóbanos/Los Almendros; Bahía Jiquilisco
Year: 2008; Counts: 26–100, 26–100, >100, respectively
SWOT Contacts: Michael Liles, Mauricio Vasquez, Wilfredo Lopez, Georgina Mariona, and Johanna Segovia

MÉXICO
Data Source: Gaos, A. Personal Communication. 2009.
Nesting Beach: Islas Tres Marías
Year: 2009; Count: Unquantified

Data Source: Peña de Niz, A. Personal communication, 2009.
Nesting Beach: Playa Rosa/Costa Careyes
Year: 2009; Count: 11–25

Data Source: Gaos, A. Personal communication, 2009.
Nesting Beach: Punta Mita
Year: 2009; Count: 11–25

NICARAGUA
Data Source: Muurmans, M. Personal communication, 2009.
Nesting Beach: Cosigüina
Year: 2009; Count: 11–25

Data Source: Otterstrom, S. Personal communication, 2009.
Nesting Beaches: El Remanso, La Flor, Ostional
Year: 2009; Counts: 11–25 per beach

Data Source: Urteaga, J. Personal communication, 2009.
Nesting Beaches: Isla Juan Venado
Year: 2009; Counts: 11–25

Data Source: Gaos, A. Personal communication, 2009.
Nesting Beach: Majagual
Year: 2009; Counts: 11–25

Data Source: Manzanares, L. Personal communication, 2009.
Nesting Beach: Estero Padre Ramos
Year: 2009; Count: 26–100

PANAMÁ
Data Source: Rodríguez, J., Ruíz, A., and Obrego, M. Comunicación personal, 2010.
Nesting Beach: Isla Cébaco, Isla Gobernad
Year: 2004; Count: Unquantified

Data Source: Rodríguez, J., Ruíz, A., and Obrego, M. Comunicación personal, 2010.
Nesting Beach: Isla Santa Catalina
Year: 2009; Count: Unquantified

Plate 9. Olive Ridley, Lepidochelys olivacea

COLOMBIA
Data Source: Amorocho, D. 2008. Informe del Taller Estandarización de Metodologías en Investigación y Monitoreo para la Conservación de Tortugas Marinas en Colombia. Convenio MAVDT-World Wildlife Fund.
Nesting Beaches: La Cuevita and Parque Nacional Natural Sanquianga
Year: 2007; Counts: 1–1,000 and unquantified, respectively
SWOT Contact: Diego Amorocho

Data Source: Barrientos-Muñoz, K. G., and Ramírez-Gallego, C. Personal communication, 2010; SWOT Database Online.
Site: El Valle, Choco
Year: 2008; Count: 1–1,000
SWOT Contact: Karla Barrientos-Muñoz

Data Source: Payan, L. F. 2009. Fortalecimiento del programa de monitoreo de tortugas marinas CIMAD—UAESPNN en el Parque Nacional Natural Gorgona. Unpublished report.
Nesting Beach: Playa Palmeras, Parque Nacional Natural Gorgona
Year: 2009; Count: 1–1,000
SWOT Contact: Diego Amorocho

COSTA RICA
Data Source: Fonseca, L. G., Murillo, G. A., Lenín, G., Spínola, R. M., and Valverde, R. A. 2009. Downward but stable trend in the abundance of arribada olive ridley sea turtles (*Lepidochelys olivacea*) at Nancite Beach, Costa Rica (1971–2007). Chelonian Conservation and Biology 8 (1): 19–27.
Nesting Beach: Nancite
Year: 2007; Count: 10,000–100,000

Data Source: Valverde, R. 2007. Global assessment of arribada olive ridley sea turtles. Final Report for the USFWS, MTCA award 6-G014.
Nesting Beach: Ostional National Wildlife Refuge
Year: 2007; Count: 100,000–1,000,000
SWOT Contact: Roldán Valverde

Data Source: Abreu-Grobois, F. A. 2010. Olive ridley nesting at Playa Nosara, Costa Rica: Personal communication. In: SWOT Report—The State of the World's Sea Turtles, vol. 5.
Nesting Beach: Playa Nosara
Year: 2009; Count: Unquantified
SWOT Contact: Alberto Abreu-Grobois

Data Source: Arauz, R., Viejobueno, M. S., Sunyer, S. P., and Naranjo, I. 2009. Conservación e Investigación de Tortugas Marinas en el Pacífico de Costa Rica (Punta Banco, Refugio Nacional de Vida Silvestre Caletas-Arío, San Miguel, Corozalito). Presentado a las autoridades del Area Conservación Tempisque (ACT) del Ministerio de Ambiente, Energía y Telecomunicaciones (MINAET).
Nesting Beaches: Caletas, Corozalito, Punta Banco, and San Miguel
Year: 2008; Counts: 1,000–10,000, 1,000–10,000, 1–1,000, and 1–1,000, respectively
SWOT Contact: Randall Arauz and Sandra Viejobueno

Data Source: Malaver, M., and Chacón, D. 2009. Informe Península de Osa Temporada 2008. Unpublished report.

Nesting Beach: Playa Carate, Rio Oro; Sirena and Corcovado
Year: 2008; Count: 1,000–10,000, 1–1,000, respectively
SWOT Contact: Didiher Chacón Chaverri

ECUADOR
Data Source: (1) Baquero, A., Muñoz, J. P., and Peña, M. 2010. Olive ridley
nesting in Ecuador: Personal communication. In: SWOT Report—The State of
the World's Sea Turtles, vol. 5. (2) Peña, M., Baquero, A., Muñoz, J., Puebla,
F., Macías, J., and Chalen, X. 2009. El Parque Nacional Machalilla: zona crítica
de anidación para la tortuga carey (*Eretmochelys imbricata*) y verde (*Chelonia
mydas*) en el Ecuador y el Pacífico Oriental. Temporadas 2007–2009. Memoria
del III Simposio Regional sobre Tortugas Marinas del Pacífico Suroriental.
Nesting Beaches: Montañita, Bahía Drake, Puerto López, Mompiche, San Lorenzo, Las Tunas, and Same
Year: 2008; Count: 1–1,000 per beach
SWOT Contact: Equilibrio Azul

Data Source: Muñoz Pérez, J., Valle, C. A., Baquero Gallegos, A., and Anhalzer
Anderson, G. 2009. Nueva playa de anidación para Lepidochelys olivacea: Por-
tete, Ecuador. Simposio Regional Santa Elena 2009. Ecuador.
Nesting Beach: Portete
Year: 2008; Count: 1–1,000
SWOT Contact: Equilibrio Azul

Data Source: Herrera, M., Coello, D., and Flores, C. 2009. Notas Preliminares:
Cabo San Lorenzo, Su Importancia Como Área de Reproducción de Tortugas
Marinas en el Ecuador. Unpublished report.
Nesting Beaches: Las Piñas, El Abra, San Lorenzo
Year: 2007; Count: 1–1,000 per beach
SWOT Contacts: Daniel Rios, Dialhy Coello, and Marco Herrera

EL SALVADOR
Data Source: Vasquez, M., Liles, M., Lopez, W., Mariona, G., and Segovia, J.
2008. Sea Turtle Research and Conservation, El Salvador. Technical Report.
FUNZEL, El Salvador.
Nesting Beaches: 7 beaches in Ahuachapan Department; 17 beaches in La Liber-
tad Department; 5 beaches in La Paz Department; 9 beaches in La Union De-
partment; 1 beach in San Vincente Department; 9 beaches in Sonsonate Depart-
ment; and 6 beaches in Usulutan Department
Year: 2008; Counts: 1–1,000, 1,000–10,000, 1–1,000, 1–1,000, 1–1,000, 1,000–
10,000, and 1,000–10,000, respectively
SWOT Contacts: Michael Liles, Mauricio Vasquez, Wilfredo Lopez, Georgina
Mariona, and Johanna Segovia

GUATEMALA
Data Source: Muccio, C. 2010. Olive ridley nesting in Guatemala: Personal communication. In: SWOT Report—The State of the World's Sea Turtles, vol. 5.
Nesting Beach: Hawaii
Year: 2008; Count: 1,000–10,000
SWOT Contact: Colum Muccio

HONDURAS
Data Source: Dunbar, S. G., Salinas, L., and Castellanos, S. 2010. Activities of the Protective Turtle Ecology Center for Training, Outreach, and Research (ProTECTOR) on Olive Ridley (Lepidochelys olivacea) in Punta Raton, Honduras. Annual Report of the 2008–2009 Nesting Seasons.
Nesting Beaches: El Muro, La Punta, La Puntilla, Don Walther, El Muerto, El Tiburon, La Cooperativa, and Palo Pique
Year: 2009; Counts: 1–1,000 per beach
SWOT Contact: Stephen Dunbar

Data Source: Dunbar, S. G., and Salinas, L. 2008. Activities of the Protective Turtle Ecology Center for Training, Outreach, and Research (ProTECTOR) on Olive Ridley (Lepidochelys olivacea) in Punta Raton, Honduras. Annual Report of the 2007–2008 Nesting Seasons. Unpublished report.
Nesting Beaches: Buquete, El Patio, La Playa, La Playa North, and La Playa South
Year: 2008; Counts: 1–1,000 per beach
SWOT Contact: Stephen Dunbar

MÉXICO
Data Source: Arista de la Rosa, E. 2009. Informe Final El Barril, Parque San Lorenzo. Unpublished report, CONANP, México.
Nesting Beach: El Barril
Year: 2009; Count: 1–1,000
SWOT Contacts: Elizabeth Arista de la Rosa and *Raquel Briseño*

Data Source: Everardo Melendez, M. 2009. Informe final Parque Nacional Bahía de Loreto. Unpublished report, CONANP, México.
Nesting Beach: Loreto
Year: 2009; Count: 1–1,000
SWOT Contacts: Mariano Everardo Melendez and *Raquel Briseño*

Data Source: Abreu-Grobois, A. Personal communication, 2010. In: SWOT Report—The State of the World's Sea Turtles, vol. 5.
Nesting Beaches: Bahía de Los Ángeles, Boca de Tomates, Boca del Cielo, Cachan de Echeverria, Chuquiapan, Cuixmala, El Mármol, Estación Biológica Majahuas, Hotelito Desconocido, Isla Pajaritos, José Maria Morelos, La Cruz de Huanacaste, La Encrucijada, La Gloria, La Placita de Morelos, La Ticla, Las Guasimas, Magdalena, Motín de Oro Michoacán, Peñas Lázaro Cardenas, Playa

Diamante, Playa Larga/ San Andrés, San Francisco, Solera de Agua, Tecuan, Teopa, and Todos Santos (Baja California)
Year: 2009; Count: Unquantified
SWOT Contact: Alberto Abreu

Data Source: Oceguera Camacho, K. 2009. Reporte Temporada 2009 Anidacion de Tortugas. Unpublished report.
Nesting Beach: San Juan de los Planes
Year: 2009; Count: 1–1,000
SWOT Contacts: Karen Oceguera Camacho and *Raquel Briseño*

Data Source: López-Castro, M. C., and Rocha-Olivares, A. 2005. The panmixia paradigm of eastern Pacific olive ridley turtles revised: consequences for their conservation and evolutionary biology. Molecular Ecology 14:3325–3334.
Nesting Beach: Las Tinajas, Punta Arena, and Punta Colorada, Baja California Sur
Years: 2002–2003; Count: Unquantified

Data Source: Murrieta Rosas, J. L. 2009. Informe final. Unpublished report, Patronato Cabo del Este, A.C., México.
Nesting Beach: Los Barriles
Year: 2009; Count: 1–1,000
SWOT Contacts: José Luis Murrieta Rosas and *Raquel Briseño*

Data Source: Rangel González, Z. 2009. Informe Final Parque Nacional Cabo Pulmo. Unpublished report, CONANP.
Nesting Beach: Parque Nacional Cabo Pulmo (Miramar, Barracas, Cabo Pulmo, Frailes)
Year: 2009; Count: 1–1,000
SWOT Contacts: Zuemy Rangel González and *Raquel Briseño*

Data Source: Tiburcio Pintos, G. 2009. Informe final. Red para la proteccion de la Tortuga Marina en el Municipio de los Cabos, Ayto Los Cabos, México. Unpublished report.
Nesting Beaches: Faro Viejo/Estero San Jose; and San Jose/Frailes
Year: 2009; Counts: 1–1,000, 1,000–10,000, respectively
SWOT Contacts: Graciela Tiburcio Pintos and *Raquel Briseño*

Data Source: Gonzalez Payan, E., et al. 2009. Informe anual. Unpublished report, ASUPMATOMA, A.C., México.
Nesting Beaches: El Suspiro and San Cristobal (both Baja California Sur)
Year: 2009; Counts: 1–1,000 per beach
SWOT Contacts: Elizabeth González Payan and Laura Sarti

Data Source: Ramirez Cruz, C. Personal communication, 2010. In: SWOT Report—The State of the World's Sea Turtles, vol. 5.
Nesting Beach: Los Esteros/Pescadero
Year: 2009; Count: 1–1,000

SWOT Contacts: Carlos Ramirez Cruz and *Raquel Briseño*

Data Source: Programa Nacional para la Conservación de las Tortugas Marinas (PNCTM), CONANP. Personal communication, 2010. In: SWOT Report—The State of the World's Sea Turtles, vol. 5.
Nesting Beaches: Bahía de Chacahua, Ceuta, and Mexiquillo
Year: 2009; Counts: 1,000–10,000, 1–1,000, and 1–1,000, respectively

Data Source: (1) Programa Nacional para la Conservación de las Tortugas Marinas, (PNCTM), CONANP. Personal communication, 2010. In: SWOT Report—The State of the World's Sea Turtles, vol. 5. (2) Valverde, R. 2007. Global assessment of arribada olive ridley sea turtles. Final Report for the USFWS, MTCA award 6–G014.
Nesting Beach: Santuario Playa de Escobilla
Year: 2003–2009; Count: > 1,000,000 clutches on average per year
SWOT Contact: Laura Sarti

Data Source: Diaz Millán, V. 2009. Informe CONANP, México. Unpublished report.
Nesting Beach: Meseta de Cacaxtla
Year: 2009; Count: 1–1,000
SWOT Contacts: Victorio Diaz Millan and *Raquel Briseño*

Data Source: Rios, D., and Programa Nacional para la Conservación de las Tortugas Marinas (PNCTM), CONANP. Personal communication, 2010. In: SWOT Report—The State of the World's Sea Turtles, vol. 5.
Nesting Beach: El Verde Camacho
Year: 2009; Count: 1,000–10,000
SWOT Contact: Laura Sarti

Data Source: Barron Hernandez, J. A. 2009. Informe final. Unpublished report, Acuario de Mazatlán, Sinaloa, México.
Nesting Beach: Playa Mazatlán
Year: 2009; Count: 1–1,000
SWOT Contacts: José Barron Hernandez and *Raquel Briseño*

Data Source: Erendira Gonzalez, D. 2009. Informe temporada 2009. México. Unpublished report.
Nesting Beach: Isla de la Piedra, Estrella del Mar
Year: 2009; Count: 1,000–10,000
SWOT Contacts: Diego Erendira González and *Raquel Briseño*

Data Source: Aguilar, H. 2009. Resultados de Conservación de Tortugas Marinas en Playa Caimanero, El Rosario, México. Unpublished report.
Nesting Beach: Playa Caimanero, El Rosario
Year: 2009; Count: 1,000–10,000
SWOT Contacts: Hector Contreras Aguilar and *Raquel Briseño*

Data Source: Peña Aldrete, V., and Grupo Ecologista de Nayarit, A.C. Personal communication, 2010. In: SWOT Report—The State of the World's Sea Turtles, vol. 5.
Nesting Beach: El Naranjo
Year: 2009; Count: 1–1,000
SWOT Contacts: Vicente Peña Aldrete and *Raquel Briseño*

Data Source: Tena Espinoza, M., and M. Nuñez Bautista. 2009. Informe Anual. Unpublished report, Campamento Tortuguero Playa Chila, A.C., México.
Nesting Beach: Playa Boca de Chila
Year: 2009; Count: 1,000–10,000
SWOT Contacts: Marco Tena Espinoza and *Raquel Briseño*

Data Source: Flores, M., and Programa Nacional para la Conservación de las Tortugas Marinas (PNCTM), CONANP. Personal communication, 2010. In: SWOT Report—The State of the World's Sea Turtles, vol. 5.
Nesting Beaches: Nuevo Vallarta; and Platanitos (both Nayarit)
Year: 2009; Counts: 1,000–10,000
SWOT Contact: Laura Sarti

Data Source: Llamas González, I. 2009. Informe final. Unpublished report, UDG Preparatoria Regional de Puerto Vallarta, México.
Nesting Beach: Mayto
Year: 2009; Count: 1,000–10,000
SWOT Contacts: Israel Llamas González and *Raquel Briseño*

Data Source: Pérez, A., and Programa Nacional para la Conservación de las Tortugas Marinas (PNCTM), CONANP. Personal communication, 2010. In: SWOT Report—The State of the World's Sea Turtles, vol. 5.
Nesting Beach: Mismaloya (El Playón section)
Year: 2009; Count: 1,000–10,000
SWOT Contact: Laura Sarti

Data Source: Martínez, C., and Programa Nacional para la Conservación de las Tortugas Marinas (PNCTM), CONANP. Personal communication, 2010. In: SWOT Report—The State of the World's Sea Turtles, vol. 5.
Nesting Beach: Chalacatepec
Year: 2009; Count: 1,000–10,000
SWOT Contact: Laura Sarti

Data Source: Abreu-Grobois, F. A., and Plotkin, P. 2009. Lepidochelys olivacea. In IUCN Red List of Threatened Species, version 2009.2.
Nesting Beaches: Cuyutlan, Colima; and Maruata-Colola, Michoacán
Years: 1999–2003; Counts: 1,000–10,000 per beach
Nesting Beach: Piedra de Tlalcoyunque, Guerrero
Year: 1997; Count: 1,000–10,000

Data Source: Hernández, A., and Programa Nacional para la Conservación de las

Tortugas Marinas (PNCTM), CONANP. Personal communication, 2010. In:
SWOT Report—The State of the World's Sea Turtles, vol. 5.
Nesting Beach: El Chupadero
Year: 2009; Count: 1,000–10,000
SWOT Contact: Laura Sarti

Data Source: CMT, and Programa Nacional para la Conservación de las Tortu-
gas Marinas (PNCTM), CONANP. Personal communication, 2010. In: SWOT
Report—The State of the World's Sea Turtles, vol. 5.
Nesting Beach: Ixtapilla
Years: 2008–2009; Count: 10,000–100,000
Nesting Beach: Morro Ayuta
Year: 2003–2009; Count: 10,000–100,000
SWOT Contact: Laura Sarti

Data Sources: (1) Abreu-Grobois, A. Personal communication, 2010. In: SWOT
Report—The State of the World's Sea Turtles, vol. 5. (2) Sarti, L., and Pro-
grama Nacional de Tortugas Marinas-CONANP. Personal communication,
2010. In: SWOT Report—The State of the World's Sea Turtles, vol. 5.
Nesting Beaches: La Zacatoza and San Juan Chacahua
Year: 2009; Count: Unquantified
SWOT Contacts: Alberto Abreu and Laura Sarti

Data Source: Ocampo, E., and Programa Nacional para la Conservación de las
Tortugas Marinas (PNCTM), CONANP. Personal communication, 2010. In:
SWOT Report—The State of the World's Sea Turtles, vol. 5.
Nesting Beach: Tierra Colorada
Year: 2009; Count: 1–1,000
SWOT Contact: Laura Sarti

Data Source: Kutzari, A. C., and Programa Nacional de Tortugas Marinas,
CONANP. Personal communication, 2010. In: SWOT Report—The State of
the World's Sea Turtles, vol. 5.
Nesting Beach: Cahuitan
Year: 2007; Count: 1,000–10,000
SWOT Contact: Laura Sarti

Data Source: Tavera, A., and Programa Nacional para la Conservación de las
Tortugas Marinas (PNCTM), CONANP. Personal communication, 2010. In:
SWOT Report—The State of the World's Sea Turtles, vol. 5.
Nesting Beach: Barra de la Cruz
Year: 2009; Count: 1,000–10,000 clutches
SWOT Contact: Laura Sarti

Data Source: Neri, S., and Programa Nacional para la Conservación de las Tortugas Marinas (PNCTM), CONANP. Personal communication, 2010. In:
SWOT Report—The State of the World's Sea Turtles, vol. 5.

Nesting Beach: Puerto Arista, Chiapas
Year: 2007; Count: 1,000–10,000
SWOT Contact: Laura Sarti

NICARAGUA
Data Source: Torres, P., Chávez, M., and Salmeron, L. 2009. Informe Proyecto de Conservación de Tortuga Tora (Dermochelys coriacea) en Playa Salamina, Villa El Carmen (Departamento de Managua), Nicaragua. Temporada 2008–2009. Unpublished report.
Nesting Beach: Isla Juan Venado
Year: 2008; Count: Unquantified
SWOT Contacts: Daniel Rios and Perla Torres Gago

Data Source: Chávez, M., and Salmeron, L. 2009. Informe Técnico del Proyecto de Conservación de Tortuga Tora (D. coriacea) en Playa Salamina, Villa El Carmen. Managua-Nicaragua. Temporada 2008–2009. Unpublished report.
Nesting Beach: Salamina
Year: 2008; Count: 1–1,000
SWOT Contacts: Daniel Rios and Perla Torres Gago

Data Source: Cornelius, S. 1982. Status of sea turtles along the Pacific coast of middle America. In Bjorndal, K. A. (Ed.), Biology and Conservation of Sea Turtles. Smithsonian Institution Press, Washington, DC.
Nesting Beaches: Masachapa, Pochomil, and Boquita
Year: 1982; Count: Unquantified

Data Source: Torres, P. 2009. Informe Proyecto de Conservación de Tortuga Tora (Dermochelys coriacea) en el Refugio de Vida Silvestre Río Escalante-Chacocente, Nicaragua. Temporada 2008–2009. Unpublished report.
Nesting Beach: Veracruz de Acayo
Year: 2008; Count: 1,000–10,000
SWOT Contacts: Daniel Rios and Perla Torres Gago

Data Source: Delegación MARENA-Rivas. 2009. Informe de Monitoreo de Tortuga Paslama (Lepidochelys olivacea) en el RVS La Flor (Departamento de Rivas, Nicaragua). Temporada 2008–2009. Unpublished report.
Nesting Beach: La Flor
Year: 2008; Count: 10,000–100,000
SWOT Contact: Daniel Rios and Perla Torres Gago

Data Source: Arana, J., and Torres, P. 2009. Informe de Monitoreo de Tortuga Paslama (Lepidochelys olivacea) en Playa Arribada del RVS Río Escalante-Chacocente (Departamento de Carazo, Nicaragua). Temporada 2008–2009. Unpublished report.
Nesting Beach: Chacocente
Year: 2008; Count: 10,000–100,000
SWOT Contact: Daniel Rios and Perla Torres Gago

PANAMÁ

Data Source: MarViva. Personal communication, 2010. In: SWOT Report—The State of the World's Sea Turtles, vol. 5.
Nesting Beach: Malena
Year: 2009; Count: Unquantified
SWOT Contacts: Jacinto Rodríguez, Argelis Ruiz, Marino Abrego, Carlos Peralta, and Harold Chacón

Data Source: Rodríguez, J., Ruiz, A., Abrego, M., Peralta, C., and Chacón, H. Personal communication, 2010. In: SWOT Report—The State of the World's Sea Turtles, vol. 5.
Nesting Beaches: Isla Taborcillo and Morrillo
Year: 2009; Count: Unquantified
SWOT Contact: Jacinto Rodríguez, Argelis Ruiz, Marino Abrego, Carlos Peralta, and Harold Chacón

Data Source: Testimonio de Moradores de la Comunidad. Personal communication, 2010. In: SWOT Report—The State of the World's Sea Turtles, vol. 5.
Nesting Beach: Cambutal
Year: 2009; Count: Unquantified
SWOT Contacts: Jacinto Rodríguez, Argelis Ruiz, Marino Abrego, Carlos Peralta, and Harold Chacón

Data Source: Ruiz, A., Rodríguez, J., and Abrego, M. Personal communication, 2010. In: SWOT Report—The State of the World's Sea Turtles, vol. 5.
Nesting Beach: Playa Marinera
Year: 2008; Count: 10,000–100,000
SWOT Contact: Jacinto Rodríguez, Argelis Ruiz, and Marino Abrego

Data Source: Ruiz, A., Rodríguez, J., and Abrego, M. 2010. Observaciones de hembras anidantes, rastros y nidos: Personal communication. In: SWOT Report—The State of the World's Sea Turtles, vol. 5.
Nesting Beach: Guánico Abajo
Year: 2009; Count: Unquantified
SWOT Contacts: Jacinto Rodríguez, Argelis Ruiz, Marino Abrego, Carlos Peralta, and Harold Chacón

Data Source: Rodríguez, J., and Trejos, J. 2010. Observación de caparazón, testimonio de moradores. Personal communication. In: SWOT Report—The State of the World's Sea Turtles, vol. 5.
Nesting Beach: La Concepción (La Yeguada)
Year: 2009; Count: Unquantified
SWOT Contacts: Jacinto Rodríguez, Argelis Ruiz, Marino Abrego, Carlos Peralta, and Harold Chacón

Data Source: (1) Plotkin, P. T. 2007. Olive Ridley Sea Turtle (Lepidochelys olivacea) Five-Year Review: Summary and Evaluation. U.S. National Marine Fish-

eries Service and U.S. Fish and Wildlife Service, Jacksonville, FL. (2) Evans, K. E., and Vargas, A. R. 1998. Sea turtle egg commercialization in Isla de Cañas, Panamá. Proceedings of the Sixteenth Annual Symposium on Sea Turtle Biology and Conservation, Hilton Head, South Carolina. NOAA Technical Memorandum NMFS-SEFSC-412.
Nesting Beach: Isla Cañas
Year: 2007; Count: 10,000–100,000

PERÚ
Data Source: Hays-Brown, C., and Brown, W. M. 1982. Status of sea turtles in the southeastern Pacific: Emphasis on Peru. In Bjorndal, K. A. (Ed.), Biology and Conservation of Sea Turtles. Smithsonian Institution Press, Washington, DC.
Nesting Beach: Punta Malpelo
Year: 1982; Count: 1–1,000

Data Source: Kelez. S., Velez-Zuazo, X., Angulo, F., and Manrique, C. 2009. Olive ridley *Lepidochelys olivacea* nesting in Peru: The southernmost records in the Eastern Pacific. Marine Turtle Newsletter 126:5–9.
Nesting Beach: Caleta Grau
Year: 2000; Count: 1–1,000
Nesting Beach: El Ñuro, Nueva Esperanza
Year: 2009; Count: 1–1,000

Data Source: Wester, J. H., Kelez, S., Velez-Zuazo, X. 2010. Tortugas marinas en el norte de Perú: Nuevo limite sur de anidación de las tortuga verde *Chelonia mydas* y golfina *Lepidochelys olivacea* en el Pacifico Este. Unpublished report.
Nesting Beach: Playa Bomba, Piura
Year: 2009; Count: 1–1,000

About the Editors

Jeffrey A. Seminoff leads the Marine Turtle Ecology and Assessment Program at the U.S. National Marine Fisheries Service's Southwest Fisheries Science Center (La Jolla, CA, USA). Since 1992 Jeffrey has been involved in ecological research and conservation of sea turtles in México and Central America. He received his Ph.D. from the University of Arizona in 2000 and was a postdoctoral fellow at the Archie Carr Center for Sea Turtle Research at the University of Florida from 2000 to 2002. Jeffrey's current research uses innovative approaches such as stable isotope analyses, biotelemetry, animal-borne imagery, and aerial surveys to elucidate the life history of sea turtles throughout the Pacific Ocean. His research has been featured in numerous scientific journals, magazines, and news outlets, as well as on the Discovery Channel, Animal Planet, PBS, and National Geographic Explorer. Jeffrey is adjunct faculty at Indiana-Purdue University and University of Florida, is an active member of the IUCN Marine Turtle Specialist Group, and is deeply involved with efforts by the U.S. Fish and Wildlife Service and U.S. National Marine Fisheries Service to update marine turtle status assessments for the U.S. Endangered Species Act. He currently serves as the U.S. delegate for the Scientific Committee of the Inter-American Convention for the Protection and Conservation of Sea Turtles. Jeff lives with his wife, Jennifer, and his young children, Quin and Graeson, in San Diego, California, along with an assortment of pets, including George, a 55-kg tortoise.

Bryan P. Wallace is the director of science for the Marine Flagship Species Program, Global Marine Division, Conservation International, and is also adjunct assistant professor in the Division of Marine Science and Conservation at the Duke University Marine Laboratory. Wallace earned his B.S. in biology at the University of Dayton (2000) and his Ph.D. from Drexel University (2005) and was a postdoctoral researcher at the Duke University Marine Lab (2005–2007). His research focuses on how to apply insights from animal physiology and behavior to pertinent conservation issues and has included biogeography, movements and behavior, nest and egg physiology, and fisheries bycatch. Working in Latin America for most of his career, Wallace is fluent in Spanish and is a regional co-vice chair for the East Pacific in the IUCN Marine Turtle Specialist Group. He is also leading efforts to evaluate the conservation status of regional management units of sea turtles globally through the Marine Turtle Specialist Group of the IUCN Species Survival Commission, and through his role as the chair of the Scientific Advisory Board of SWOT (State of the World's Sea Turtles). He has coauthored many peer-reviewed papers and blogs, has consulted on a children's book, and is an editor of the State of the World's Sea Turtles—SWOT Report. He lives in Washington, DC, with his wife, Jessica, their son, Leander, and their two dogs, Lupe and Sancho.

About the Contributors

F. Alberto Abreu-Grobois waded innocently into the world of sea turtles back in 1989 when he offered to help Raquel Briseño organize a national information center for the species. Since then he was hooked and found he could put to good use his expertise in population genetics. His interests as a research scientist at the Unidad Mazatán del Instituto de Ciencias del Mar y Limnología (UNAM) in that field have focused on Mexican sea turtle species. He has mentored many postgraduate students at home and abroad studying the various species. His commitment have included organizing the first sea turtle symposium to be held outside of the United States, chairing the IUCN Marine Turtle Specialist Group (MTSG), serving as president of the Consultative Committee of the Inter-American Convention for the Conservation and Protection of Sea Turtles, coauthoring the current Red List assessment of olive ridleys, and collaborating in the evaluation of the conservation status of Mexican hawksbills. Alberto is currently the MTSG regional vice chair for the Greater Caribbean. His free time is spent as a full-time father and husband.

Joanna Alfaro-Shigueto began working as a fisheries researcher in Perú after graduating with a degree in biology from Ricardo Palma University, Lima. Joanna spent considerable time working in small fishing villages, where she learned the importance of collaboration among communities, researchers, and government in dealing with conservation matters. In 2001 she became president of Pro Delphinus, a Peruvian nongovernmental organization whose priority is to encourage young Peruvian researchers to continue working in conservation locally. Joanna has authored and collaborated on a number of peer-reviewed publications in collaboration with the international research community. She is also currently working toward her PhD degree at University of Exeter, United Kingdom.

Javier Alvarado Díaz received his undergraduate degree from the Universidad Autónoma de Nuevo León and his M.Sc. from the University of Missouri, USA. He is a member of the National System of Investigators of México and is a research professor at the Universidad Michoacana de San Nicolás de Hidalgo (UMSNH). He is also the head of the Department of Zoology of the Natural Resources Research Institute and president of the State Advisory Board for the Recovery of Sea Turtles in Michoacán. Javier was the founder of the Aquatic Biology Laboratory and the Ecological Recovery of Sea Turtle Populations on the coast of Michoacán Project at UMSNH. He has published several articles about reproductive biology of sea turtles in particular and herpetology generally in national and international journals.

Diego Amorocho is a Colombian biologist who earned his M.Sc. in environmental science from the Australian National University and his Ph.D. in biological sciences from Monash University (Australia). He was awarded with the Molly Holman Medal for best doctoral thesis for the faculty of science from Monash University in 2009, and received the Whitley Award in 2010 for his contributions to wildlife conservation in his native Colombia. Diego has been active in sea turtle research and conservation in Colombia for more than twenty-five years, with experience specifically in the sea turtle field research along the Pacific and Caribbean coasts of Colombia; regional species conservation and environmental management policy design; coordination of sea turtle biology and conservation training courses in Colombia (1995, 1998, 1999, 2000, 2001, 2004, 2009), including international invitations to present similar workshops; and production and distribution of educational material concerning sea turtles and wildlife conservation management. He is a member and current regional vice cochair for the East Pacific Region of the IUCN Species Survival Commission's Marine Turtle Specialist Group. In addition to his native Spanish, he speaks English and Portuguese. He has authored several book chapters, scientific articles, and seminar papers and presented posters at national and international forums. He has supervised eight undergraduate marine turtle biologists and ecologists, who have conducted their bachelor's thesis research on marine turtles in the San Andrés Archipelago, the central Caribbean coast, and Gorgona National Park in the Colombian Pacific. Diego is the current executive director for the Research Center for Environmental Management and Development (CIMAD) based in Cali, Colombia.

Randall Arauz is the founder and president of the Costa Rican nongovernmental organization Programa Restauración de Tortugas Marinas, better known for its acronym, PRETOMA. he graduated with a B.Sc. in biology from the University of Costa Rica in 1988. Randall worked as the manager of the Green Turtle Station in Tortuguero in 1991 and was the director of Las Baulas National Park in 1992. He was hired as the Central American Director of the San Francisco–based Turtle Island Restoration Network in 1994, a position he still holds, and moved on to found PRETOMA in 1997. His work has focused on the impact of industrial fisheries on sea turtles, mainly shrimp trawling and longlining. After he obtained perhaps the first footage of shark finning while monitoring the activities of a longliner in late 1997, Randall became involved in shark conservation as well. He is currently working to protect four nesting beaches in Costa Rica and studying sea turtle and shark migrations with acoustic and satellite telemetry. Through PRETOMA, Randall has led campaigns for responsible shrimp fishing and against shark finning that have had global impacts. He has also served as technical consultant to the Ministry of Environment and official delegate at important international meetings such as the United Nations Law of the Sea and the Convention of Migratory Species. He is currently the Central American cochair of the IUCN Shark Specialist Group and member of the IUCN Marine Turtle Specialist Group. His work has received acknowledgements by Conservation International (2004), Shark Project Germany (2005), Protect the Sharks of Netherlands (2006), and the Duke of Edinburgh of the United Kingdom (2007). He has received major international awards such as the prestigious Whitley Award of the United Kingdom (2004) and the Goldman Prize of the United States (2010).

Andres Baquero Gallegos graduated in 1998 in ecology from the Universidad San Francisco de Quito (Ecuador). After completing his undergraduate studies, he began working on conservation projects in Parque Nacional Machalilla, the most important protected area on the coast of Ecuador. During this time, he started projects dealing with fisheries and marine conservation involving local communities. He continued his studies at University of California, San Diego, obtaining his M.S. in marine science in 2006 with a thesis on telemetry of sharks. After returning to Ecuador, he cofounded Equilibrio Azul, a nongovernmental organization that has since developed several important marine research and conservation projects involving fisheries, seabirds, sea turtles, and elasmobranchs. He is the director of Equilibrio Azul, and is a member of the faculty at the University of San Francisco de Quito.

Ana Rebeca Barragán studied biology and earned her M.Sc. in animal biology from the National Autonomous University of México, participating as a student in the university's Sea Turtle Conservation Project in 1988. Since 1992 she has been involved in the conservation of the leatherback turtle in the Mexican Pacific, doing field research and protection activities in Mexiquillo and Tierra Colorada, two of the oldest camps devoted to the recovery of that species in México. In 1997 Ana coordinated the Leatherback Conservation Project in Tortuguero, Costa Rica, and from 1998 to 2000 she was principal investigator of the Leatherback Project at Sandy Point National Wildlife Refuge, U.S. Virgin Islands, where she lead the monitoring and research of that important population and trained hundreds of Earthwatch volunteers. When not in field season in the Caribbean, Ana was part of the research team that monitored the nesting trend of the leatherback in the eastern Pacific, participating in aerial surveys in México and Central America from 1996 to 2006. In 2000 Ana became the coordinator of the leatherback conservation project in Playa Cahuitán, the most recently discovered of the Mexican index beaches, as part of Proyecto Laúd. There she worked with the local communities to raise awareness on the importance of the conservation of sea turtles and involved them in the beach protection activities. Under her leadership the community of Cahuitán formed a local committee for the protection of sea turtles, which in 2004 joined the Network for Oaxacan Wetlands and is working in collaboration with the biologists of Proyecto Laúd to this date. Ana is currently part of the technical team of the National Sea Turtle Conservation Program in México, at the National Commission for Protected Areas in SEMARNAT, where she helps develop the national strategies for the recovery of important sea turtle populations and gives technical advice to regional conservation groups like the Leatherback Community Network and the Network for Oaxacan Wetlands.

Scott R. Benson is a research fishery biologist with the NOAA—U.S. National Marine Fisheries Service and a research affiliate with Moss Landing Marine Laboratories. He received his M.S. in ecosystem studies at San Jose State University's Moss Landing Marine Laboratory in 2002. Stationed in Moss Landing, California, on Monterey Bay, Scott's research projects have included integrated studies of marine mammals, seabirds, and leatherback turtles, with emphasis on abundance, distribution, foraging ecology, and oceanographic patterns influencing the occurrence of these species.

Raquel Briseño-Dueñas is a marine biologist at the Centro Interdisciplinario en Ciencias Marinas, Instituto Politécnico Naciona, in Mazatlán, México. She has an M.Sc. in fisheries biology from the Universidad Autónoma de Sinaloa and an international diploma in coastal management from the joint program of Centro Universitario de la Costa Sur/Universidad de Guadalajara (México), Dalhousie University (Canadá), and Universidad de La Habana (Cuba). She is recognized as one of the pioneers in conservation biology research in México. Throughout her career, she has worked on conservation of sea turtles and their coastal and marine habitats in association with local communities for more than thirty years, offering support and technical supervision. In 2004, she completed a report on the first complete inventory of sea turtle nesting beaches in México for the Comisión Nacional de la Biodiversidad. She is the coordinator of the Marine Turtle Information Bank, based in the Unidad Mazatlán del Instituto de Ciencias del Mar y Limnología de la Universidad Nacional Autónoma de México, and she is a regional co-vice chair for the East Pacific region of the IUCN Marine Turtle Specialist Group. In these functions and as a member of the advisory group to the Grupo Tortuguero de las Californias, she advocates a conservation model that integrates participation of local groups from public, private, and social sectors. She has published eight book chapters and seven articles in scientific journals and has made numerous contributions in other communications media. She has participated in national international forums in her professional specialty areas and has directed master's and honors theses. For the past twenty-four years she has served as an adviser to the Environmental Education Department of the Mazatlán Aquarium, and she has written eight scripts for dance and music performances with environmental themes. She also is president of Amigos del Museo Arqueológico de Mazatlán, an organization whose aim is to help to maintain the national and historical heritage of northwest México.

Didiher Chacón Chaverri graduated from the Universidad Nacional de Costa Rica with an M.Sc. in marine biology and an M.Sc. in marine and coastal sciences. He is the Coordinator of Wider Caribbean Sea Turtle Conservation Network (WIDECAST) for Latin America and is the president of WIDECAST-Costa Rica. He is a well-respected technical adviser on issues dealing with management of marine and coastal resources, including sea turtles, for government agencies and conservation organization, such as CITES, IUCN, Wildlife Conservation Society, Fauna and Flora International, and NOAA, among others. He created the Central American Regional Network for Sea Turtle Conservation, is a member of the IUCN Marine Turtle Specialist Group, and has won several awards recognizing his efforts for marine and coastal conservation, including the prestigious Whitley Award.

Carlos Delgado Trejo received his M.Sc. from the Universidad Michoacana de San Nicolás de Hidalgo (UMSNH). He is a researcher and professor in the Natural Resource Research Institute (INIRENA) at UMSNH, where he runs the Marine Turtle Biology and Conservation Laboratory. He is the director of the Ecological Recovery of Sea Turtle Populations on the coast for the Michoacán project at UMSNH and is the coordinator of graduate studies for INIRENA. He is currently working on his doctoral degree in the Institute of Geophysics at the Universidad Nacional Autónoma de México.

Peter H. Dutton is a senior research biologist and leader of the Marine Turtle Genetics Program at the U.S. National Marine Fisheries Service's Southwest Fisheries Science Center in La Jolla, California. He is a member of the IUCN Marine Turtle Specialist Group. He received his B.Sc. in biology from Stirling University in Scotland, his M.Sc. in ecology from San Diego State University, and his Ph.D. in zoology from Texas A&M University. He has twenty-nine years of experience with sea turtle conservation in countries throughout the Caribbean, the Americas, and Melanesia and has published more than sixty scientific articles and book chapters related to the biology and conservation of marine organisms. His research interests include the evolution, phylogeography, ecology, and conservation biology of marine turtles. He uses genetics and satellite telemetry as tools to study the life history, migration, and habitat use of sea turtles.

Christina Fahy has worked for the Southwest Regional Office, U.S. National Marine Fisheries Service (NMFS), since 1998 primarily as a fisheries biologist and more specifically as the sea turtle recovery coordinator for the U.S. West Coast. She specializes in the analysis of impacts of threats to sea turtles and marine mammals, including fisheries, Navy activities, and other anthropogenic activities. Her fieldwork experience includes conducting aerial and vessel surveys for marine mammal stock abundance, pinniped capture/tagging, and sea turtle capture/tagging. Prior to working for NMFS, she worked on commercial fishing vessels in Alaska and the Pacific Northwest as a fisheries observer and in both areas as a research diver.

Jack Frazier is a researcher at the Smithsonian Conservation Biology Institute, in Virginia, USA. He received a D.Phil. in zoology from Oxford University and subsequently served as a professor at Universidad National in Heredia, Costa Rica, and at México's Centro de Investigación y de Estudios Avanzados del Instituto Politécnico Nacional. Most of his career has been spent in the tropics of the developing world, carrying out research, professional training, and environmental education in Africa, Asia,, and Latin America. Jack has published more than 100 articles and has served as chief editor, guest editor, editorial adviser, editorial committee member, academic coordinator, and academic council member in various countries. He serves as chair of the advisory committee of the Indian Ocean South East Asia Marine Turtle Memorandum of Understanding for the Indian Ocean and South East Asia and until recently was also chair of the Scientific Committee of the Inter-American Sea Turtle treaty. In recent years he has been exploring ways to integrate biological science with other disciplines and human endeavors, taking into account social and policy issues that are routinely ignored or avoided in the natural sciences. Much of his work has been *ad honorem*.

Alexander R. Gaos is the cofounder and regional director of the Eastern Pacific Hawksbill Initiative, an international collaborative effort to understand the biology, identify priority habitat, mitigate direct threats, and promote recovery of hawksbill turtles in the eastern Pacific Ocean. Prior to his current role, he managed a variety of research programs involving sea turtle nesting beach conservation, in-water monitoring, and fisheries bycatch reduction in countries such as México, Nicaragua, and Costa Rica. Having spent much of his childhood abroad, Alexander has an adept ability to work with an amalgam of people from an array of cultures and backgrounds. He earned a B.A. in environmental studies from California State University, East Bay, an M.S. in conservation ecology from San Diego State University (SDSU), and is currently a Ph.D. student with the Joint Doctoral Program in Ecology at the University of California, Davis and SDSU. While Alexander currently lives with his wife, Ingrid Yañez, and young son, Joaquinn, in San Diego, California, their work leads them on journeys throughout Latin America, including a recent 10,000-mile sojourn from the United States to Costa Rica and back via pickup truck.

Martin Hall has been principal scientist and head of the Tuna-Dolphin Program of the Inter-American Tropical Tuna Commission (IATTC) since 1984. The program succeeded in reducing dolphin mortality in the tuna purse seine fishery of the eastern Pacific to less than 1 percent of the initial figures, without reducing the productivity of the fishery. Keys to the success were the implementation of an observer program to diagnose the causes of mortality and a fishers education program to disseminate information on the solutions to the problems identified, together with the widespread adoption of improved gear and procedures. Martin has also been directly involved in developing and implementing the international agreements that address the tuna-dolphin issue. In 2003, IATTC received a request from the government of Ecuador to assist it in the development of a program to mitigate sea turtle bycatch by vessels that fish for tunas and mahi-mahis with longline gear. This program was developed in cooperation with the World Wildlife Fund, NOAA, national fisheries agencies, and local and international conservation groups and is currently under way in most countries of the Pacific coast of America, from Perú to México. More recently, he became involved in the coordination of the global efforts to reduce bycatch in the fishery for tunas associated with floating objects. His publications center on bycatch issues in general and the strategies and approaches to implement successful mitigation programs. He has presented papers at numerous scientific and management conferences and organized well over a hundred workshops for fishers, on bycatch problems and solutions. He got his first degree from the University of Buenos Aires, Argentina, and his Ph.D. from the University of Washington.

Mark Helvey is the assistant regional administrator for sustainable fisheries at NOAA's U.S. National Marine Fisheries Service, Southwest Region. He holds B.S. and M.S. degrees from the University of California, Santa Barbara and the University of Arizona, respectively, and an M.B.A. from California State University, Long Beach. Mark is involved in domestic and international fishery management issues as they relate to highly migratory and coastal pelagic species. His present interests revolve around seeking sustainable fishing opportunities for U.S. West Coast fishermen.

Shaleyla Kelez is a Peruvian biologist who received her B.S. in 2000 from the Universidad Nacional Agraria La Molina (Lima). Currently she is a Ph.D. candidate at the Nicholas School of the Environment at Duke University. From 1999 to 2005, Shaleyla worked in several nongovernmental agencies conducting research on sea turtles and bycatch. She worked mainly in Perú but also in Costa Rica and Puerto Rico. In Perú, she also did research on penguins and marine protected areas. At Duke, Shaleyla was a research assistant for Project GloBAL of the Duke Center for Marine Conservation for three years. In 2009 she cofounded the Peruvian nongovernmental organization ecOceánica, which aims to conserve marine ecosystems in the southeast Pacific through scientific research, community involvement, and sustainable development. Shaleyla wants to see economic activities such as fisheries develop sustainably while ensuring the preservation of species like sea turtles.

David Maldonado Díaz works as sustainable fisheries manager with the Grupo Tortuguero's Proyecto Caguama in Baja California, México, where he has partnered directly with fishermen to enhance fisheries selectivity and livelihoods since 2004. David holds a degree in marine biology from the Universidad Autonoma de Baja California Sur and carried out his graduate work focused on natural resource management at the Instituto Nacional Politécnico's Centro Interdisciplinario de Ciencias Marinas Marine Sciences. He has more than fifteen years experience in administration and management of natural resources and natural protected areas in northwestern México, including six years as subdirector of the Parque Nacional de Bahía de Loreto. David also works as an independent consultant with fishing cooperatives and federations to help secure their permits, rights, and opportunities.

Jeffrey C. Mangel is a biologist with Pro Delphinus. He holds an M.S. in environmental science from the SUNY College of Environmental Science and Forestry (1997) and a master's degree in environmental management from Duke University (2003). Prior to attending Duke he resided in Hawaii, where he participated in numerous research and conservation projects including studies of hawksbill turtles, monk seals, coral reef ecology, and seabird conservation. He also worked on sea turtle nesting beach projects in Costa Rica and Suriname. In 2003 he received a Fulbright Award to study marine otters in southern Perú. In 2005 he worked as a resource management liaison with the NOAA-Pacific Islands Regional Office, Office of Protected Resources. Jeffrey now resides in Perú, where he is involved in all aspects of Pro Delphinus research and conservation projects with sea turtles, seabirds, marine mammals, and sharks, as well as with artisanal fishery monitoring and bycatch mitigation. His work has resulted in numerous peer-reviewed publications, reports, book chapters, and conference presentations. Jeffrey is also currently working toward a Ph.D. in biological sciences at the University of Exeter, Penryn, United Kingdom.

MariLuz Parga is head of conservation medicine at the nongovernmental organization Submon, based in Barcelona, Spain. She qualified as a veterinary doctor at the University of Leon (Spain) in 1996 and since then has specialized in wildlife health, working in a number of wild animal rescue centers, exotic animal clinics, and zoological gardens, both in Spain and in the United Kingdom. She completed an M.Sc. in wild animal health at the Royal Veterinary College (University of London) and the Royal Zoological Society (London Zoo) in 1999 and did a three-year residency at the Royal School for Veterinary Studies (Edinburgh University), specializing in exotic and wild animal medicine and surgery. Back in Spain in 2003, she was the head veterinarian at the Rescue Centre for Marine Animals in Barcelona, where she first got involved with sea turtles, collaborating with the University of Barcelona in research related to diagnostic imaging, blood analysis, and physiology of loggerheads. She is a member of the IUCN Wildlife Health Specialist Group. Currently, her main work is related to longline bycatch reduction of sea turtles, and she has been involved in different projects related to improving handling and dehooking techniques of sea turtles, assessing hook-related lesions and postcapture mortality, and training fishermen and fishery observers.

S. Hoyt Peckham has directed Proyecto Caguama since 2002, partnering with fishermen, conservationists, and government in Pacific México, Hawaii, and Japan to augment fisheries sustainability and reduce bycatch of endangered megafauna. Previously, Hoyt codirected the Center for Cetacean Research and Conservation, a U.S.-based nonprofit that combines research and education to effect marine conservation in Polynesia and the Caribbean. Hoyt is currently based in Baja California Sur, where he serves as director of international relations for the Grupo Tortuguero de las Californias, A.C., and as a member of the IUCN's Marine Turtle Specialist Group. Hoyt earned his Ph.D. in ecology and evolutionary biology at the

University of California, Santa Cruz, and double-majored in biology and English literature at Bowdoin College. Hoyt is an accomplished commercial captain and underwater cameraman.

Rotney Piedra Chacón is a biologist and a master's student in marine and coastal sciences at the National University of Costa Rica. He is married to fellow biologist Elizabeth Vélez, who is also passionate about sea turtles, and they have two beautiful daughters, Tamara and Samantha. Piedra was born in Ciudad Quesada, an agricultural and ranching region located far from the sea. In his first two years of study, it seemed that his training would be in pharmacy. However, during his third year, he took several biology courses that made his academic training take a turn. His passion for sea turtles began in 1994 when he first visited the Parque Nacional Marino Las Baulas in Costa Rica and had his first encounter with a leatherback turtle in a night full of excitement and curiosity. Since then, he has been connected to this special protected area. Piedra and Vélez, along with several friends, colleagues, and volunteers, protect and carry out research on leatherback turtles in the Playa Langosta, a nesting beach of the park. In 1998, Rotney became the director of the Parque Nacional Marino las Baulas, with consolidation of existing lands within park boundaries as a main objective of this protected area as a sanctuary for leatherbacks and other species. He has always been aware that success depends on the participation and integration of a significant number of key stakeholders in the management of the area—government, community, biologists, volunteers, nongovernmental organizations, private entrepreneurs—a difficult but not impossible integration. He is currently a member of the IUCN Marine Turtle Specialist Group. He has decided to dedicate his professional life to the protection and conservation of sea turtles.

Pamela T. Plotkin received her B.S. from Penn State University and earned her M.S. and Ph.D. degrees from Texas A&M University. She was a Dean John A. Knauss Marine Policy Fellow in the U.S. National Marine Fisheries Service Office of Protected Resources, a postdoctoral fellow at Drexel University, an assistant professor at the University of Delaware, and a senior conservation scientist with the Center for Marine Conservation and is currently a research administrator at Cornell University. Her area of expertise is in the conservation biology of sea turtles. Much of her research over the past twenty years documents the behavior and ecology of sea turtles and the effects of human activities on populations.

Peter C. H. Pritchard has a B.A. (with honors) and M.A. in chemistry and biochemistry from Oxford University and a Ph.D. in zoology from the University of Florida, where he studied sea turtle biology with Dr. Archie Carr. After four years of work with the World Wildlife Fund, he became an officer of the Florida Audubon Society in 1973, where he held various positions, including those of assistant executive director, senior vice president, and acting president. Since 1998 he has been director of the Chelonian Research Institute in Oviedo, Florida, and is an adjunct professor of biology at Florida Atlantic University and the University of Central Florida. He is best known as an authority on the biology and conservation of turtles and tortoises. Both before and after receiving his doctorate in 1969, he has undertaken extensive fieldwork with turtles in all continents and many remote islands, and he has established a permanent field station for turtle conservation in northwestern Guyana. Three species of turtle are named after him—a snake-necked turtle from New Guinea, a pond turtle from northern Burma, and a giant fossil side neck turtle from Colombia. He has been recognized as a "Champion of the Wild" by the Discovery television channel and as a "Hero of the Planet" by TIME Magazine. In 2001, he was declared "Floridian of the Year" by the Orlando Sentinel newspaper. He has written ten books: *Living Turtles of the World* (1967), *Marine Turtles of Micronesia* (1977), *Encyclopedia of Turtles* (1979), *Turtles of Venezuela* (1984), *The Alligator Snapping Turtle* (1990 and 2007), *Galapagos Tortoises: Nomenclatural and Survival Status* (1996), *Cleopatra the Turtle Girl* (2007), *Tales from the Thébaide* (2007), and *The Pinta Tortoise: Globalization and the Extinction of Island Species* (2005). Peter is a popular lecturer and speaker, at home and overseas, both on topics relating to his specialty and on general topics relating to endangered species conservation, travel, and philosophical subjects.

Vincent S. Saba received his Ph.D. in marine ecology in 2009 from Virginia Institute of Marine Sciences. His general research interests are in the fields of marine ecosystems and climate variability, fisheries oceanography, sea turtle biology, ecosystem modeling, and ocean color variability. He is particularly interested in the trophic cascading dynamics within marine ecosystems between lower trophic primary producers and higher trophic organisms through both large- and small-scale climate variability. This includes both bottom-up factors such as resource availability and top-down factors, including anthropogenic climate warming and fisheries. He is currently a research oceanographer at Princeton University's Atmospheric and Oceanic Sciences Program.

Yonat Swimmer has been a fisheries biologist working at NOAA's Pacific Islands Fisheries Science Center for twelve years. Yonat received her B.S. from the University of California, Santa Cruz and her M.S. and Ph.D. from the University of Michigan. Her primary expertise is leading research aimed to reduce the incidental capture of sea turtles in fisheries, primarily longline fishing gear. Yonat's area of research has taken advantage of her strong background in sea turtle and tuna physiology and behavior, as well as a strong interest in marine conservation. She has worked extensively with sea turtles for more than twenty years and enjoys the challenges of conserving marine species and habitats in an ever-changing world.

José Urteaga works as Fauna and Flora International's Nicaragua country coordinator, where he supports the sea turtle conservation program, the development of a marine program, and land and fresh water conservation on Biosphere Reserve of Ometepe Island. Jose graduated with a B.Sc. in biological sciences in 2000 and a postgraduate diploma in sea fisheries in 2001 from the Universidad Nacional de Mar del Plata, Argentina. Since joining Fauna and Flora International in June 2002, his role has involved the development of comprehensive Sea Turtle Conservation Program in the Pacific coast of Nicaragua that at present contributes to the protection of more than 40 km of coast in four different sites, protecting per year more than 50,000 sea turtle nests from the threat of illegal egg poaching activities. José and his team, working closely with several local organizations, enhance the survival rate of sea turtle and hatchlings by relocating the turtles' eggs to hatcheries, ensuring nesting beach protection, providing environmental education to the local communities as well to national audiences, and helping them to develop alternative livelihoods that do not rely on unsustainable use of resources. José was announced as a 2010 Emerging Explorer of the

National Geographic and in 2005 as a Conservation Hero of the Disney Foundation. He is member of the IUCN Marine Turtle Specialist Group and Nicaraguan cocoordinator of the Wider Caribbean Sea Turtle Conservation Network.

Ingrid L. Yañez is the cofounder of the Eastern Pacific Hawksbill Initiative, an international collaborative effort to understand the biology, identify priority habitat, mitigate direct threats, and promote recovery of hawksbill turtles in the eastern Pacific Ocean. She has nearly a decade's worth of experience working with sea turtles at nesting beaches and foraging grounds in community-based and remote settings. She also has extensive experience working with a variety of wildlife species throughout the coastal and inland regions of Central and South America, with a particular focus on the Peruvian Amazon, where for several years she led expeditions along the regions main rivers and numerous tributaries. Ingrid earned a B.S. in biology from the University of Cayetano Heredia in Perú and currently lives with her husband, Alexander Gaos, and son, Joaquinn, in San Diego, California.

Patricia Zárate is a marine biologist who graduated from Universidad Catolica del Norte, Chile. She dedicated the first years of her professional career to the study and evaluation of management areas along central Chile working closely with fishermen associations and the Instituto de Fomento Pesquero. In 2000, she moved to the Galápagos Islands, Ecuador, and became part of the staff at the Charles Darwin Foundation, where she was the coordinator and principal investigator of the sea turtle research leading projects at the nesting beaches and foraging grounds for eight years. Her work has greatly increased the understanding of the ecology of sea turtles in the equatorial waters of the Galápagos and provided important, up-to-date information to managers of the national park. Patricia has also been deeply involved in shark research and bycatch issues and served for several years as Ecuador delegate for the Scientific Committee of the Inter-American Convention for the Protection and Conservation of Sea Turtles and the Permanent Commission for the South Pacific, CITES, and as technical adviser at the Galápagos Marine Reserve management meetings. She is a member of the IUCN Marine Turtle Specialist Group and a Ph.D. student at the Archie Carr Center for Sea Turtle Research and Department of Biology of University of Florida (USA). Patricia's doctoral dissertation will cover reproductive and foraging biology of the green turtle *Chelonia mydas* through applications of stable isotopes and genetic analysis to elucidate the ecological role and connectivity of this species in the Galápagos Islands and within the eastern Pacific Ocean.

Index